Engineering Materials and Processes

Series Editor
Professsor Brain Derby, Professor of Material Science
Manchester Science Centre, Grosvenor Street, Manchester, M1 7HS, UK

Bruno Predel · Michael Hoch · Monte Pool

Phase Diagrams and Heterogeneous Equilibria

A Practical Introduction

With Technical Cooperation of Felicitas Predel

With 270 Figures

 Springer

Prof. em. Dr. Dr. h.c. Bruno Predel
Haugstr. 26
D-70563 Stuttgart
Germany
predel@mf.mpg.de

Prof. em. Dr. Michael Hoch
5300 Hamilton Av., Apt. 1706
Cincinnati, OH 45224-3165
USA
niklah@fuse.net

Prof. em. Dr. Monte Pool
University of Cincinnati
Dept. of Chemical and Materials Engineering
PO Box 210012
Cincinnati, OH 45221-0012
USA
monte.pool@uc.edu

and
University of Cincinnati
Dept. of Chemical and Materials
Engineering
P.O. Box 210012
Cincinnati, OH 45221-0012
USA

With Technical Cooperation of Felicitas Predel
Max-Planck-Institut für Metallforschung, Stuttgart

Original German Edition published by Steinkopff Verlag Darmstadt, 1992

Library of Congress Cataloging-in-Publication Data
Predel, Bruno.
Phase diagrams and heterogeneous equilibria : a practical introduction / Bruno Predel, Michael Hoch, Monte Pool.
p. cm. – (Engineering materials and processes) Includes bibliographical references and index.
ISBN 3-540-14011-5 (alk. paper)
1. Phase diagrams. 2. Phase rule and equilibrium. 3. Thermodynamics. 4. Chemistry, Metallurgic. I. Hoch, M.J.R. (Michael J. R.), 1936 – II. Pool, Monte. III. Title. IV. Series.
QD503.P72 2003
541'.363–dc22

ISBN 3-540-14011-5 Springer Berlin Heidelberg New York

Springer. Part of Springer Science+Business Media
springeronline.com

© Springer-Verlag Berlin Heidelberg 2004
Printed in Germany

Typesetting: medionet AG, Berlin, Germany
Cover Design: Erich Kirchner, Springer Heidelberg, Germany

Printed on acid-free paper   62/3020/M   5 4 3 2 1 0

Preface

Since J.W. Gibbs in 1878 succeeded comprehensively in establishing the basic principles for an understanding of equilibria in heterogeneous systems, numerous books concerning constitution diagrams have been written, some of them providing a formal treatment of phase equilibria down to the small detail. The purpose of the present book is to provide an introduction to the practical applications of phase diagrams. In the first instance it is intended for students of chemistry, metallurgy, mineralogy and materials science, but also for engineers and students of science and engineering disciplines concerned with materials. To facilitate the start of an involvement with heterogeneous equilibria, reactions and dynamic equilibria will be treated first, since these are familiar to chemists and metallurgists.

Of course, a description of phase equilibria is not possible without a minimum of formalism. The formalistic description, however, will be made lighter by clear explanations of experimental methods used to determine the constitution of a system, by application examples, as well as by discussing realistic cases from chemistry, metallurgy, materials science and mineralogy. By this, the necessity of the knowledge of phase diagrams can be shown. On the other hand a practical exercise is possible.

The physical and energetic background to phase equilibria will also be treated. In so doing, the principles of thermodynamics of mixtures will be discussed and the correlation between energetics and constitution demonstrated. In this way, the more qualitative framework which often surrounds the teaching of constitution will be surmounted and the interested reader will be provided with a tool enabling him to make quantitative predictions also concerning phenomenological energetics and the structural and physical factors governing an individual system. It will also be possible to make predictions concerning phase equilibria for systems for which experimental results can only be obtained with difficulty.

From the standpoint of practical application, the treatment of nucleation of phase transitions, the production and stability of technologically important metastable phases and attempts to understand metallic glasses will also be discussed. There is currently a large-scale technologically motivated research effort in this area which is providing a broader and deeper understanding. A short survey of the most important facts will be presented.

Finally, a condensed presentation of the thermodynamics and constitution of polymer systems is included.

The book "Heterogene Gleichgewichte" though printed in 1982 is still very actual. We tried to make it a thermodynamic treatment for all materials: cover ceramics, organic materials, polymers and aqueous solutions (for geologists).

To upgrade the book, we introduced two new solution models, which permit the calculation of enthalpy of mixing and of phase diagrams in ternary, quaternary, quinary and larger systems from binary data alone. This is important in practical applications, where after calculations a few experimental measurements are sufficient to check the results. We also deal in greater detail with second order transitions in metals and polymers. For the use of ceramicists, we especially described the phase rule in ternary systems. In the aqueous solutions we show, how solubilities of several salts in water can be calculated, again using only binary data. Last but not least, we show that the same thermodynamic formulas used for metals and ceramic materials can be applied to organic materials, polymers and aqueous solutions.

No other text to our knowledge covers all these areas.

For critical review of the text, we are grateful to several experts in the field, especially Prof. Dr. Dr. h.c. mult G. Petzow, Prof. Dr. Dr. h.c. W. Gust, Prof. Dr. F. Sommer, Prof. Dr. W. Funke, Dr. I. Arpshofen and T. Gödecke. Mrs. G. Kümmerling and Dr. I. Arpshofen have prepared drawings.

Stuttgart and Cincinnati,   *Bruno Predel*
Summer 2003   *Michael Hoch, Monte Pool*

Contents

List of Symbols

Latin Letters

at-% A	atomic percent of component A
a_A	thermodynamic activity of component A
C_p	molar heat capacity at constant pressure
C_V	molar heat capacity at constant volume
c	concentration, common
v	vapor
e	binary eutectic point
E	ternary eutectic point
E_Q	quaternary eutectic point
F	free energy per g-atom (molar free energy)
G	Gibbs energy per g-atom (molar Gibbs energy)
H	heat content per g-atom (molar enthalpy)
L	liquid, melt
l	lamellar spacing in an eutectic or in a lamellar precipitation
M_A	atomic weight of component A
n	number of moles, common
p	pressure, common; in T-x phase diagrams: binary peritectic point; ternary peritectic point
p_A	partial vapor pressure of component A
$p_A{}^0$	vapor pressure of pure component A
p_K	critical pressure
Q	molar heat per g-atom
Q_F	activation energy of the freezing reaction
Q_M	activation energy of the melting reaction
R	gas constant
R_A	rate of reaction; common
R^A	rate of precipitation
R_E	rate of the eutectic reaction
R^K	net rate of crystallization
R^F	rate of freezing
R^M	rate of melting
S	a) entropy per g-atom (molar entropy)
	b) used as an index: solid

T	temperature in K; common
T_A	melting temperature of component A
T_A^0	boiling temperature of component A
T_e, T_E	eutectic temperature
T_F	freezing temperature; common
T_g	temperature of equilibrium
T^G	temperature of glass transformation
u	a) internal energy per g-atom; (molar energy)
	b) in phase diagrams: ternary transition equilibrium
V	molar volume
V_A	partial volume of component A in g-atom
wt-% A	weight percent of component A
x_A	mol fraction (atomic fraction) of component A
x_e, x_E	eutectic concentration
z	coordination number

Greek Letters

$\alpha, \beta, \gamma, \varepsilon, \mu$	in phase diagrams: solid phases
ν^L	oscillation frequency of atoms in a liquid
ν^S	oscillation frequency of atoms in a solid
σ	specific grain boundary energy

Differential Quantities

ΔF	variation of the molar Gibbs energy due to formation of a mixed phase (integral molar Gibbs energy of mixing)
ΔG	changes in the molar Gibbs energy
	a) due to formation of a mixed phase (integral molar Gibbs energy of mixing)
	b) by a phase transformation
ΔG_A^F	$G_A^L - G_A^S$ = molar Gibbs energy of freezing of component A
	G_A^L, G_A^S Gibbs energy of liquid, respectively solid A
ΔG^{ex}	Integral molar excess Gibbs energy
ΔG_A	Partial molar Gibbs energy of mixing of component A
ΔG_i	Ideal integral molar Gibbs Energy of mixing
$\Delta G_{A(i)}$	Ideal partial molar Gibbs Energy of mixing of component A
ΔH	change in molar enthalpy due to the formation of a mixed phase (integral molar enthalpy of mixing or of formation)
ΔH_A^F	molar enthalpy of melting of component A
ΔH^U	molar enthalpy of transformation
ΔH^v	molar enthalpy of evaporation
ΔS	change in molar entropy due to the formation of a mixed phase (integral molar entropy of mixing)
ΔS_A^F	molar entropy of melting of component A
ΔS^{ex}	Integral molar excess entropy

ΔS_i Ideal integral entropy of mixing
ΔS_A Partial molar entropy of mixing of component A
$\Delta S_{A(i)}$ Ideal Partial molar entropy of mixing of component A
ΔT difference in temperature, general
ΔV change in molar volume due to the formation of a mixed phase (integral molar volume of mixing)
ΔV^F change of molar volume due to melting in cm^3/g-atom (molar volume of melting)
ΔV^U change of volume due to transformation (molar transformation volume)
ΔV_A Partial molar volume of mixing of component A

Annotation

If not mentioned otherwise, the extensive state functions were referred to 1 mol, for instance F, G, H, S and all differential values marked with Δ (except ΔT). For mixtures with common content, these values are identical with those related to 1 g-atom.

    Care should be taken, if molecules or molecule like species were considered, which have an exactly defined stoichiometry. In such cases the values related to 1 mol have to be converted into values related to 1 g-atom. This is done by dividing the molar value by the number of atoms in the formula unit, for, as well known, 1 mol consists of $6.023 * 10^{23}$ molecules, whereas 1 g-atom contains $6.023 * 10^{23}$ atoms.

    As an example: Melting enthalpy of water is $\Delta H^F_{H_2O} = 6.007$ kJ/mol. This corresponds to $\Delta H^F_{H_2O} = 2002$ kJ/g-atom.

CHAPTER 1

Fundamental Facts and Concepts

1.1
General

Experience teaches us, that under given external conditions a material is in a certain state. This state is defined by certain properties. At 200 K the compound H_2O is solid, at 300 K liquid and at 400 K gaseous. Properties, which are characteristic for a state of a material, are called intensive properties. Among them is the molar volume. As one can see from Fig. 1.1, the molar volume of water is different at 200, 300 and 400 K.

Even without a change in phase, a variation of the external condition, here the temperature, can change the state of the system. The molar volume at 300 K is different from that at 277 K. As is well known, water has its highest density at

Fig. 1.1
Molar volume of water as a function of temperature at normal pressure [1]

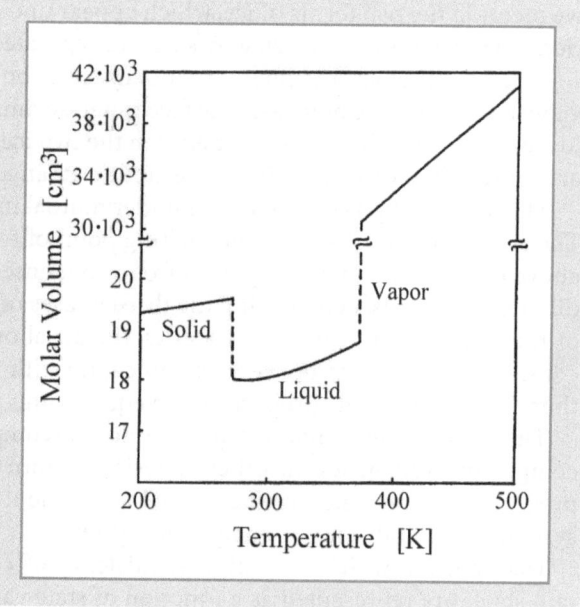

277 K and differs in this respect, clearly, from water at 300 K; this is of great importance for the heat balance of the oceans.

Quantities, which fix a certain state, are called state variables. In addition to the temperature T, they are the pressure p and the composition c. Only in special cases, which are not discussed here, can other state variables be significant.

Whether a change of state really takes place or not during the time of observation due to the change in the external conditions, depends on the kinetics of the process, which reorders the atoms or molecules of the material, as for instance ice → water. In solids, especially at low temperatures, the rate of reaction is so slow, that long times are needed to attain the required state. If one has attained the state corresponding to the given p, T and c, then no change will occur during further waiting. This is the equilibrium state, or stable state of the material for the chosen state variables. As long as the equilibrium state is not reached, the material is in a metastable state. In technology, metastable states can, under certain conditions, be of greater importance than stable states. The equilibrium state is independent of the path the system takes to reach it. Water at 300 K has the same properties, independently if it was obtained by melting ice of 200 K, or condensing water vapor of 400 K. The state variables, the properties of the material in equilibrium, are independent of the path used to attain them.

One and the same material can appear in different forms, called phases. Phases are in a macroscopic sense homogeneous. Water, ice and water vapor are different phases. The concept of phase is not always connected to a form of aggregation. A solid material can, depending on the state variables, appear in different crystal structures (modifications), which are also defined as phases, as they are, as required by their definition, macroscopically homogeneous. For example we mention the two forms of ice, which appear under two different (p, T) conditions. Two phases existing side by side, are separated by a phase boundary.

When melting ice, the phase boundary is the boundary surface ice-water, at vaporization it is the boundary surface water-steam, and when two forms of solid ice coexist it is the boundary between the two ice surfaces. The phase boundary represents a discontinuity in the atomic arrangement of the material.

The totality of phases, which are under mutual interaction, is called a system. The system water consists at the melting point of two condensed phases, a little above or below the melting point of one condensed phase. Other phases from different materials, neighboring the three phases of water, for example the glass of the container in which the water is, the air above the water surface are not considered, as they do not react significantly with the material of interest, and thus, do not change its properties and equilibrium state.

The system water contains one material, one component. It is defined as a one-component system. Adding other materials to pure water, which may change the properties of the water, creates a two-component system, a three-component system, in general: a multi-component system.

Diagrams, in which the range of existence of certain phases and thus, certain states are represented as a function of state variables, are called phase dia-

Fig. 1.2
Schematic phase diagram of
water in the neighborhood
of the triple point t

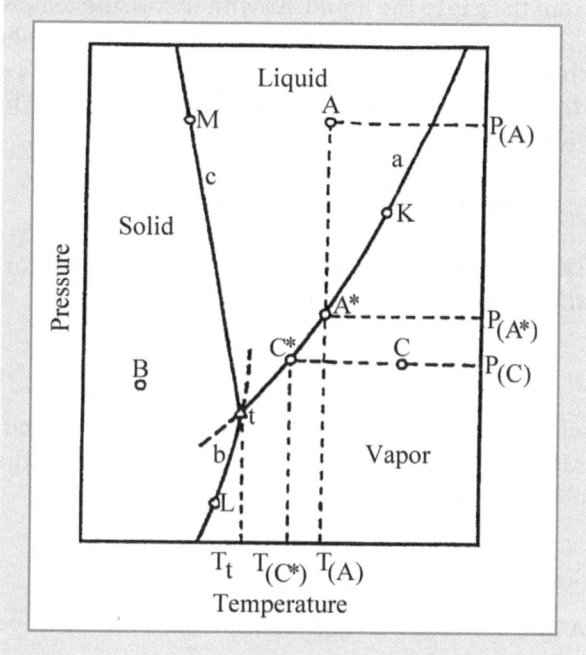

grams. Figure 1.2 represents schematically the basic outlines of the water phase diagram. In this case we have only one component, thus, the composition does not appear as a variable. The state of water is uniquely determined by the pressure p and temperature T.

1.2
Vaporization Equilibrium

First, let us look at the behavior of atoms during vaporization of a liquid. The molecules in a liquid have very different kinetic energies; there is a certain kinetic energy distribution. The average value of this energy depends on the temperature.

For a molecule to transfer from the liquid phase into the gas phase, one has to overcome the interatomic forces in the liquid. This is possible for single molecules, when their kinetic energy lies above a certain value, the enthalpy of vaporization. They can pass the phase boundary liquid-gas, and escape the attractive forces of the liquid. Outside of the liquid phase the molecules encounter the container walls, and develop a certain pressure. They can reach the boundary liquid-gas again, and become part of the liquid. The vaporization enthalpy (condensation enthalpy) used up during vaporization is then recovered.

At a certain temperature T a dynamic equilibrium is reached, where per unit time, the same number of molecules pass from the liquid into the gas, as pass from the gas to the liquid. As with increasing temperature the kinetic energy of the molecules increases, the vapor pressure of the liquid also increases with increasing temperature. If in the case of ΔH_W^V is the enthalpy required to vaporize 1 mol of H_2O, then the following relationship holds for the vapor pressure p_W of water:

$$P_W = A \cdot exp\left(-\frac{\Delta H_W^V}{RT}\right) \tag{1.1a}$$

The relationship described in Eq. (1.1a) is valid for all vaporization equilibria, and is expressed in general

$$p = A \cdot exp\left(-\frac{\Delta H^V}{RT}\right) \tag{1.1b}$$

where A is a constant, property of the material, and R the general gas constant. ΔH^V is the molar enthalpy of vaporization, and p the vapor pressure. Taking the natural logarithm of Eq. (1.1b) we obtain

$$\log p_W = \log A - \frac{\Delta H_W^V}{R} \cdot \frac{1}{T} \tag{1.2}$$

A plot of ln p_W vs $1/T$ gives a straight line, the slope of which permits the evaluation of ΔH_W^V.

In Fig. 1.2 the vapor pressure curve of the melt is represented schematically by lines.

Similar is the vaporization of ice. The enthalpy of vaporization of ice ΔH_I^V is, however, larger than that of water. The difference

$$\Delta H^F = \Delta H^V - \Delta H_W^V \tag{1.3}$$

corresponds to the enthalpy needed to melt 1 mol of ice. Thus, at a fixed temperature the slope of the vapor pressure curve of ice is greater than that of water (see Fig. 1.2, curve b). There is a point of intersection t of the vapor pressure curves. Only at the temperature of this point are the two vapor pressure curves above water and ice equal. Above the temperature of T_t the vapor pressure of water is less than that of ice, below T_t the vapor pressure of ice is less than that of water. The effects of this situation are shown in the experiment illustrated in Fig. 1.3.

Ice and water are in a closed container, with a common gas phase, but otherwise separated. Care is taken, so that no difference in temperature exist between ice and water. If one chooses the temperature of the system as $T > T_t$, a dynamic vaporization equilibrium cannot be sustained. As $p_W < p_i$, ice will vaporize and condense in the boat containing water. At $T < T_t$ the reverse occurs. Only at $T = T_t$ is there equilibrium. Thus, the following trivial conclusion appears: below temperature T_t – in addition to the gas phase in equilibrium with the condensed phase – only ice is the stable phase, above T_t only water is stable. Only at T_t can both condensed phases coexist.

Fig. 1.3
Illustration of stability of
coexisting phases. p_W and p_i
are vapor pressures of water
and ice, respectively

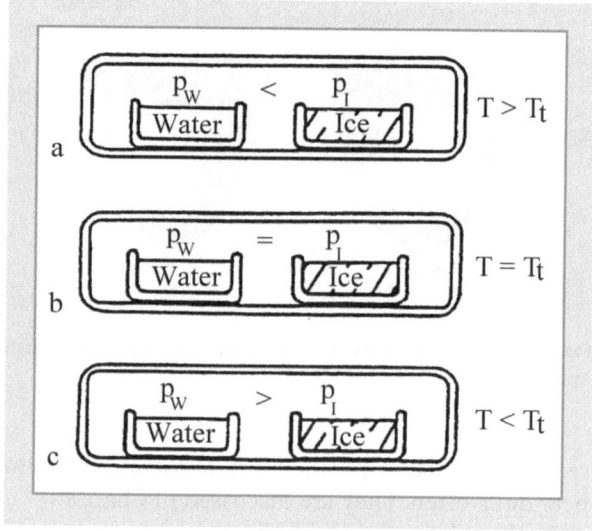

The vapor pressure p is in this thought experiment a visible indicator for the
energetic background of the stability of the phases and the equilibria relationships. During distillation at $T > T_t$ (Fig. 1.3a) the free energy of the system decreases. Its value depends on the relationship of the vapor pressures P_I and P_W,
and for n moles:

$$\Delta F = -n \cdot RT \cdot \ln \frac{p}{p_W} \tag{1.4}$$

After the distillations process the total free energy is lower than before. This is
really the only criterion for the stability of phases. Under preset conditions (T, v
constant) the phase (or the totality of coexisting phases) is the stable one, which
has the lowest free energy.

During phase changes, occurring at constant pressure, we have to consider
the change in energy due to possible change in volume, as explained in Chap. 6.
The change in free energy ΔF is replaced by the change in Gibbs energy, ΔG.

The energetic situation to attain equilibrium is equivalent to the conditions
in a mechanical system, as shown in Fig. 1.4. If the ball K is at the bottom of the
container, we have, as is well known, a stable equilibrium (lowest potential energy). In case **a**, which if this situation may be realized by high friction forces
(analogous: kinetic hindrance of the reaction), we have an unstable system (no
minimum in potential energy). Eliminating this hindrance (analogous at chemical systems is the increase in temperature) allows the transition to a stable situation and the system will lose energy. Similarly for case **c**, where the ball is in a
side minimum, by expending a small amount of potential energy the threshold
S can be overcome, and the minimum in potential energy of the system can be

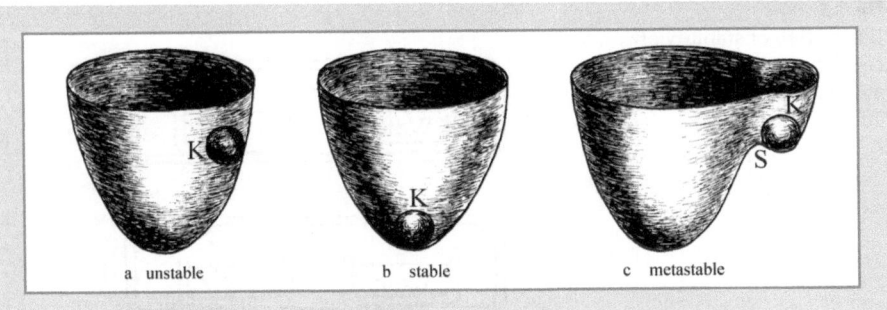

a   unstable   b   stable   c   metastable

Fig. 1.4 Mechanical model to explain conditions of equilibrium. **a** unstable, **b** stable, **c** metastable

reached. The case **c** is called a metastable equilibrium. Metastable situations appear quite often. They are discussed in Chap. 8.

In the schematic phase diagram for water in Fig. 1.2, line a represents the boundary between the region of existence of water and steam, line b the boundary between the region of existence of ice and steam. At values of (p, T) which belong to point A ($p_{(A)}$, $T_{(A)}$), only one phase (water) exists, at B only ice and at C only gas. At K steam and water are in equilibrium, at L ice and steam coexist. If we decrease the pressure starting at A and at constant temperature T_A, when curve a is reached, boiling starts. The pressure stays constant, at $p_{(A*)}$, until all the water is vaporized, as in this case at a fixed temperature of $T_{(A)}$, water and gas can only coexist at a pressure $p_{(A*)}$ (two phase equilibrium). During this boiling process, the vaporization enthalpy is extracted from the surroundings. Only after all the water is vaporized, can the vapor pressure decreased further.

Similarly, starting from C, and decreasing the temperature at constant pressure p_C, for instance, by shutting down the furnace, which contains the container with the vapor. As soon as curve a is reached at C^*, condensation begins. The temperature stays constant at $T_{(C*)}$, because at fixed pressure $p_{(C)}$ two phases can only coexist at the temperature of $T_{(C*)}$. During condensation heat (enthalpy of condensation) will be liberated. As soon as all the gas is condensed, the temperature can be decreased further.

The line a in the p-T diagram represents in general conditions for two-phase equilibria, in the special case of point t, three phases are in equilibrium: ice, water and steam. This is the melting point of water, under its own vapor pressure. This point, where three phases coexist is called the triple point. In the case of water it is at p = 610 Pa and T = 273.1575 K (4.58 Torr, 0.0075 °C)

References

Citations
[1] R.C. Weast (Editor), "Handbook of Chemistry and Physics", 49. Edition, The Chemical Rubber Co., Cleveland, Ohio (1969)

General References
A.M. Alper (Editor), "Phase Diagrams, Materials Science and Technology", Vol. 1, Academic Press, New York (1970)

P. Gordon, "Principles of Phase Diagrams in Materials Systems", McGraw-Hill Book Comp., New York (1968)

W. Paul and D.M. Warschauer (Editors), "Solids Under Pressure", McGraw-Hill Book Comp., New York (1963)

R.G. Ross and D.A. Greenwood, "Liquid Metals and Vapours under Pressure", in "Progress in Materials Science", Editors: B. Chalmers and W. Hume-Rothery, Vol. 14. No. 4, Pergamon Press, London (1969)

CHAPTER 2

Phase Equilibria in One-Component Systems

2.1
General

On melting a substance heat is absorbed in order to transform the crystal into the melt, in which the interatomic interaction is weaker. In addition, the degree of order is lower than in the solid body. The melting process is therefore connected with an absorption of heat (heat of melting or enthalpy of fusion) as well as with a volume change. This implies that the melting equilibria are determined by an interaction of variables of state, that is, pressure and temperature. The direction of the effect of an external pressure is given by Le Chatelier's principle of the smallest force (1884). According to this principle, an equilibrium is always shifted into the direction leading towards a reduction of a coercion imposed from without (of the given change of the state variables). If, on melting, the volume of a substance is increasing, a pressure rise will result in an increase of the melting temperature, as on solidification of the melt a solid body is formed which is able to counterbalance the external force.

Water melts under a reduction of volume. Therefore, its melting point is decreasing with increasing pressure. In general, the basic relationship is expressed by the Clausius-Clapeyron equation:

$$dp/dT = \Delta H^F/(T^F x \Delta V^F)$$

Here, ΔH^F is the melting enthalpy per mol, T^F the absolute temperature of fusion, and ΔV^F the volume change per mol which is connected with the melting process. In Fig. 1.2, the melting pressure curve (c) of ice is drawn in starting at the triple point t. It represents the entirety of the points in the p-T diagram at which water and ice are in equilibrium.

2.2
Transformation Equilibria in the Solid State

As is well known the atomic structure of a solid body is determined by the nature and the strength of the interatomic, resp. intermolecular interactions. Just as by certain changes of the variables of state a transition from the crystalline

long range ordered structure of the solid body into the only short range ordered arrangement of the melt can be effected, for many substances there exists the possibility to attain, by the variation of the pressure and the temperature, a transformation of the crystal structure into a different lattice with other
symmetry relationships as well as other distances between the atoms or molecules.

This phenomenon has been discovered by Mitscherlich in 1821, and has been
called "polymorphism". The transformation of a solid phase into another solid
phase is proceeding, as a rule, by diffusion processes within the phase boundary or in its immediate neighborhood. In the case of the coexistence of two solid phases, at the phase boundary there is a dynamic equilibrium in which an
equal number of atoms or molecules is diffusing from phase 1 to phase 2 as in
the opposite direction, per unit area and time. In principle, analogous circum-
stances are prevailing as on evaporation and on melting. Therefore, the principles of phase equilibria are analogous to those of melting and of evaporation.
In the phase diagram, distinct phase areas have to be attributed to the different
solid phases. Along the different p-T-curves separating these areas, two phases
are in equilibrium. Phase transformations between stable solid phases (equilibrium phases) are called "enantiotropic" transformations. In principle, they can
be carried out in a reversible manner.

In addition, monotropic transformations may occur in the course of which a
metastable solid phase is transformed into a stable one. Such transformations
are not reversible. They always proceed in the direction of the stable phase. A
stable phase cannot be in equilibrium with a metastable one. As the phase diagram is an equilibrium diagram, monotropic transformations cannot be drawn
in. However, it is usual to indicate, for the purpose of a quick orientation, the
probable areas of existence of metastable phases.

Similar to the melting equilibria, also the transformation equilibria are connected with volume changes. Therefore, the transformation temperature is dependent on pressure. The pressure-transformation temperature curve is described, similar to the pressure-melting temperature curve, by the Clausius-
Clapeyron equation. For the dependence of the transformation temperature on
the pressure the transformation enthalpy ΔH^T as well as the volume change on
transformation ΔV^T, are responsible.

As an example of the phase diagram of a mono-component system exhibit-
ing phase transformations in the solid state, in Fig. 2.1 a part of the phase diagram of water is reproduced. In the region up to $3 \cdot 10^6$ kPa (approx. $3 \cdot 10^4$ at)
ice is occurring in seven stable modifications. At (p,T) values complying with
the drawn in curves, the phases belonging to the adjacent areas are in equilibrium. It ought to be remarked that the transformations which can be expected to
happen, at a variation of p or T, on passing an equilibrium curve according to
the Clausius-Clapeyron equation, as e.g. V/VI, are setting on only with hesita-
tion. Until nuclei of the new phase are formed which are able to grow, the initial
phase, of course, is remaining in existance, in the metastable state.

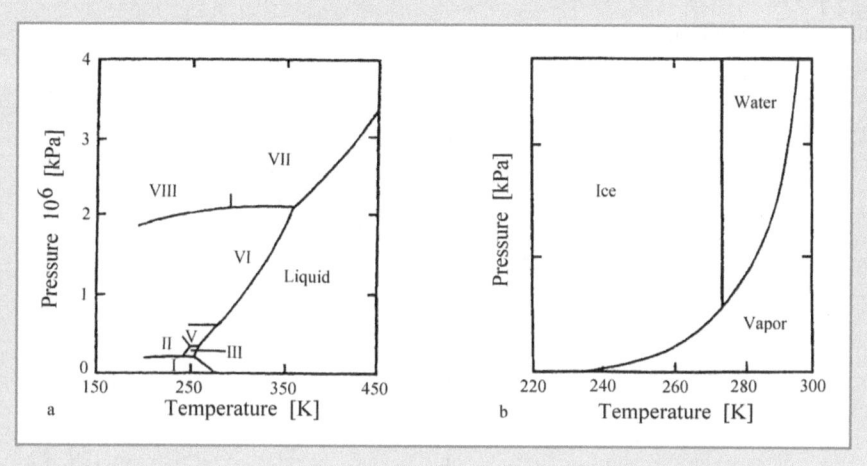

Fig. 2.1 Phase diagram of water. **a** Phase equilibria in the solid state after [1], **b** Phase equilibria at pressures < 3 kPa [2]

Furthermore, it is remarkable that the melting points of the modifications designated as III, V, VI and VII are increasing with increasing pressure, while the melting point of ordinary ice (I) is decreasing. The high pressure modifications have a smaller volume than water, whereas the ice I forming at lower pressures by the freezing of water has a higher molar volume than water. As, in all cases, the sign of the melting enthalpy ΔH^F remains the same, according to the Clausius-Clapeyron equation a change of the sign of ΔV^F results in a change of the sign of dp/dT. At 4.10^6 kPa, the melting point of ice is already reaching 470 K (approx. +200 °C). Finally, it may be mentioned that the modification called ice IV is metastable and therefore is not shown in the equilibrium diagram (Fig. 2.1a).

Also metals which have relatively simple crystal structures can occur in different modifications. Already at atmospheric pressure, at different temperatures they exhibit different crystal structures. Titanium, e.g. at 1155 K (882 °C) is changing from the low temperature modification with hexagonal closed packed structure into the high temperature modification having body centered cubic structure. Uranium is occurring in three, manganese in four different modifications. Of peculiar importance is the polymorphism of iron. Below 1184 K (911 °C), iron is present in a body centered cubic structure (*a*). Between 1184 K and 1665 K (1392 °C) it has a face centered cubic structure (γ) and above 1665 K (1392 °C), again the bcc structure (α) is stable. At high pressure, finally, a hexagonal closed packed modification (ε) is occurring. The phase diagram of iron is reproduced in Fig. 2.2.

Fig. 2.2
Phase diagram of iron after
[2]

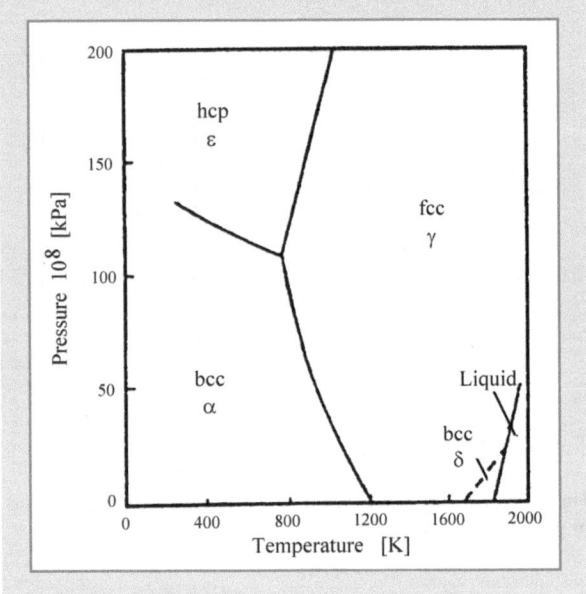

2.3
Monotropic Transformations

Monotropic transformations are transitions of metastable phases into stable ones. Metastable phases are not formed by equilibrium reactions. They are growing, e.g. by condensation from the vapor phase, by solidification from the melt in solid state reaction processes or are resulting as products of chemical reactions, if, for some reason nuclei of the stable modification which are able to grow cannot be formed.

In Fig. 2.3 the vapor pressure curves of a melt and of the corresponding stable solid body α, are depicted. The equilibrium point of fusion may be denoted as T_g. A metastable phase of this system, α_{met}, has to have a higher vapor pressure than α: $P\alpha_{met} - P\alpha = \Delta p$. On an isothermic distillation, α_{met} changes over into α; in doing so the expansion work determined by the ratio of $P\alpha_{met}/P\alpha$ represents the diminuation of the Gibbs energy connected with this process (comp. with Eq. (1.4)). If one starts with a melt at $T > T_g$, by lowering the temperature the crystallization of the equilibrium phase may fail to take place, as this is often observed. The temperature of the melt may sink below T_g. If the nucleation of the α_{met} modification proceeds more rapidly than that of α at T_{met}, crystallization of the metastable phase takes place. On heating it is transitioning into α. The temperature at which this monotropic transformation occurs is not as well defined as an equilibrium temperature but primarily depends on the mobility of the atoms or molecules in the α_{met} lattice as well as on the probability of the formation, out of the α_{met} lattice, of nuclei which are able to grow. This is no transformation point. The temperature range of this transformation, as experience shows, is considerably dependent on the heating rate.

In real systems, often several metastable phases can be observed. In general, on non-equilibrium reactions the phase characterized by the lowest stability (that is, with the highest Gibbs energy) is formed. Usually, this phase is not transitioning immediately into the stable phase. On the contrary, a gradual reduction of the Gibbs energy takes place so that the metastable phase is formed which is the next one in the order of stability, and so on, until eventually by such single steps the stable phase is obtained. (Ostwald's rule of steps). This course is due to the fact that metastable phases which are only a trifling different from each other in regard to their energetics, in general are also showing only little differences as to their atomic arrangements. Thus, the probability for the spontaneous formation of a volume part (nucleus) which has the crystal structure of the respective phase with the higher stability and is able to grow under reduction of the Gibbs energy of the system is higher than at considerable structural differences as they may exist regarding the stable phase.

From Fig. 2.3 can also be seen that the metastable modification, due to its higher vapor pressure as compared with the stable modification, obviously

Fig. 2.3
Schematic phase diagram

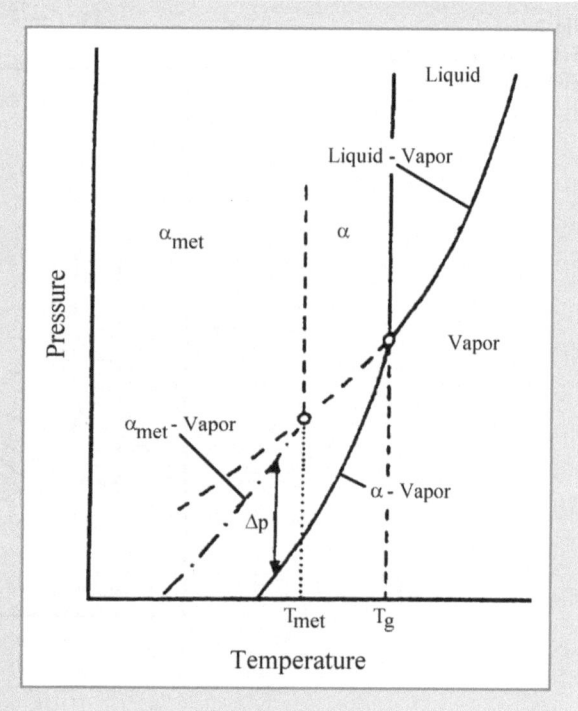

exhibits a lower melting temperature. This is fundamentally valid. Metastable phases always have a lower melting point than the corresponding stable phases of the same system.

In Fig. 2.4, the phase diagram of silicon dioxide is schematically represented. Several stable as well as metastable phases are occurring. Usually, only some of these phase transformations are taking their course without being disturbed. Among these is, e.g. the transformation of α-quartz. Already on the solidification of the melt, difficulties regarding nucleation occur, so that it is easily possible to avoid crystallization at all.

Here, the viscosity of the SiO_2 melt is rapidly enhanced at decreasing temperatures so that solidification into a "silicate glass" takes place. From the metastable SiO_2 glass, according to Ostwald's step rule, at temperatures of about 1070 K α-quartz is formed which is stable in this temperature range. Rather, as the first product of transformation the metastable α-tridymite is formed. As is well known, the formation of (metastable) glasses in silicate systems is of considerable technological importance.

A typical case of a monotropic transformation, furthermore, is present in phosphorus. The violet modification is stable under normal conditions while the white one is metastable. The white modification can be transformed, on heating, into the violet one but not the violet into the white one.

Fig. 2.4
Schematic phase diagram of
SiO_2 after [3]

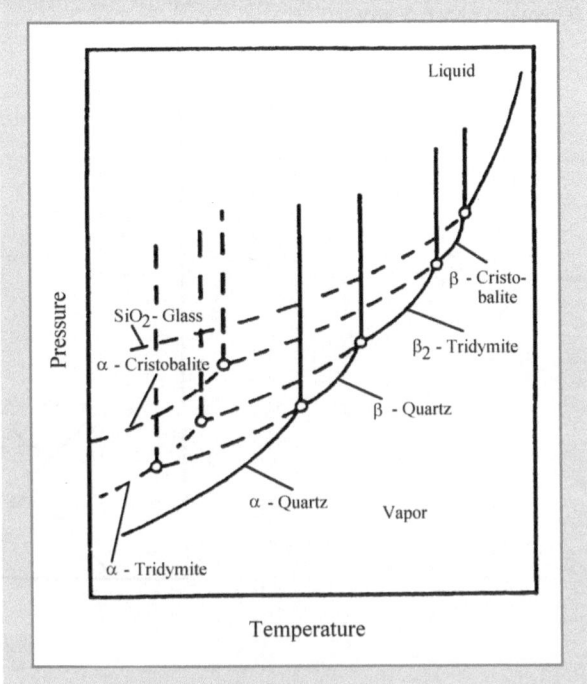

The formation of nuclei of stable modifications can often be sped up using catalysts. As an example, the transformation of white phosphorus to violet phosphorus can be accelerated by adding small amounts of violet iodine. To speed up nucleation in SiO_2-systems small additions of Na_2WO_4 may be used.

References

Citations

[1] J.C. Bailar, Jr., H.J. Emeleus, Sir Ronald Nyholm, A.F. Rotman Dickenson (Editors), "Comprehensive Inorganic Chemistry", Vol. 2, Pergamon Press, Oxford (1973)

[2] P. Gordon, "Principles of Phase Diagrams in Materials Systems", McGraw-Hill Book Comp., New York (1968)

[3] R. Vogel, "Die heterogenen Gleichgewichte", 2. Edition, Akademische Verlagsgesellschaft, Leipzig (1959)

General References

W. Paul and D.M. Warschauer (Editors), "Solids under Pressure", McGraw-Hill Book Comp., New York (1963)

R.G. Ross and D.A. Greenwood, "Liquid Metals and Vapours under Pressure", in "Progress in Materials Science", Editors: B. Chalmers and W. Hume-Rothery, Vol. 14, No. 4, Pergamon Press, London (1969)

CHAPTER 3

Phase Equilibria in Two-Component Systems under Exclusion of the Gas Phase

3.1
Definition of the Composition

The composition appears as a state variable in multi-component systems. For the description of atomic situations, its definition as atomic fraction or mol fraction x is especially useful. The atom fraction of one component A of a multi-component system is given by

$$x_A = \frac{n_A}{\sum_i n_i} \tag{3.1}$$

where n_A is the number of g-atoms (or mols) of A and $\sum_i n_i$ the sum of the g-atoms (or mols) of all components. In addition

$$x_A + x_B + x_C + \ldots = 1 \tag{3.2}$$

In a two-component system one only needs the concentration of one-component, because

$$x_A = 1 - x_B \tag{3.3}$$

It is possible to indicate the concentration in atomic-percent (atomic-%). Here

$$\text{at.-\% } A = x_A * 100 \tag{3.4}$$

To prepare defined mixtures of several components the specific materials are weighed. Therefore first the knowledge of the weight percents of mixtures are needed. Weight (wt.)-% and atomic (at.)-% can be converted easily into one another, considering the atomic and molecular-weights. For a two-component system we have

$$\text{wt.-\% A} = \frac{100 \cdot (\text{atom} - \%A) \cdot M_A}{(\text{atom} - \%A) \cdot M_A + (\text{atom} - \%B) \cdot M_B} \tag{3.5}$$

$$\text{atom} - \%A = \frac{100}{1 + \dfrac{100 - (\text{wt.-\%A})}{\text{wt.-\%A}} \cdot \dfrac{M_A}{M_B}} \tag{3.6}$$

M_A is the atomic weight of A and M_B is the atomic weight of B.

For the representation of phase equilibria the concentration is represented in at.-%, less useful and less often in wt.-%.

3.2
Partial Reactions of the Solid-Liquid Transition

The melting of a material occurs empirically in the measure as heat (enthalpy) is supplied. The quantity of solid and liquid phase, which are in equilibrium at a certain temperature T^F and at a certain moment, depends on the fraction of the total enthalpy which has been transferred up to this moment. The rate of the melting process depends on the speed of the enthalpy transfer.

This rate, which can be measured directly, can be understood as the difference of two rates, which determine the dynamic melting rate. This is the melting process, where the atoms or molecules of the solid phase pass the phase boundary and reach the liquid phase, as well the process of solidification, where parts of the liquid transfer to the solid. For a particle to perform the step of a reaction, it must possess sufficient high energy, the activation energy of the melting-process, Q^M, or the activation energy of the solidification-process, Q^F.

Other factors are of importance. The higher the vibration frequency v^S of the particle in the lattice of the solid, the higher the number of possibilities per second that an atom in the solid can leave the solid phase. Similarly for the back reaction, the partial process of melting (v^L = vibration frequency of the particle in the melt).

In addition, the atom, which has the energy to escape, must also have a favorable direction, so that it has a sufficiently high velocity component perpendicular to the solid surface, to overcome the attractive range of the atoms in the solid. This special condition can be considered by using geometrical factors while describing the velocities of the partial reactions: G^M for the process of melting, G^F for the process of solidification.

Furthermore, a particle which escaped from one phase, must find an acceptable place to be inserted. It must then give up some of its energy, to avoid returning through the activated state to the starting phase. The probability, that an insertion takes place, is expressed by the accomodation coefficient: A^M for the melting process, A^F for the solidification process.

Finally the velocity of the fusion reaction, R^M is proportional to N^S the number of atoms or molecules which are per unit surface area of the phase boundary are located on the solid side. The velocity of the solidification reaction, R^F is proportional to N^L the number of atoms or molecules which are per unit surface area of the phase boundary on the melt side.

The velocity of the partial reactions, which are described by the number of atoms, which pass per second and cm^2 across the boundary is according to [1].

$$R^M = N^S \cdot A^M \cdot G^M \cdot v^S \cdot \exp\left(-\frac{Q^M}{RT}\right) \qquad (3.7)$$

$$R^F = N^L \cdot A^F \cdot G^F \cdot v^L \cdot \exp\left(-\frac{Q^F}{RT}\right) \qquad (3.8)$$

The temperature dependence of the partial reactions for the case of copper is represented in Fig. 3.1.

It must be emphasized, that with thermally activated processes, which reach a dynamic equilibrium, the difference in the activation energy of the forward and backward reaction corresponds to the reaction enthalpy of the total process. In the case of solid-liquid equilibrium we have (Fig. 3.2)

Fig. 3.1
Rates of partial reactions of freezing, R^F, and of melting, R^M, for copper as a function of temperature [1]

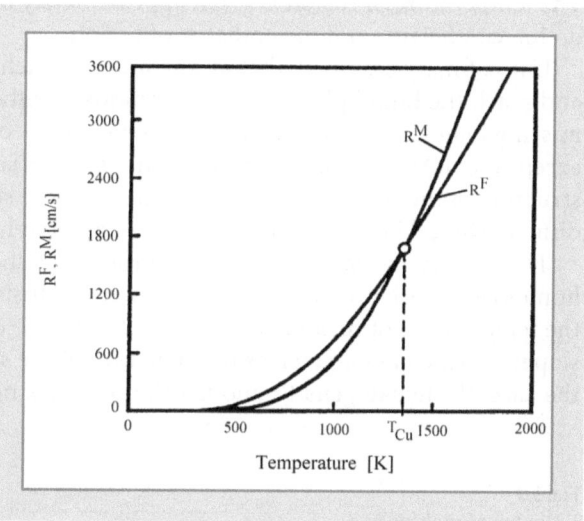

Fig. 3.2
Relation between activation energy of the partial processes of the solid-liquid equilibrium and the enthalpy of melting, ΔH^F

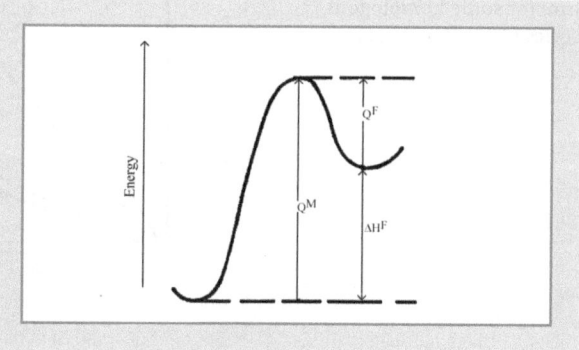

$$\Delta H^F = Q^M - Q^F \tag{3.9}$$

where ΔH^F is the molar enthalpy of fusion of the material under consideration. According to Eq. (3.9) there is $Q^M > Q^F$. The shape of the $R^F - T$- and $R^M - T$-
curve is different. The curves intersect at the equilibrium temperature T_g. Here
the forward and reverse reactions have the same rate. For this case ($R^M = R^F$) it
follows from Eqs. (3.7)–(3.9)

$$\frac{\Delta H^F}{RT_g} = \ln \frac{A^M \cdot G^M \cdot N^S \cdot v^S}{A^F \cdot G^F \cdot N^L \cdot v^L} \tag{3.10}$$

For a plane surface, $G^M \sim G^F$. Furthermore, in first approximation for simple
materials such as metals, $N^S \sim N^L$ and $v^S \sim v^L$. Equation (3.10) simplifies to

$$\frac{\Delta H^F}{RT_g} \approx \ln \frac{A^M}{A^F} \tag{3.11}$$

The temperature of fusion is given approximately by the ratio of the accommodation coefficients and the enthalpy of fusion.

If it is finally assumed, that all the atoms, which come from the solid phase
and reach the liquid phase, will accommodate without difficulty because of the
missing long range order in the liquid phase, thus, one can take $A^M \sim 1$. With the
exception of ΔH^F the fusion temperature is essentially dependent on the crystal
structure of the solid phase. More complicated the structure of the solid is, more
difficult should be the accommodation of an arriving atom or molecule.

This affects, in reality, the melting process, as shown in Fig. 3.3, in which for
homologous series of materials the enthalpy of fusion is plotted as a function of
the temperature of fusion. Based on Eq. (3.11) one can expect straight lines, the
slopes of which are greater, as the structure of the solid is more complicated. In
the case of the halogens F_2 and Cl_2 the particles not only have to arrive at the

Fig. 3.3
Enthalpy of melting as a
function of melting temperature for some homologous
series of substances [1]

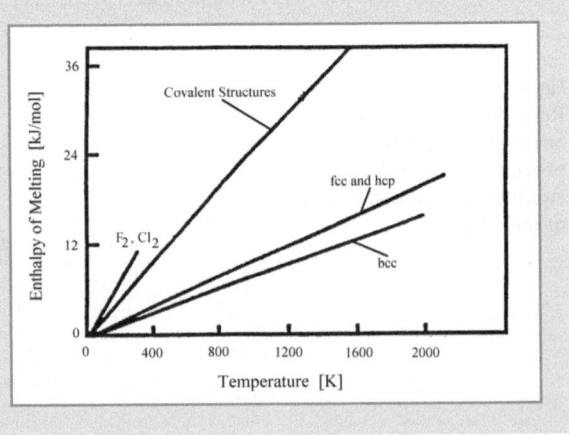

correct place on the lattice, but also in the right direction of the dumbbell molecule, to be inserted.

The difference in the structural change during fusion for the materials in the different groups can be easily recognized from a thermodynamic point of view. The ratio

$$\frac{\Delta H^F}{T_g} = \Delta S^F \qquad (3.12)$$

which can be obtained from the slope of the lines in Fig. 3.3, represent the entropy of fusion. For metals with densest sphere packing, or cubic body centered structure it is of an order of magnitude $\Delta S^F \sim 8.5$ J/g-atom K for the halogens $\Delta S^F \sim 20.4$ J/g-atom K. The different entropies of fusion of real metals and halogens indicate a different change in structure during melting of the two groups.

3.3
Process of Fusion in a Two-Component System

In a two-component system at the solid-liquid equilibrium one has to consider the forward and reverse reactions of both components. Here the reaction velocity of the partial reaction of component i is proportional to the concentration of this component in the phase, from which it comes. Is x_i^S the concentration (mol fraction) of the component i in the solid, and x_i^L in the liquid, then in analogy to Eqs. (3.7) and (3.8)

$$R_i^N = A_i^M \cdot G_i^M \cdot v_i^S \cdot N^S \cdot x_i^S \cdot \exp\left(-\frac{Q_i^M}{RT}\right) \qquad (3.13)$$

$$R_i^F = A_i^F \cdot G_i^F \cdot v_i^L \cdot N^L \cdot x_i^L \cdot \exp\left(-\frac{Q_i^L}{RT}\right) \qquad (3.14)$$

At equilibrium, $R_i^M = R_i^F$ must occur. Thus,

$$\frac{x_i^S \cdot N^S \cdot A_i^M \cdot G_i^M \cdot v_i^S}{x_i^L \cdot N^L \cdot A_i^F \cdot G_i^F \cdot v_i^L} = \exp\left[\frac{Q_i^M - Q_i^F}{RT}\right] \qquad (3.15)$$

Simplifying with $N^S \sim N^L$, $G_i^M \sim G_i^F$ and $v_i^S \sim v_i^L$ we have

$$\ln \frac{x_i^S \cdot A_i^M}{x_i^L \cdot A_i^S} = \frac{\Delta H_i}{RT} \qquad (3.16)$$

where ΔH_i is the enthalpy of fusion of component i.

For the temperature of fusion of the pure component we have, since $x_i^S = x_i^L = 1$ (see Eq. (3.11))

$$\ln \frac{A_i^M}{A_i^S} = \frac{\Delta H_i}{RT_{g,i}} \tag{3.17}$$

Subtracting (3.17) from (3.16)

$$\ln \frac{x_i^S}{x_i^L} = \frac{\Delta H_i}{R} \left[\frac{1}{T} - \frac{1}{T_{g,i}} \right] \tag{3.18}$$

It follows, that under the condition $x_i^S/x_i^L \neq 1$, equilibrium only occurs at a temperature $T \neq T_{g,i}$. Whether this temperature is higher or lower than $T_{g,i}$, depends if x_i^S/x_i^L is larger or smaller than 1.

For a two-component system A – B, transforming Eq. (3.18) for small concentrations of x_B^L and x_B^S one obtains

$$\frac{\Delta H_A}{R(T_{g,A})^2} = \frac{x_B^L}{\Delta T_A} - \frac{x_B^S}{\Delta T_A} \tag{3.19}$$

with

$$\Delta T_A = T_A - T \tag{3.20}$$

and

$$\ln (1-x) \approx - x \text{ for } x \ll 1 \tag{3.21}$$

T_A is the temperature of fusion, and ΔH_A is the enthalpy of fusion of the pure component A.

Through addition of a second component to a material A the partial reaction velocity of the two partial reactions R_A^M and R_A^F are slowed.

Figure 3.4 shows, that one has a melting point lowering if R_A^F is decreased more than R_A^M. This is the case, if $x_A^S > x_A^L$, which is to be expected in metals, when the atomic radii of the components differ greatly, and no tendency is present to form a compound with a separate lattice. The introduction of atoms too large or too small into a base lattice (A) is connected with an elastic distortion. This requires an expenditure of energy during insertion of B-atoms. If the B-atom leaves during fusion of the solid solution, the lattice distortion energy is liberated, and is, in addition to the vibrational energy of the atom, available as part of the total activation energy. Thus, at a given temperature T, a larger number of B atoms of the solid solution are in a position, to obtain the activation energy for a partial step of the fusion process, as would be the case without a lattice distortion. With a large dilution of the B atoms in the solid solution, this does not apply to the A atoms, though the local distortion is present in the vicinity of each B atom, but not in the vicinity of all, in large concentration present A atoms.

In the melt, where the long range order is absent, the difference in atomic radii does influence the energy relationships to such a great extent. Considering all four partial reactions it follows, that B atoms, present in small concentrations,

Fig. 3.4
Illustration of the freezing
point depression by addition
of a second substance to a
solvent

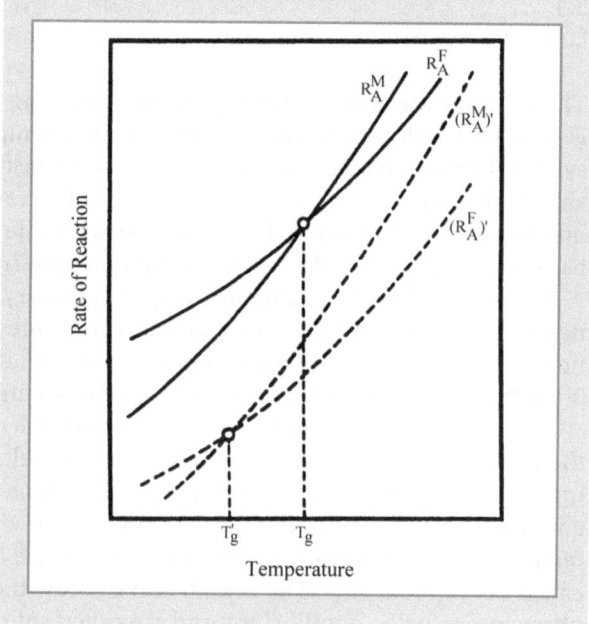

Fig 3.5
Possible melting equilibria
in a binary system at low
concentrations of the component B

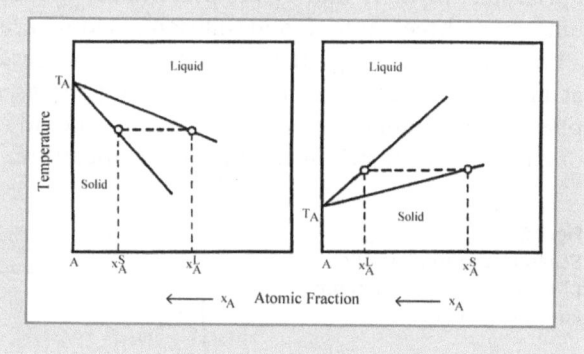

and causing lattice distortion energy, will enrich in the liquid phase. An equilibrium is attained, where $x_B^L > x_B^S$ and correspondingly $x_A^L < x_A^S$.

If during introduction of a B atom the interatomic interactions are not weakened, as in the case of lattice distortions, but for example enforced through charge transfer between different type of atoms, then we have $x_B^L < x_B^S$ and $x_A^L > x_A^S$. A melting point increase occurs. Both cases are represented in Fig. 3.5. The amount of melting point change is represented by Eq. (3.19). ΔT_A is determined by, in addition to $x_B^L \sim x_B^S$ by the enthalpy of fusion and the temperature of fusion of the principal component A. A detailed description follows in Chap. 6.

3.4
Eutectic System

The relationship contained in Eq. (3.19), describing the change in melting point is valid for small concentrations of the second component in a two-component system. Through changes in the energetic relationships at intermediate compositions, the relationship between x and T can be very complicated. For a certain number of binary systems, the relationships for dilute solutions already give the basic relationships for the fusion equilibria at intermediate compositions.

Let us consider a two-component system (binary system), where the components A and B have different crystal structures. Due to the structural differences no mixed crystal system (solid solution) can extend in the concentration range 0–100 %. Let us assume that the molten components are completely miscible.

Equation (3.19) is valid for the fusion equilibria of both components close to the temperatures of fusion T_A and T_B. If both melting points are lowered during addition of the other component, one obtains in the simplest case fusion equilibria represented in Fig. 3.6. The lines a and a′, which represent at a given temperature the concentration of the liquid phase in the fusion equilibrium, is called the liquidus line. The lines b and b′ show the concentration of the solid phase in the fusion equilibrium, and are called solidus lines. At an intermediate concentration the liquidus lines intersect. At the intersection both fusion equilibria, starting at T_A and T_B are present. For the identification of the equilibrium, where three phases (solid solution α, Liquid L, and solid solution β) coexist, a line is drawn through the concentrations of these three phases. A line, which at a given temperature connects coexisting phases, is called a tie line. The three phase equilibrium between two solids and a liquid is called eutectic equilibrium or eutectic for short. T_e is the eutectic temperature.

Fig. 3.6
Schematic presentation of phase equilibria in a simple eutectic system

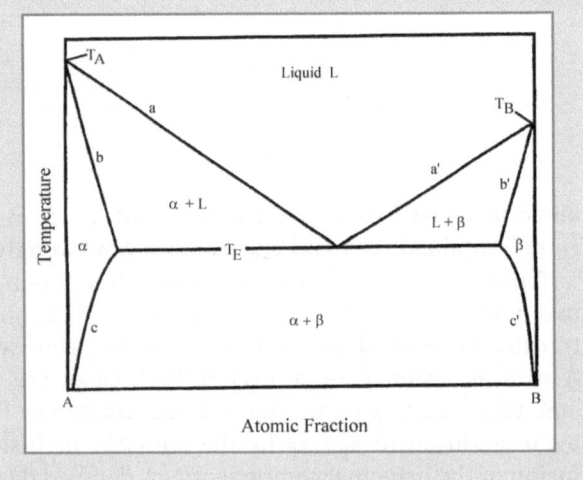

Below T_e everything is solid. At very low concentration of B we have a α-solid solution, at very low concentration of A we have a β-solid solution. The solubilities of A in B and of B in A are temperature dependent. Up to T_e, they increase with increasing temperature. The corresponding solubility lines are c and c'. They naturally intersect at T_e the solidus lines b and b', because at this temperature the solid solutions α and β have the same concentration as the solution equilibrium α-β as well as the fusion equilibria melt-α and melt-β.

The region of existence of the various phases is shown in Fig. 3.6.

3.5
Eutectic Real Systems

Figure 3.7 shows as an example the phase diagram lead-silver. The eutectic point is at high lead concentrations. Generally, the eutectic point is on the side of the lower melting component. Let us remark, that the liquidus curve, which starts at the melting point of silver, does not descend linearly, or with a constant curvature to the eutectic point, but shows an inflection point. This is caused by a special energetic behavior in the middle of the diagram, and cannot be deduced from Eq. (3.19) which is valid for the limiting case of small concentrations.

In addition, the solidus concentration on the silver side in the system does not increase with decreasing temperature, again due to individual energetic reasons, as this would be expected based on the limiting relationship Eq. (3.19). A "returning" or retrograde solidus curve is present. Finally, let us mention, that the solubility of silver in lead is so small, that it cannot be drawn in Fig. 3.7. At the eutectic temperature only 0.19 at.-% Ag can be dissolved in solid lead.

Fig. 3.7
Phase diagram Ag – Pb [2]

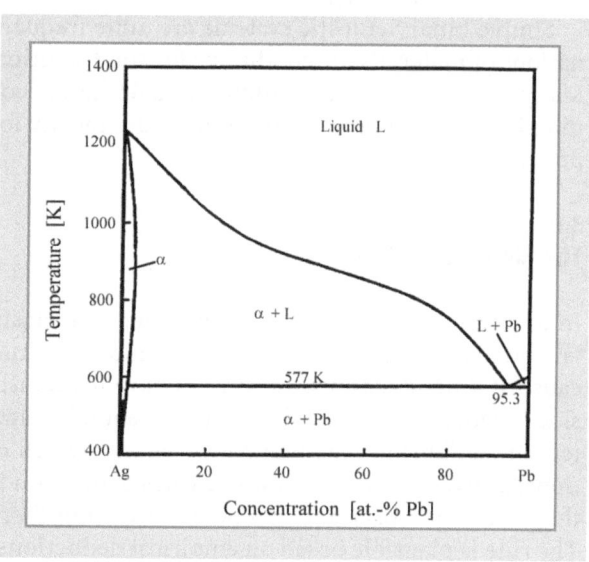

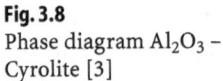

Fig. 3.8
Phase diagram Al_2O_3 –
Cyrolite [3]

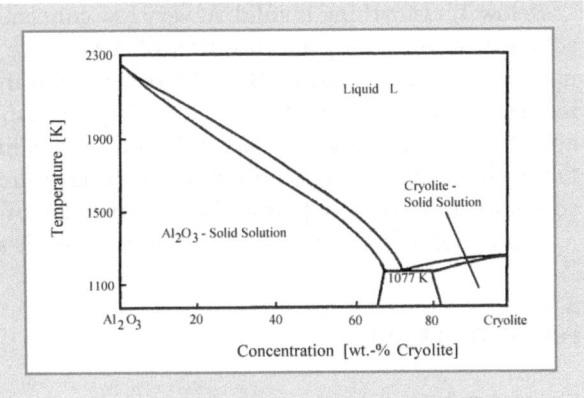

The fusion equilibria in the system silver-lead gained certain importance in winning silver from raw lead. In the "Pattinson" method, the very rich liquid lead alloy is cooled, as lead with very little silver crystallizes. The lead crystals are removed with a screen. Finally the eutectic melt, at 577 K (304 °C) remains in which the silver is enriched to 4.7 at.-%. This liquid, undergoes an oxidation process, where in a puddling-furnace the lead is transformed into litharge (PbO) and removed as a liquid. Liquid silver, with over 95 at.-% silver remains.

The eutectic melt consisting of an oxide and a salt is used during aluminum winning. Figure 3.8 shows the essential part of the system cryolite (Na_3AlF_6) – argillaceous earth (Al_2O_3). For the liquid electrolysis the lowest melting mixture is used, so that the Al_2O_3 dissolved in the cryolite can be electrolyzed < 1250 K. Other salts (such as CaF_2) are added to lower the melting point and increase the electrical conductivity.

Simple binary eutectic systems are quite frequent, because for many combinations of materials the conditions for solid solution formation are not fulfilled. Many more binary combinations of materials are completely miscible in the liquid phase. In metals, eutectic systems are mostly found with low melting components.

3.6
The Gibbs Phase Rule

In any system, in principle, any number of metastable phases are imaginable, and often can also be realized. If an equilibrium situation cannot be attained, because of kinetic reasons, one cannot have relationships between phases present side by side, in a system which is not in equilibrium. On the other hand, in a system in equilibrium, the number of phases, which coexist under certain conditions, is given by strict rules. The rules depend on the energetic background of the phase equilibria, and can be derived from thermodynamic considerations. The rule is plausible based on empirical deductions. It is the Gibbs phase rule.

The state of a substance is described by equations of state, the simplest of which is the one for ideal gases:

$$p \cdot V = R \cdot T \tag{3.22}$$

Here V is the molar volume. For real gases, liquids and solids similar equations of states apply, but they are much more complicated. To fix the state of a single phase system, one needs only two of the three state variable p, V and T. The third is fixed by the equation of state. According to J.W. Gibbs (1839–1903) the state variables which can be chosen freely (for example p and T; then in a single phase system V is fixed) are called "degrees of freedom of the equilibrium" or simply degrees of freedom F. Based on our experience, we find that in the presence of a single stable phase in a one-component system, the number of degrees of freedom is $F = 2$.

For a two phase equilibrium, for instance at the melting point on a pure substance, $F = 2$. The fusion temperature is a well defined quantity in a one-component system at a fixed pressure. It can only be changed, if the pressure is changed. In a one-component system with a two-phase equilibrium one can only dispose freely of one state variable. The system has only one degree of freedom. This is valid for all two-phase equilibria in a one-component system (vaporization, solidification, transformation). In general the rule is

$$P + F = 3 \tag{3.23}$$

Where P is the number of phases, and F the number of freedoms.

In a two-component system, melting at constant pressure does not occur at constant temperature. Liquid and solid can be in equilibrium at different temperatures, though the concentration in the phases is different. The concentration x appears as another state variable besides p, T and V. Even with p constant, T and x can be changed by finite amounts, without one of the two phases disappearing or a new one appearing. T and p can be fixed simultaneously. Of course, then the concentrations in the liquid and solid are fixed. For a two-component system the rule, which is expressed by Eq. (3.23) for a one-component system, has to be expanded. For a two-component system it is

$$P + F = 4 \tag{3.24}$$

The difference between Eqs. (3.23) and (3.24) is conditioned by the fact, that the number of components which constitute the system has changed by 1. One can write in general

$$P + F = B + 2 \tag{3.25}$$

where B is the number of independent components.

The particular advantage of the phase rule is in the possibility, to check very confusing phase diagrams for consistency. It provides an indispensable method of control in the experimental determination of phase equilibria.

3.7
Application of the Phase Rule

For further illustration of the Gibbs phase rule, and to introduce further concepts, the phase diagram cadmium-zinc, shown in Fig. 3.9, is considered. The following cases are to be distinguished:

1. Melting point of zinc

   $P = 2$; $B = 1$; $F = 0$ at $p =$ constant

At constant pressure, for a pure substance the melting point is a substance property. Invariant equilibrium.

2. Liquidus line

   $P = 2$; $B = 2$; $F = 1$ at $p =$ constant

When T is fixed, the concentration of the phases is given. Univariant equilibrium.

3. Single phase melting

   $P = 2$; $B = 2$; $F = 2$ at $p =$ constant

Both temperature and concentration can be changed by limited amounts, without a change in the number of phases. Bivariant equilibrium.

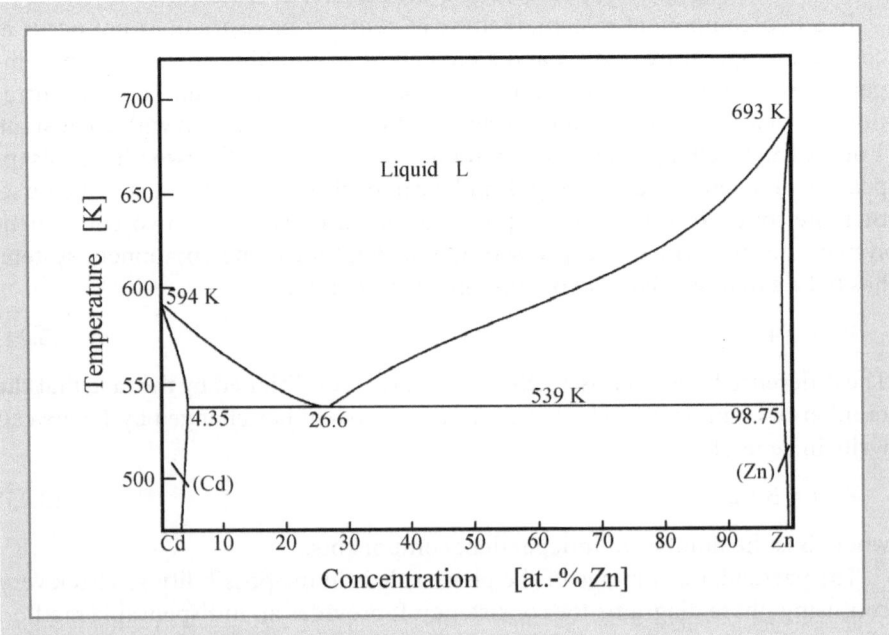

Fig. 3.9 Phase diagram Cd – Zn [4]

4. Solid-liquid equilibrium

$P = 2; B = 2; F = 1$ at p = constant

Solid solution of zinc and liquid of a certain concentration are in equilibrium. Raising the temperature changes the concentrations of the solid solution and the melt. Univariant equilibrium.

5. Eutectic equilibrium

$P = 3; B = 2; F = 0$ at p = constant

At the eutectic point zinc solid solution, cadmium solid solution and liquid, each with its fixed composition are in equilibrium. Invariant equilibrium.

6. Solubility equilibrium in the solid state

$P = 2; B = 2; F = 1$ at p = constant

At a given temperature zinc solid solution and cadmium solid solution each with its fixed composition are in equilibrium. This is similar to case 4: Univariant equilibrium.

3.8
The Lever Rule

From a phase diagram of a binary system one can obtain immediately the type of the coexisting phases and their concentration. Based on a simple relationship, one can also give the ratio of the amounts of the phases in equilibrium.

At a fixed temperature let the mixture of two substances A and B with a total concentration x_B consist of coexistent phases α and β (see Fig. 3.10). Also assume

Fig. 3.10
Lever rule to determine the ratio of masses of coexisting phases in a binary system

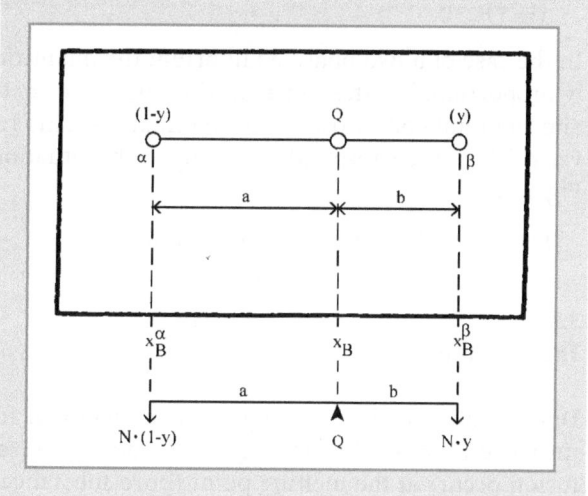

N the number of atoms (or molecules) in the total mixture Q
y the fraction of the N atoms, which are contained in β
(1-y) the fraction of the N atoms, which are contained in α.
We have

$$N * (1-y) + N * y = N \qquad (3.26)$$

We take

n_B^Q the number of B atoms in the alloy Q
n_B^α the number of B atoms in the phase α
n_B^β the number of B atoms in the phase β

The mol fractions of the total mixture and in the phases in equilibrium are:

$$x_B = \frac{n_B^Q}{N}; \quad x_B^\alpha = \frac{n_B^\alpha}{N(1-y)}; \quad x_B^\beta = \frac{n_B^\beta}{N \cdot y} \qquad (3.27)$$

Furthermore

$$n_B^Q = n_B^\alpha + n_B^\beta \qquad (3.28)$$

With Eq. (3.27) it follows

$$x_B = x_B^\alpha (1-y) + x_B^\beta * y \qquad (3.29)$$

Transformation gives

$$y = \frac{x_B - x_B^\alpha}{x_B^\beta - x_B^\alpha} \qquad (3.30)$$

If, according to Fig. 3.10 we set $(x_B - x_B^\alpha) = a$ and $(x_B^\beta - x_B) = b$, after further transformation

$$\frac{y}{(1-y)} = \frac{a}{b} \qquad (3.31)$$

In the case of a two phase equilibrium the quantities of the phases are inversely proportional to the concentration difference between the phase in question and the total concentration of the mixture. This relationship is called the "lever rule", as it corresponds formally to the situation in a mechanical lever (see Fig. 3.10)

Load $\cdot$ Loadlever = Force $\cdot$ Forcelever.

3.9
Thermal Analysis

During solidification of a liquid the enthalpy of fusion is liberated. If no disturbances, caused by non-equilibria occur, the delivery of the enthalpy of fusion occurs at the melting point (pure substance) or in the liquidus-solidus-

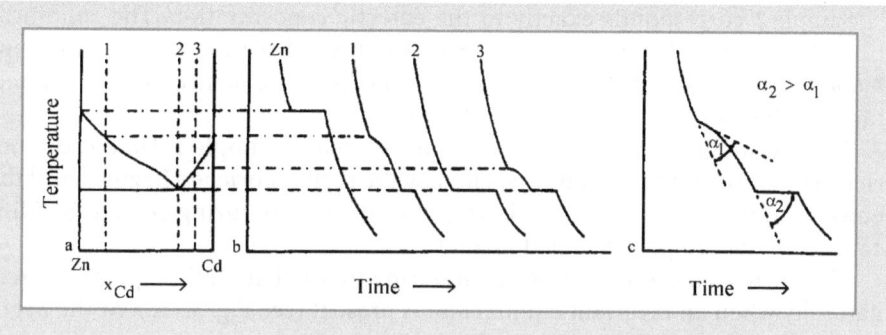

Fig. 3.11 Cooling curves of a simple thermal analyses of Cadmium-Zinc alloys. **a** Schematic phase diagram of the Cd – Zn system (solubility in the solid state is neglected), **b** cooling curves, **c** explanation of the sharpness of kinks in a cooling curve

region (multi-component system). This fact is used in "thermal analysis" to study fusion equilibria. In the simplest situation, the substance is melted in a suitable container, and then cooled. The temperature of the substance is registered as a function of time during cooling. The results are illustrated for the system cadmium-zinc (see Fig. 3.11).

The specimen to be investigated is cooled from a temperature far above the temperature of fusion. Normally, this occurs according to Newton's cooling law, if the system is left alone. If the specimen is pure zinc, when the solidification point of this metal is reached, crystallization occurs. The enthalpy of fusion is liberated, with a velocity dictated by the experimental heat transfer conditions, to keep the temperature constant. Liquid and solid zinc can only coexist at the temperature of fusion of zinc. If the sample temperature is registered as a function of time, an "arrest" shows up during solidification at the melting point (see Fig. 3.11b). The solidification process represents here an invariant equilibrium. Only when the whole sample is solidified, does the cooling continue.

With sample 1 (see Fig. 3.11a, b) solidification starts when the liquidus line is reached. The solidification equilibrium is, however, not bound to a constant temperature. One has an univariant equilibrium. The solid crystallizing from the liquid is almost pure zinc. Thus, the liquids content of this component decreases. As a consequence, the liquidus temperature decreases. During crystallization enthalpy is liberated, which slows the cooling rate of the sample, but an "arrest" does not occur in an univariant equilibrium as long as a two-phase equilibrium is present. Finally the liquid reaches the eutectic point, a result of the crystallization of zinc rich crystals. Now in addition to the crystallization of zinc rich crystals, the crystallization of cadmium rich crystals begins. Three phases are in equilibrium now. This corresponds to an invariant equilibrium. Accordingly the eutectic crystallization occurs at constant temperature. Zinc-rich crystals and cadmium-rich crystals precipitate side by side in such a ratio, that no change in concentration of the liquid takes place. At the temperature of the eutectic an "arrest" occurs.

Sample 2 corresponds exactly to the eutectic concentration. The solidification of the whole alloy occurs in an invariant three phase equilibrium at the eutectic temperature, with cadmium rich crystals and zinc rich crystals precipitating simultaneously.

The thermogram of sample 3 is similar to that of sample 1. Here cadmium rich crystals precipitate primarily. Thereby the melt is enriched in zinc, until the eutectic concentration is reached. Only after the now occuring eutectic solidification can the sample be cooled further.

The break in a cooling curve taken during thermal analysis appears clearest naturally when an invariant equilibrium is present (see Fig. 3.11c). At the begin of an univariant solidification equilibrium the change of direction of the cooling curve is basically smaller, because crystallization can only occur while the sample is cooled further. The angle is also dependent on the number of moles of the substance, which solidify per degree of temperature lowering, that is on the slope of the liquidus and solidus lines in the temperature range close to the liquidus temperature of the sample. Sometimes it may become difficult, to determine the liquidus temperature from a small change of the cooling rate.

The sensitivity can be increased considerably, if, together with the experimental sample a comparison sample is cooled under the same thermal conditions, but which, in the temperature range of interest, does not show a thermal effect (see Fig. 3.12). If instead of the sample temperature the temperature difference ΔT between sample and reference material is registered, a much sharper change of slope occurs at the beginning of the solidification, than with a simple thermal analysis (see Fig. 3.13b). This method is called differential thermal analysis (DTA). It has almost completely replaced the simple thermal analysis because of its advantages. To obtain the temperature of the change in slope in the ΔT-time curve the temperature of the sample must be registered separately. To simplify the evaluation, the modern commercial DTA equipment registers not ΔT as a function of time, but as a function of the sample temperature at a given cooling or heating rate.

If the cooling condition remains the same while investigating several samples with different concentrations, the eutectic arrest time t and the surface F (see Fig. 3.13), caused by the eutectic effect under ΔT-time curve, are a measure of the fraction of the total melt, which solidified eutectically. When the concentration of the sample approaches the eutectic concentration x_e, t and F increase linearly with concentration and reach their maximum value at the eutectic (see Fig. 3.14).

For a rapid orientation about an unknown eutectic system it is possible to obtain information about the concentration of the eutectic liquid through determination of arrest times t or surfaces F for a few alloys (for example a, a', b, b', Fig. 3.14). If the mutual solubility of the components in the solid state is small, cooling curves of two samples (a and b) are sufficient, because in this case the straight line a-a' passes approximately thru a", and the straight line b-b' passes approximately through b". Trivially, at the solidification of the pure components no eutectic effect appears.

Fig. 3.12
Example of an experimental device for differential thermal analyses (DTA). A crucible with sample, B crucible with inert metal, H electrical furnace, M massive inert metallic block, I_1, I_2 inlet for inert gas, Th_1, Th_2 thermal elements, Thermoelement Th_1 (a measure of temperature of the sample A), ΔE difference of voltages of Th_1 and Th_2 (a measure of the temperature difference ΔT between sample and reference sample) Ar protecting gas, E voltage

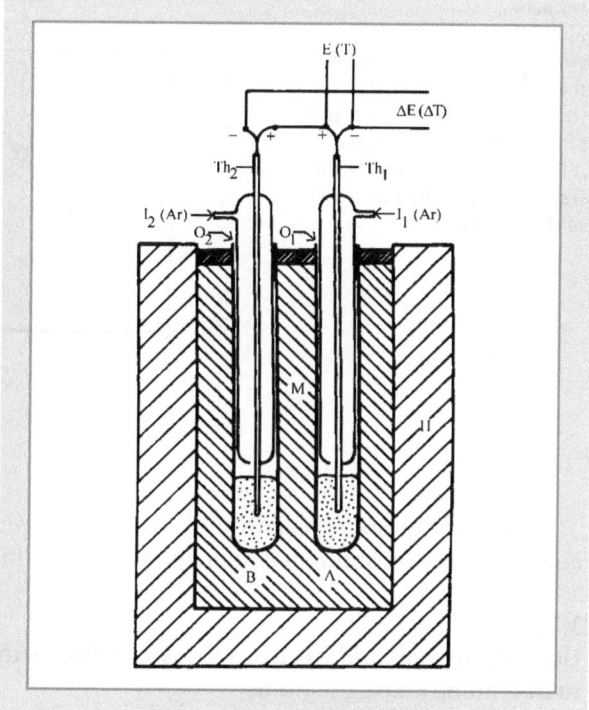

Fig. 3.13
Explanation of the recorded curve of differential thermal analyses. **a** Cooling curve of a sample A and the inert reference sample B (see Fig. 3.12) for the case of nonvariant freezing equilibria starting at T_0, **b** temperature difference between A and B as a function of time for the case of freezing shown in **a**

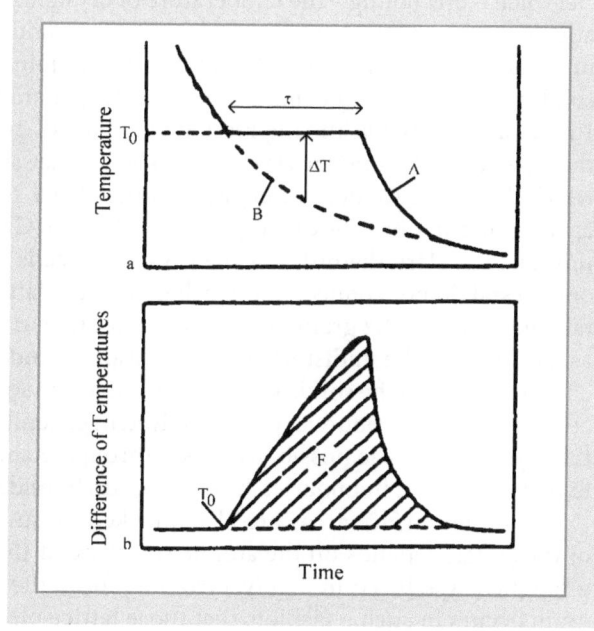

Fig. 3.14
Method to determine the
eutectic concentration from
the time of duration of
thermal arrest in the thermal
analysis or the area under
ΔT-time curve caused by the
freezing of the eutectic part
of the sample

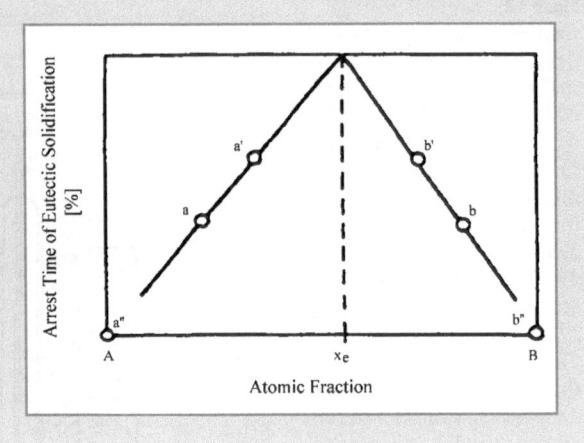

Thermal analysis, especially in the different forms of the differential thermal analysis, is the most important method to obtain fusion equilibria.

3.10
Light Microscopic and Electron Microscopic Research Methods to Determine Phase Diagrams

The crystallization of a melt of a pure substance begins with the formation of nuclei when – on cooling – the temperature of crystallization is reached. This nucleation occurs in general in various places in the liquid. The nuclei grow to crystals, until they reach each other. The single crystals, joined to form a polycrystalline solid, have in general unequal boundaries, are generally small (diameter less than 1 mm), and often give the impression of a mass of grains sticking together, thus, they are called "grains" or crystallites. The contact surfaces between two grains, which due to the random orientation of the nuclei, touch each other with difference in orientation, are called grain boundaries. Grain boundaries are surface mismatches. Here the order of the crystal lattice is disturbed. Lattice planes of one crystal do not connect undisturbed with the lattice planes of the other crystal. The atoms in the grain boundary are not tied as perfectly to their neighbors as the atoms in the undisturbed lattice. Grain boundaries are places of increased energy content in the solid. At a chemical attack by solvents, liquid or gaseous, the atoms on the grain boundary are dissolved in general faster, than in other parts of the crystal. Thus, the possibility exists, through etching, using suitable etchants, to make the grain boundaries "visible" (grain boundary etching, see Fig. 3.15).

The other surface areas are also attacked by suitable etchants. The severity of the attack depends on the atomic structure of the surface being attacked. In a sample crystallized in a polycrystalline form, the removal of the surface of a grain occurs in such a fashion, that those lattice planes are removed the fastest,

Fig. 3.15

Explanation of differences in brightness of grains due to etching of a polycrystalline sample (---- etching of grain boundaries; hatched area: etching of grain surface)

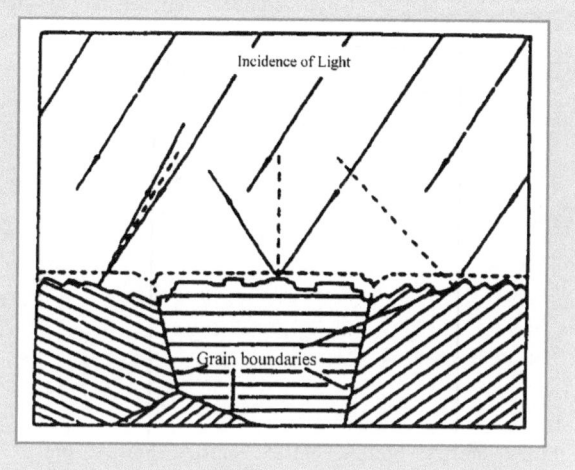

for which the solution rate is the highest. This leads to a stepwise roughing of the surface (see Fig. 3.15). Illumination of this surface with parallel light, gives reflections in correspondingly different directions, so that certain regions appear light to an observer, other as dark. This "grain boundary etching" can serve to make the crystal texture visible.

Often the removal of the surface yields a reaction of the atoms dissolved with certain parts of the etchant, giving possibly insoluble products, which precipitate in different fashion on the various crystal surfaces. This „precipitation etching" also creates different brightness of the grains. We must remark, that there are a large number of chemical and physical methods to make a structure "visible".

On zinc-covered iron objects, atmospheric attack on the zinc layer, makes it possible to see the structure with bare eyes. The grains of most metals in use and of most industrial nonmetallic materials are so small, that they can only be distinguished by the light microscope. This requires grinding of the sample to obtain a flat surface, then polish it to a bright surface, and finally etching. Figure 3.16 shows the structure of pure polycrystalline zinc, obtained using a light microscope. With metals and other non-transparent materials incident light has to be used, with a lot of minerals through light microscopy is possible. Non cubic substances deflect more or less the plane of polarized light, both with through beam, or in the case of non-transparent metals with incident light. Depending on the orientation, which the crystallites of a polished, unetched surface have towards the incident light beam, they appear lighter or darker under polarized light. This effect is illustrated in Fig. 3.16 using zinc, which crystallizes in a be close-packed hexagonal structure. This fact can, as with the etching technique, used to investigate structures.

The difference in attack, which the various planes of a crystal undergo with an etching agent, is even more pronounced with crystallites with different con-

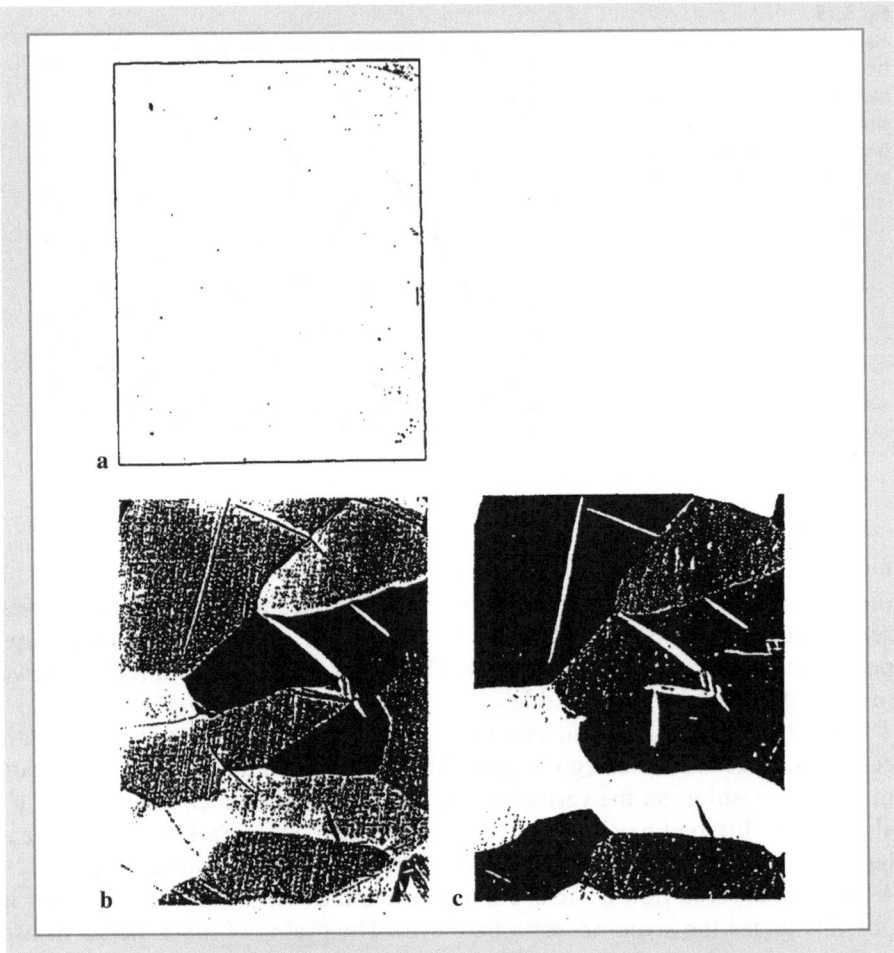

Fig. 3.16 Photos of polycrystalline zinc taken by light microscopy (source: K. Polczer), magnification: 64-times. **a** As-polished, **b** precipitation etching using a solution of potassium disulfate and sodium thiosulfate, **c** as-polished, polarized light

centrations and even more with different crystal structures. As a rule it is possible, in a simple manner, to observe the existence of different solid phases in a specimen after suitable etching. This method is used extensively to control the results of the thermal analysis, and to obtain information about predictions provided by the thermoanalytical method. As an example the structure of samples from the system Cd – Zn are reproduced in Fig. 3.17.

The solidification of an alloy with 48 at.-% zinc begins at 580 K (see Fig. 3.9). First nuclei of zinc alloys form, which with decreasing temperature grow to crystallites, until finally the eutectic temperature is reached. Now zinc and cadmium crystallize side by side (invariant equilibrium). As a rule this crystalliza-

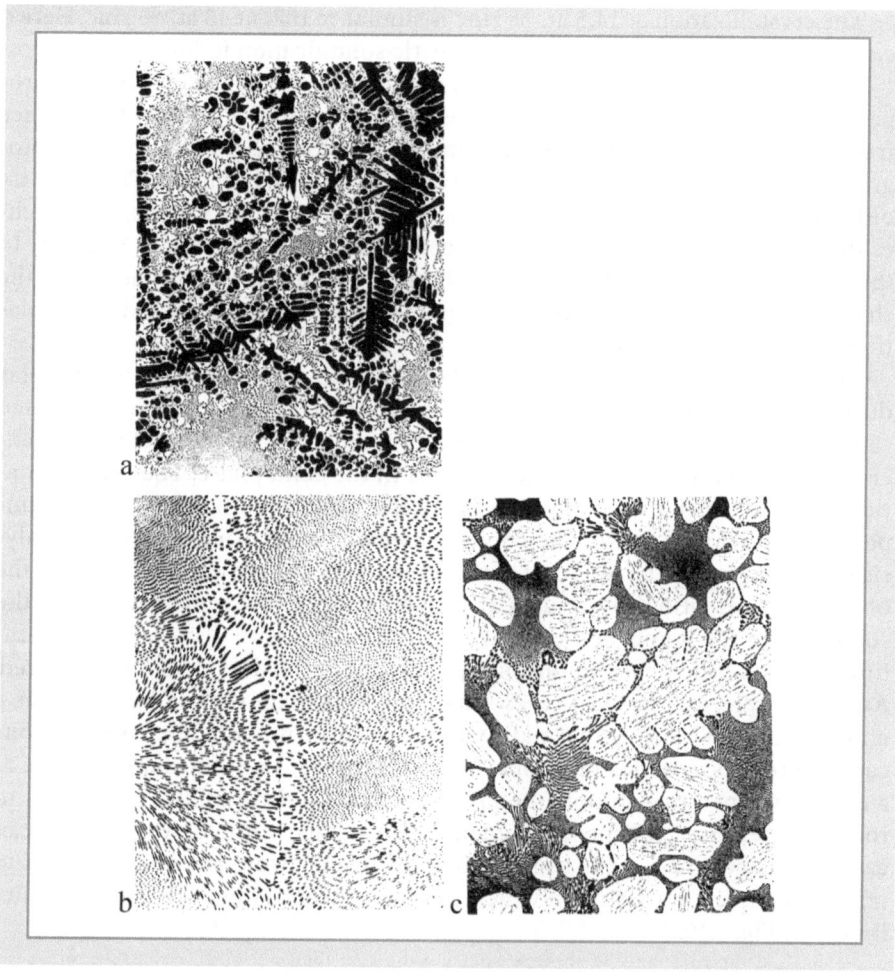

Fig. 3.17 Microstructure of cadmium-zinc alloys. **a** 48 at.-% Zn, primary zinc (black dendrites), rest eutectic; light microscopy; magnification: 40-times (source: S. Schmidt), **b** 26.5 at.-% Zn, eutectic; light microscopy; magnification: 300-times (source: E. Scheil) **c** 14.5 at.-% Zn, primary: Cd (bright areas), rest eutectic; light microscopy; magnification: 75-times (source: S. Schmidt)

tion leads to a texture, in which lamellae of zinc alloy and cadmium alloy alternate, and fills the room between the primarily formed zinc crystals.

At the eutectic composition no primary crystals of any component are present. The whole sample consists of a lamellar texture, which occurs during the eutectic crystallization, and is called "eutectic". The eutectic crystallization starts at different places in the sample and yields regions of more or less similar orientation of the lamellae. These regions finally meet each other. The boundaries are called eutectic grain boundaries.

The crystallization at 14.5 at.-% zinc is similar to that at 48 at.-% zinc. Here a cadmium alloy crystallizes first. The eutectic solidification follows.

The surface fractions of the primary crystallites and of the eutectic are proportional to the volume fraction of these structures in the sample. The surface fractions can be determined using a planimeter on a corresponding microphotograph, or directly with commercial optical-electronic equipment. The quantity of the eutectic depends linearly on the concentration of the sample, and reaches the maximum (100 %, see Fig. 3.18) at the eutectic composition. Similarly to the eutectic arrest time, the eutectic concentration can be determined from the share of the eutectic of the whole specimen in different samples, and extrapolation (to 100 % eutectic share).

The transformation solid-liquid takes place as a rule without much delay during the recording of cooling curves in a thermal analysis, and is terminated rapidly in the measure that the solidification enthalpy is removed; this is generally not the case during transformations in the solid phase. The transformation reactions often starts only at temperatures way below the equilibrium temperature, and do not finish with cooling rates which are usual in thermal analysis (0.1–10 K/min). The mobility of the atoms, which is the precondition for the phase transformation to occur, is at a given temperature much smaller in the solid than in the liquid, and decreases exponentially with decreasing temperature. To reach solid phase equilibrium, sometimes considerable time is required. As an example, it is in most cases impossible, to obtain the solubility curves c and c′ (see Fig. 3.6) using thermal analysis. However, microscopic investigations can give results. A series of samples, which were held a sufficient long time at a certain temperature to obtain equilibrium, are cooled rapidly (quenched) to room temperature, polished, etched and investigated with a microscope. It can easily be determined, if the sample contains one or two phases, which helps to answer the question, between which sample concentrations lays the solubility limit (see Fig. 3.19a).

Fig. 3.18
Part ε of eutectic in a simple eutectic system as a function of concentration

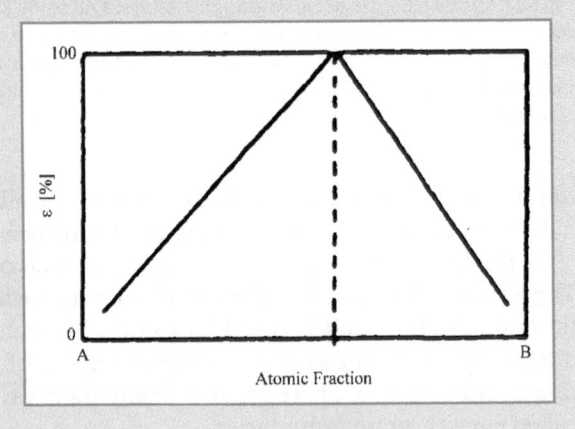

Fig. 3.19
a Determination of solubility equilibria by qualitative analysis of the microstructure ○ = one-phase alloy; ◐ = two-phase alloy,
b Determination of solubility limits by quantitative analysis of the microstructure

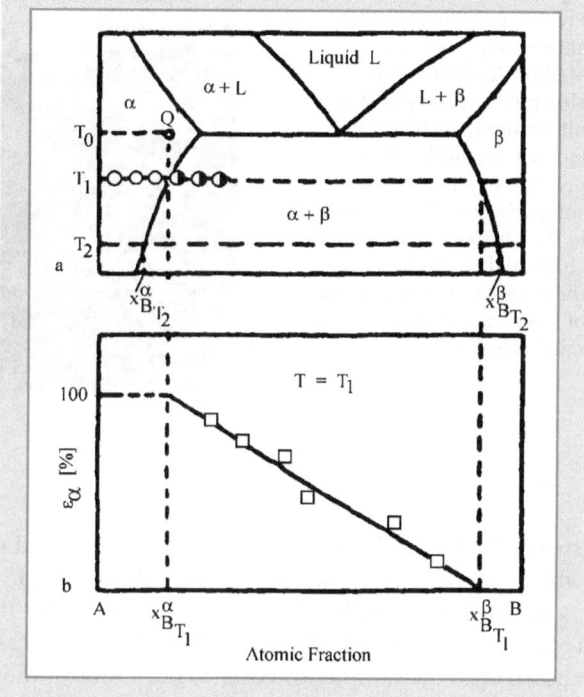

A second possibility exists, based on a quantitative determination of the percentage of one-phase particles in the volume of the alloy. The value of ε is almost equal to the fraction of one-phase grains of the microstructure visible on the surface of the sample. In Fig. 3.19b ε is plotted as a function of concentration. There results a straight line, which can be extrapolated up to the values 0 % ε and 100 % ε. Thus, the solubilitiy limits x_B^α and x_B^β were gained at the temperature T_1, at which the alloy has been equilibrated.

The grain structure of solids can be so fine, that the resolution of the light microscope is insufficient. In this case, useful results can be obtained through electron microscopic methods. Examples of this are precipitation structures, when a solid solution (one phase) Q with the concentration x_B^α (see Fig. 3.19), is rapidly cooled from a temperature T_0, at which it is stable, to a temperature T_2, where it should decompose into phases α + β. The precipitation of phase β from the α solution can occur very slowly, start at extremely large number of sites in the starting structure, and thus, leads to a very fine texture. For an electron microscopic investigation the sample is thinned down, so that an electron beam can pass through it. Figure 3.20 shows an example for a precipitated structure using transmission electron microscopy.

Also the surface scanning electron microscope can be used for structure investigations. This is an equipment, operating with incident light, and having a

Fig. 3.20
Precipitation in a Co-Cu
alloy with 2 at.-% Co.
Method: Transmission
electron microscopy; Mag-
nification: 1,300,000-times
(source: M. Rühle). The
precipitations consist of Co
solid solution. They induced
in the Cu-rich matrix me-
chanical distortions being
the reason of the formation
of a contrast in the electron
microscopic image

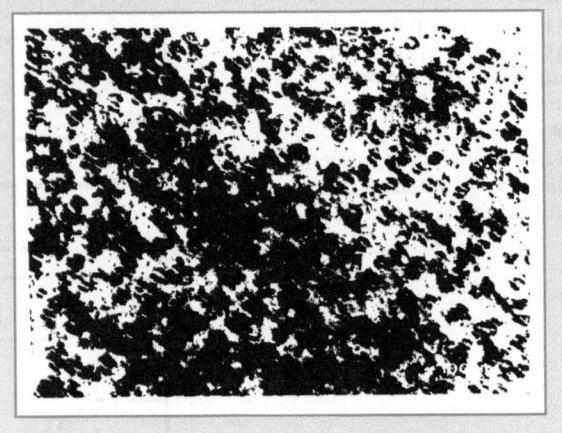

strong depth resolution, and thus, very well suited to investigate surface reliefs, which can occur for example during solidification.

Finally, let us mention the electron microprobe. It is a combination of an electron microscope and an X-ray tube. A highly accelerated electron beam is focussed onto the sample, and excites the atoms of the sample locally to emit X-ray radiation. The electron beam can be made to scan the sample surface. The X-ray radiation is decomposed spectrometrically into the various wavelengths characteristic of the chemical elements, and used to form the image. The resulting image gives the position of the atoms emitting at the specific wavelength in the structure. Not only can a relative distribution of the various atoms in the different grains be determined, but also a quantitative determination of the concentration of the components in the structure obtained.

3.11
X-Ray Diffraction Methods

The structure of a solid can be determined with respect to the form of the various phases using X-ray diffraction methods. This is especially useful in the determination of phase diagrams, because in the solid, in addition to solid solutions of the components, other phases can appear.

In addition, solubility equilibria can be easily determined. By adding a second component, which leads to solid solution formation, the lattice parameter of a substance will in general be changed, because the different atoms or molecules in the solid solution are in the rule of different size. This is represented in Fig. 3.21 for the example of the copper-silver system. The lattice parameter of silver decreases during incorporation of the smaller copper atom, until the solubility limit is reached at a temperature, at which, after long enough tempering,

Fig. 3.21
Determination of solubility
equilibria by lattice spacing
measurements

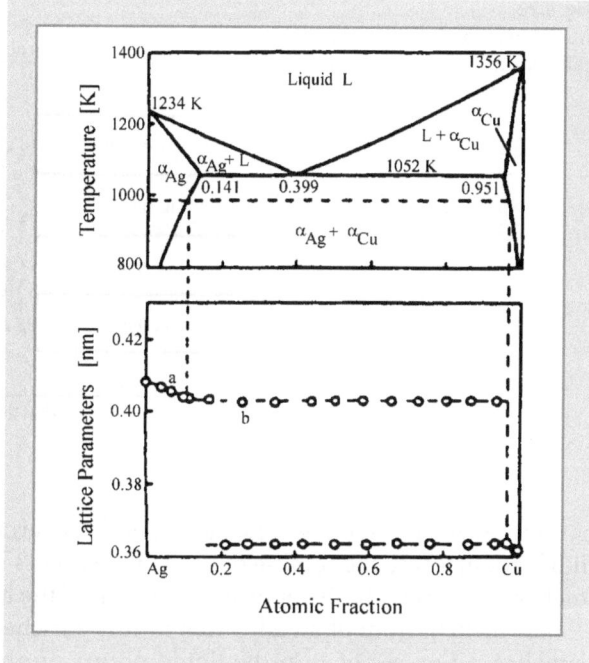

equilibrium is attained. In the two phase equilibrium between silver and copper
solid solutions, the lattice parameters of both solutions are independent of the
overall composition. The intersection of lines a and b gives the solubility limit.
The same applies to the copper side of the system.

3.12
Other Experimental Methods

In principle, all properties, which indicate changes of phases or structures, can
be applied to the determination of the structure of solids and fusion equilibria.
In reality, the following physical properties were used: electrical resistance,
magnetic properties, optical properties, hardness, thermal expansion, and some
other properties. Not all procedures are equally suitable for the study of all possible phase equilibria. For special cases some very complex research methods
have been developed.

3.13
Eutectic Crystallization

From a melt, with the eutectic composition, both solid phases being part of the
invariant equilibrium, crystallize simultaneously. The existence of the lamellar

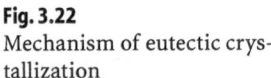
Fig. 3.22
Mechanism of eutectic crystallization

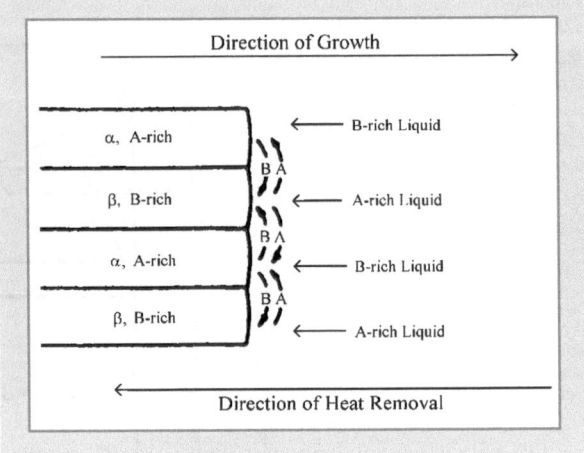

structure which often appears, has kinetic backgrounds. Under single axis solidification, where the cooling of the sample occurs in one direction only, the lamellar front grows in the opposite direction of the heat removal (see Fig. 3.22).

The melt, in front of an alloy rich in B, is enriched in A. In front of an A-rich lamellae, an excess of B in the liquid occurs. In the melt, directly next to the growth front of the lamellae, the different atoms are carried to the neighboring lamellae through diffusion, created by the concentration gradient. The excess of B atoms in the liquid, arisen in front of an A rich lamellae is decreased through diffusion to neighboring lamellae rich in B, where the liquid was depleted due to the growth of B rich lamellae. In return the excess in A atoms in front of the B rich lamellae are transported through diffusion into the region in front of A rich lamellae, where due to their growth a deficit in A atom exists. Thus, only a very short path is needed, to distribute the two types of atoms, which are present homogeneously in the liquid, to the two solid phases. The diffusion in the liquid, which represent the mechanism of transport in the liquid to effect the solidification reaction, becomes very effective because of the shortening of the diffusion path, due to the favourable arrangement of the solid phases. The rate of solidification is faster, the faster the heat removal is. At large heat removal rates, the distance between lamellae becomes small. Thus, the separation of the two components, controlled by diffusion, into the two phases can occur faster. However, the total surface area per mole created becomes larger, which represents a loss in Gibbs energy for the driving force during solidification. Under the given thermal and material conditions a certain lamellae distance occurs. According to Chadwick

$$l \approx R_E^{-\frac{1}{2}} \tag{3.32}$$

where l is the distance between lamellae and R_E the velocity of growth of the eutectic.

If one of the two phases appearing at the eutectic crystallization is transparent, the plate-like texture of the solid can immediately be recognized. An impressive example is shown in Fig. 3.23, which represents the eutectic in the system UO_2 – W. UO_2 is transparent in thin layers.

The lamellar structure degenerates at very low solidification rates. Local enlargements, discontinuity of the lamellar structure and rounding of the end of the lamellae occur (see Fig. 3.24a). Through the reduction of the total phase boundary surface the grain boundary energy of the two phase system is decreased.

Another peculiarity occurs, if in addition to the two components making up the eutectic, other materials are dissolved in the liquid. This can lead to a rod-like eutectic, where one phase is the basic matrix, which encloses the other component in form of rods (see Fig. 3.24).

The advantage of the lamellar or rod-like structure at eutectic crystallizations is in addition of the short diffusion path, that no repetitive nucleation has to occur. The lamellae or rods grow continuously. This can easily be realized, if no special individual crystal morphology exists. On the other hand, if one of the phases, formed during eutectic crystallization, has a crystal structure with a strong growth anisotropy, the ordered formation of the lamellar structure is no longer certain. As an example, the eutectic of the systems Al – Si and Pb – Sb are shown in Fig. 3.25.

Fig. 3.23
Lamellar eutectic in the UO_2 – W system. Light microscopy; dark field image in reflected light; magnification: 1000-times. The W-plates extend up to the polished surface. The second component, UO_2, is transparent (G.W. Clark, A.T. Chapman, and J.C. Gower, Metals and Ceramic Devision, Oak Ridge, National Laboratory; sponsored by the Materials Science Devision, U.S. Departement Carbide Corporation)
The color intensity between the plates is the result of the selective light absorption of UO_2. It deepens with the increasing thickness of UO_2. Where the W plates reach the surface, interference lines appear

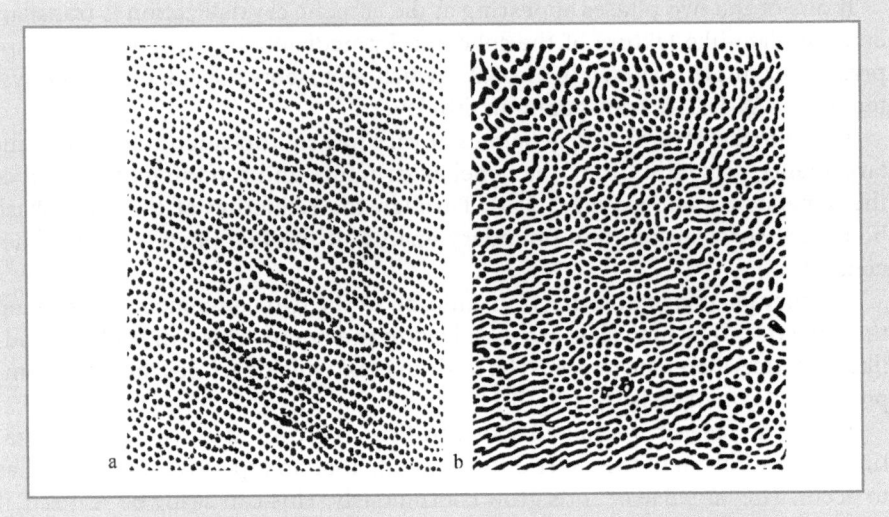

Fig. 3.24 Special shapes of eutectic structure. **a** Slowly solidified eutectic in the Co – Sn system with 20 at.-% Sn (source: M. Ellner); constituents of the eutectic structure: α-Co solid solution (dark) and γ-phase (bright); light microscopy; magnification: 1500-times, **b** rod-like eutectic in the Cd – Zn system; 26.5 at.-% Zn; light microscopy; magnification: 600-times

The structure of such an "abnormal" eutectic can also be influenced by the different nucleation probability for the crystallization of the two phases. For example if during cooling of the eutectic liquid L_E (Fig. 3.25c) below the eutectic temperature T_E only nucleation of the a-phase but not of the b-phase occurs, then the concentration of the liquid changes with the subsequent crystallization from a along line a_1, which is the extension of the liquidus line. Only when a sufficiently large undercooling occurs, for example T_U, does the crystallization of b start. The result of such a crystallization path is another ratio of a and b, than would be expected from a normal eutectic solidification.

Growth anisotropy can cause a spiral eutectic, which, however, is quite rare. Figure 3.26 reproduces a spiral eutectic in the system Zn – Mg.

A "degenerate" eutectic occurs, when the eutectic concentration lies close to one of the components, for example A. During primary crystallization of a liquid with intermediate concentration, a large quantity of B rich alloys can precipitate before the eutectic point is reached. The leftover eutectic liquid fills the small places between primary crystals.

Nuclei of the A rich phase are formed, which is unavoidable for an eutectic crystallization. The formation of nuclei of the B rich phase can be omitted. The precipitation of a B rich phase, occuring in the invariant equilibrium besides of the precipitation of A rich phase, takes place onto the surface on primary crystals, because no long diffusion paths are needed, the regions of the melt being narrow. As an example Fig. 3.27 represents such a degenerate eutectic.

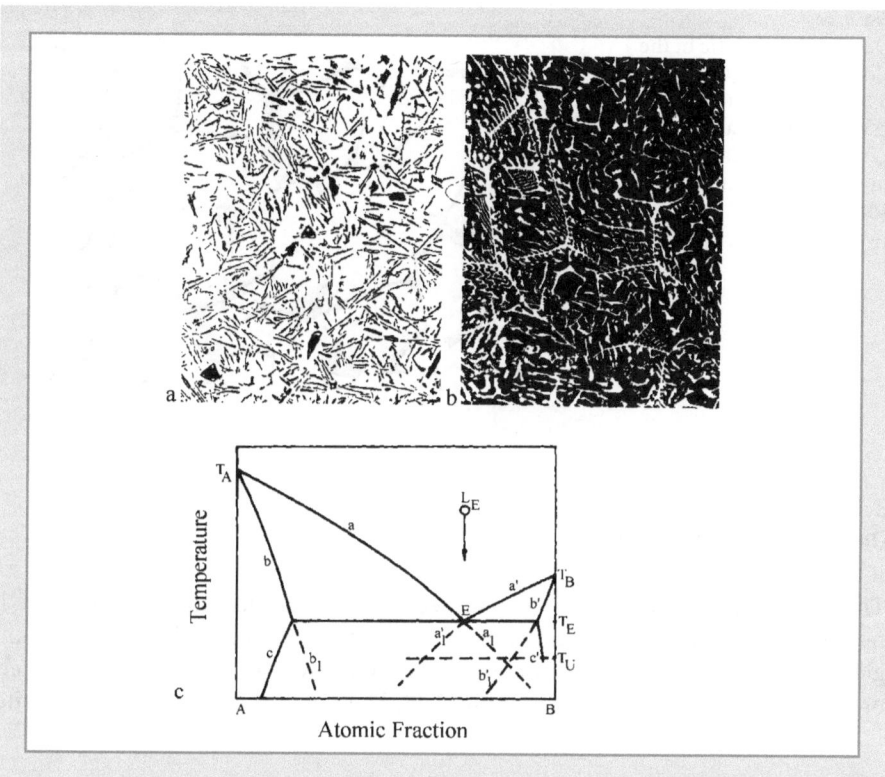

Fig. 3.25 Anomalous eutectics. **a** Al – Si system (source: Dujardin); 11.3 at.-% Si; light microscopy, **b** Pb – Sb system (source: Spengler); 17.5 at.-% Sb; light microscopy, **c** explanation of the course of solidification of an anomalous eutectic

Fig. 3.26
Spiral eutectic in the
Mg – Zn system (source:
E. Scheil). 7.7 at.-% Mg; light
microscopy; magnification:
600-times

Fig. 3.27
Degenerate eutectic in the
Cd – Sb system (Gützler).
5.5 at.-% Cd, primary
crystals, bright: antimony,
grey: Cd – Sb phase, light
microscopy, magnification:
150-times

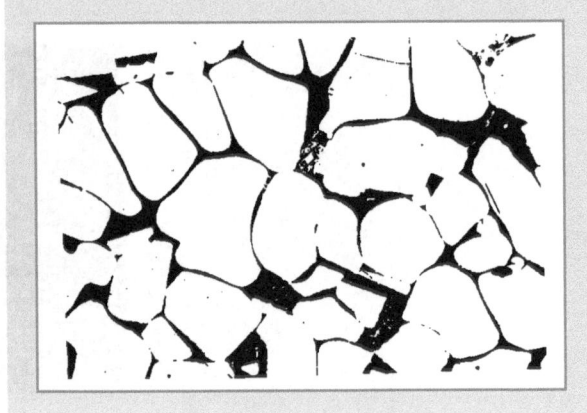

3.14
Dendritic Crystallization

The form of the crystallites in a solidified, polycrystalline two phase mixture is not always polyhedric. Sometimes only skeleton like crystals with intruding angles appear as primary precipitates, called dendrites, which can be also finely branched. An example is shown in Fig. 3.28. The point of a transparent dendrite,

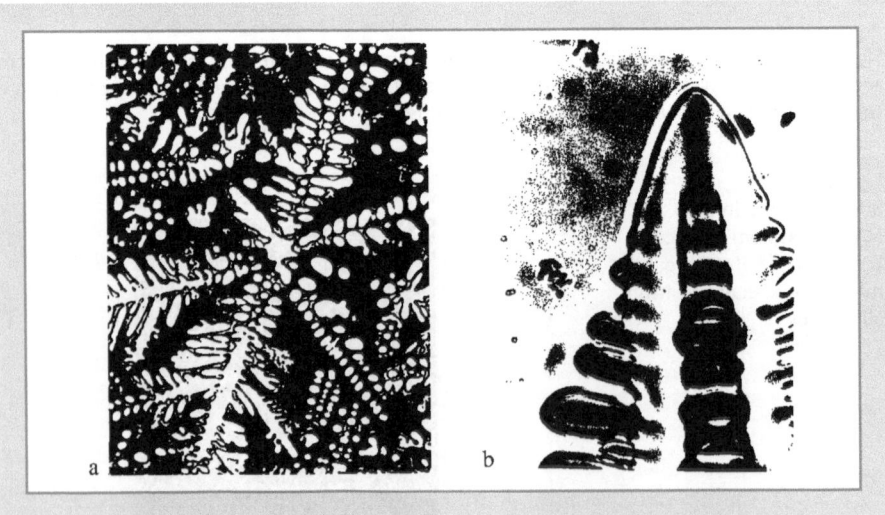

Fig. 3.28a Dendritic crystallization in the Cu – Fe – Al system (source: E. Wachter) 25 at.-% Fe, 22 at.-% Al, primary dendrites of the κ-phase, light microscopy, magnification: 750-times, **b** tip of a dendrite grown in-situ within the microscope by uni-directional freezing. Picrate acid with 0.5 at.-% naphthaline. Rate of growth: $R^K = 4 \times 10^{-1}$ cm/s; gradient of temperature: 15 K/cm. Magnification: 230-times (sources: T.A Steinmann, and W. Kurz)

as can be observed under a microscope during solidification is shown in Fig. 3.28a. The main and side branches correspond to directions with preferred growth rate. The reasons of the dendritic crystallization are the pecularities of the temperature distribution in the liquid. A strong undercooling of the liquid under the equilibrium liquidus temperature causes after nucleation first dendritic growth of the primary crystals. Only when the enthalpy of solidification, liberated through dendritic crystallization, raises the temperature to the equilibrium liquidus temperature, follows the normal crystallization of the liquid between the dendrites.

In pure materials the dendritic structure is barely visible on the polished surface, because the remaining hollow parts of the skeleton are filled with the same liquid substance, which also solidify. On the surface of solidified ultra pure samples one can see dendritic reliefs. As a result of the volume contraction during solidification, the leftover liquid is sucked into the space between dendrites. The dendrite branches can thus, protrude as elevated skeletons from the sample surface.

3.15
Simple Phase Equilibria with Complete Solubility in the Solid and Liquid Phases

Complete solution both in the liquid and solid phases can be formed in a two-component system, if the components have the same crystal structure, the difference in the size of atoms or molecules is small, and the bonding is similar. Figure 3.29 shows the simplest solidus and liquidus line combination for this case. It is assumed, that the two components behave differently with respect to the change in melting point during addition of the other component. Addition of B to A causes a melting point lowering. Addition of A to B causes a melting point increase. This happens in systems, in which only a very small change in interatomic interactions occurs when solid and liquid solutions are formed.

No invariant equilibrium exists in the whole composition range, apart from the melting equilibria of the pure components A and B. The solidification during extremely slow cooling takes the following path: Cooling the liquid Q with the starting concentration $(x_B^L)_1$ and starting temperature T_0, precipitation of solid solutions begins when the liquidus curve is reached; the composition of the solid is $(x_B^S)_1$ given by the tie line K_1. As $(x_B^S)_1 < (x_B^L)_1$ the liquid is enriched in B during crystallization. The liquidus temperature decreases in the measure as the concentration of B in the liquid increases. When the composition of the liquid reaches $(x_B^L)_2$, the temperature must simultaneously reach T_2. The composition of the phases now in solidification equilibrium is given by the tie line K_2. As the concentration of the solid solution in equilibrium $(x_B^S)_2$ is again smaller than the coexisting liquid $(x_B^L)_2$, the liquid is further enriched. The decrease of the liquidus temperature continues.

Simultaneously with the change of liquidus temperature, the liquidus and solidus compositions, the ratio of liquid to solid also changes. When reaching

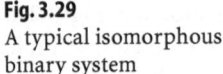

Fig. 3.29
A typical isomorphous
binary system

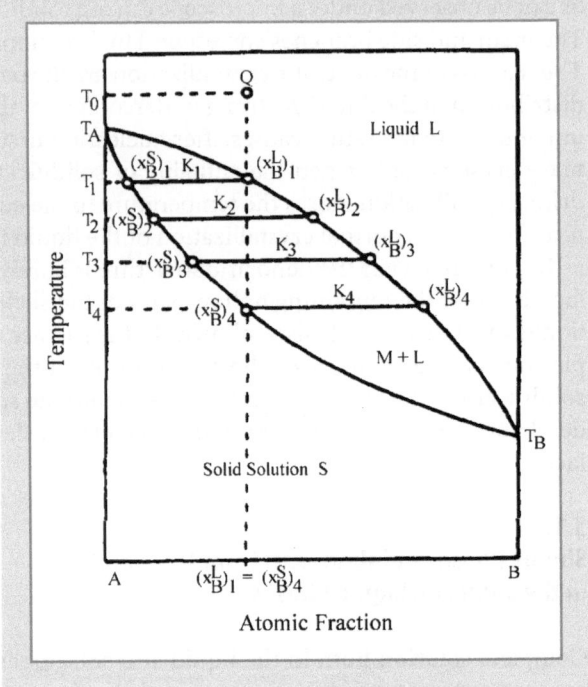

the liquidus line at T_1 during cooling a single phase liquid, the whole sample is liquid. The quantity of the coexisting solid solution with the composition $(x_B^S)_2$ is negligibly small. With decreasing temperature the quantity of the liquid solution decreases, that of the solid solution increases. The quantities are determined by the lever rule.

Let us consider the equilibrium at T_3. The overall composition of the sample is $(x_B^L)_1$, the concentration of the coexisting phases is $(x_B^S)_3$ and $(x_B^L)_3$. According to Eq. (3.31) we have

$$\frac{M_3^L}{M_3^S} = \frac{\left(x_B^L\right)_1 - \left(x_B^S\right)_3}{\left(x_B^L\right)_3 - \left(x_B^L\right)_1} \tag{3.33}$$

or in general

$$\frac{M_i^L}{M_i^S} = \frac{\left(x_B^L\right)_1 - \left(x_B^S\right)_i}{\left(x_B^L\right)_i - \left(x_B^L\right)_1} \tag{3.34}$$

where M_3^L is the quantity of the liquid, and M_3^S that of the solid in fusion equilibrium at T_3. Correspondingly, the indices i apply to any temperature T_i between T_1 and T_4.

At T_1 $(x_B^L)_1 - (x_B^S)_i$ is maximum, and $(x_B^L)_1 - (x_B^L)_i = 0$. Correspondingly, $M_i^S = 0$, and M_i^L is maximum. With decreasing temperature $(x_B^L)_1 - (x_B^S)_i$ decreases, and $(x_B^L)_1 - (x_B^L)_i$ increases. Likewise the ratio M_i^L/M_i^S decreases, until it becomes 0 at T_4. Here we have $(x_B^L)_1 - (x_B^S)_i = (x_B^L)_1 - (x_B^S)_4 = 0$ and $(x_B^L)_i - (x_B^L)_1 = (x_B^L)_4 - (x_B^L)_1$ is maximum. The whole sample now is solid.

This course is only guaranteed with complete equilibration. This is only possible with extremely slow cooling rates. After formation, the first solid solutions have the composition $(x_B^S)_1$, which corresponds to the equilibrium with the liquid at T_1. For example, with a high cooling rate, temperature T_2 is reached rapidly. Now a solid solution with the concentration $(x_B^S)_2$ precipitates, and surrounds the solid precipitated before, with $(x_B^S)_1 < (x_B^S)_2$. The equilibrium for the whole system is only reached, when the whole solid, which is in equilibrium with the liquid having the concentration $(x_B^L)_2$, has the same concentration $(x_B^S)_2$ through the whole solid, inside and out. To obtain this, B atoms from the liquid must diffuse into the interior of the crystal, to increase the concentration there to $(x_B^S)_2$. A relatively long time is needed to attain this, as one deals with a solid state reaction. With rapid cooling, no homogenization of the solid can be achieved. This means, that at T_2, though on the boundary liquid-solid the equilibrium concentrations $(x_B^S)_2$ and $(x_B^L)_2$ are present, thus, the equilibrium conditions are satisfied locally, the whole system, however, is not in equilibrium. Too few B atoms are present in the interior of the solid. They are still in the liquid (see Fig. 3.30). More and more B atoms (atoms with the lower melting point) are enriched in the liquid during continued cooling, thus, when T_4 is reached, and where the solidification in equilibrium should be finished, liquid is still present. But correspondingly less solid phase is formed. The solidification of the remaining liquid occurs below T_4. Possibly the temperature T_B is reached, where the last remnants of the liquid solidify.

Fig. 3.30
Layered growth during solidification in isomorphous system

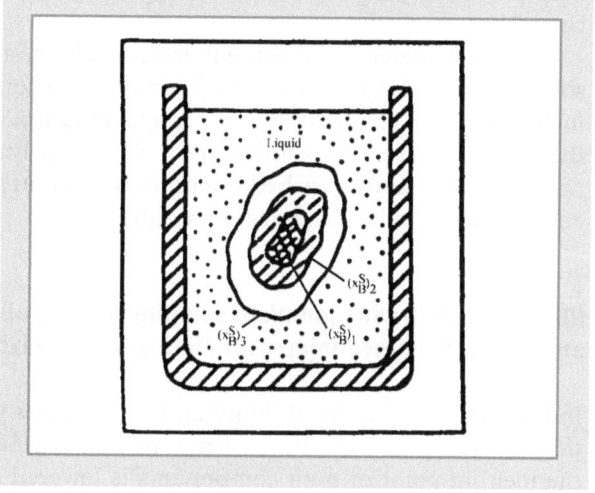

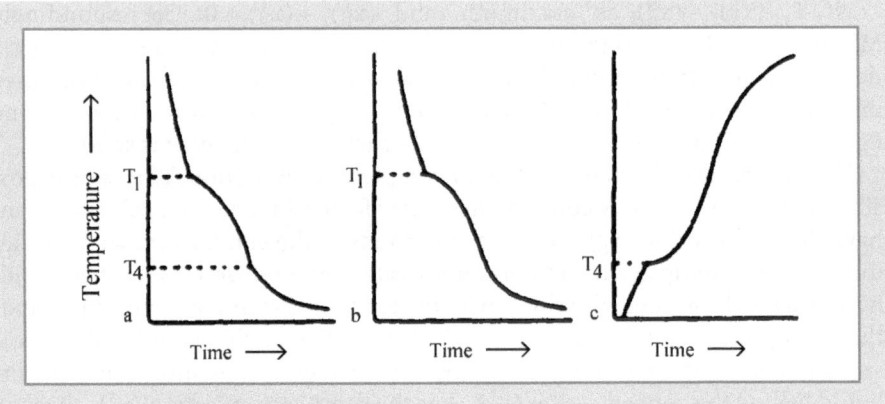

Fig. 3.31 Schematic cooling and heating curves for solidification respectively for melting of solid solution with total concentration $(x^L_B)_1 = (x^S_B)_4$ (see Fig. 3.28)

The lever rule is no longer applicable, as no total equilibrium is reached during the whole solidification process. As a product of solidification we have a solid solution with uneven distribution of the various atoms (cored crystal).

Figure 3.31 represents schematically the time-temperature curves during thermal analysis for the solidification and melting of solid solutions with the overall concentration $(x^L_B)_1 = (x^S_B)_4$ (see Fig. 3.29). If total equilibrium is reached continuously (Fig. 3.31a), a clear break point appears both at the liquidus temperature T_1 and at the solidus temperature T_4. In the absence of total equilibrium only the liquidus point can be recognized in the cooling curve. The end of the solidification is "spread out" (see Fig. 3.31b). Normally, the cooling rates in thermal analysis are generally too high, to be able to obtain equilibrium. Using this route, only the point of the liquidus can be determined experimentally. In some geological solidification processes, however, total equilibrium can be reached.

T_4 must be determined using thermal analysis, starting from a homogenized solid solution, tempered for a long period, with concentration of $(x^S_B)_4$, and the melting process recorded during heating. Figure 3.31c represents schematically the heating curve. The solidus point T_4 is clearly marked. However, now, for reasons similar and described in a non equilibrium situation during cooling, the effect of the liquidus is not clearly visible.

3.16
Phase Equilibria with Complete Solubility in the Solid and Liquid Phases and a Melting Point Minimum or Melting Point Maximum

If during the formation of liquid and solid phases with complete solubility noticeable changes in the interatomic interactions occur, then it is possible that the melting point of both components is lowered or increased by addition of the other partner. If the melting point of both components is lowered, a melting point minimum must necessarily appear. If for both components a melting point increase takes place, a melting point maximum must follow. The behavior of the liquidus and solidus lines for both cases are shown in Fig. 3.32.

The liquidus and solidus lines meet at the melting point minimum or maximum. This can be proven based on thermodynamic considerations, but can also be explained while inspecting Fig. 3.33. If the liquidus and solidus curves would not meet at the melting point maximum it would be possible, through joining

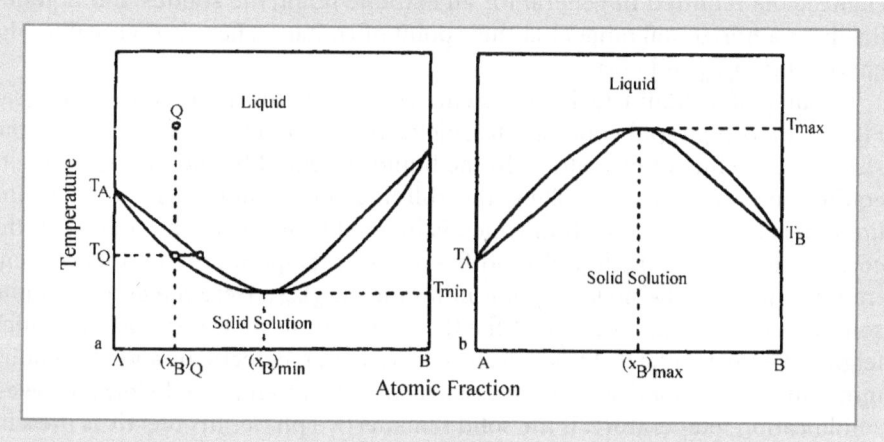

Fig. 3.32 Phase diagram with a melting point minimum or a melting point maximum with complete miscibility of the component in the solid as well as in the liquid state

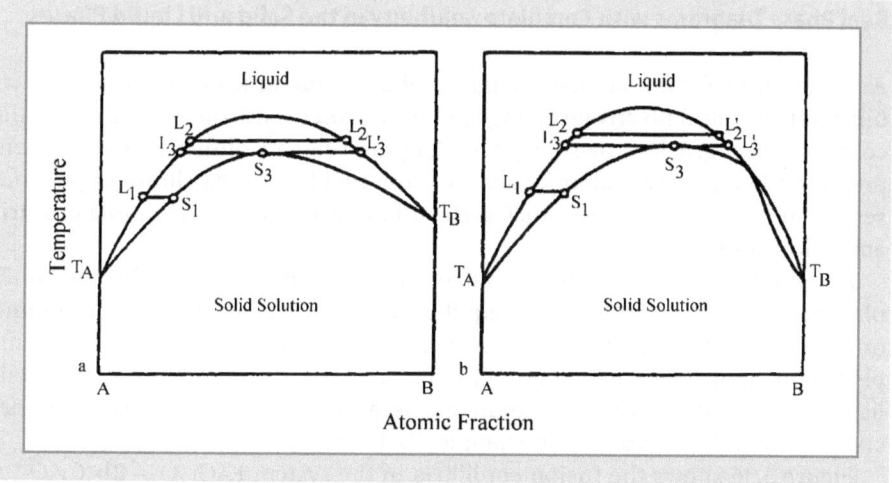

Fig. 3.33 Explanation of the necessity of contact of liquidus with solidus at the extremum of melting point

the liqudus and solidus lines to have tie lines, where two liquid phases (L_3, L_3') of different composition are in equilibrium with a solid phase (S_3). This possibility is, however, senseless, because the necessary two phase character of the fusion equilibrium is no longer maintained. Between the melting point maximum or minimum and that of a component, fusion equilibria appear, which correspond essentially to simple systems with complete miscibility in the solid and liquid phases (see Figs. 3.29 and 3.32). At the melting point maximum and minimum the univariant equilibria degenerate into invariant equilibria. The solidification occurs at a constant temperature, because the concentration can not be changed. As required in general for an extreme point, the solidus and liquidus line have a horizontal tangent at their point of contact. There is no break in the curves at $(x_B)_{min}$ and $(x_B)_{max}$.

Cooling of a liquid Q in a system with a melting point minimum (see Fig. 3.32a), using rapid cooling rates, yields the formation of layer crystals and stronger enrichment of B atoms in the liquid, as would be the case if complete equilibrium was achieved. Thus, the solidification is not finished, when the liquid reached a concentration which is in equilibrium with the solid with the composition $(x_B)_Q$. The liquid solidifies at lower temperatures, and its concentration can reach the melting point minimum, $(x_B)_{min}$. Here the leftover liquid solidifies. In the cooling curve of the thermal analysis an arrest appears, which feigns an eutectic effect. To decide, if in a system an eutectic or a melting point minimum exits, a solid, which shows such an arrest is tempered below the lowest solidification temperature. If the solid remains two phase, an eutectic is present. If it becomes single phase, a minimum melting point is present. The proof can be found microscopically or by X-ray diffraction.

3.17
Real Phase Diagrams with Complete Solubility in the Solid and Liquid Phases

As an example for a real phase diagram of a system, in which complete miscibility appears, and no strong change in interaction occurs during mixing of the components, Fig. 3.34 shows the phase diagram FeS-CrS. Similarly to the system presented here, other systems, such as sodiumfeldspar – calciumfeldspar, and several metallic systems (Ag-Au, Cu-Ni and many others) show simple univariant fusion equilibria.

Among the systems, which show an unlimited solubility in the solid and liquid phases, about 35 % have fusion equilibria without melting point minimum or maximum. In about 65 % of the cases a melting point minimum occurs. Simple fusion equilibria with a melting point maximum do not occur. If in a metallic system a melting point maximum exists, at other concentrations other special phase equilibria occur (for example Pb-Tl see Fig. 3.35).

Figure 3.36 shows the fusion equilibria in the system $K_2Cr_2O_7 - Rb_2Cr_2O_7$. A melting point minimum occurs. A melting point maximum was found in the system d-carvoxim-l-carvoxim.

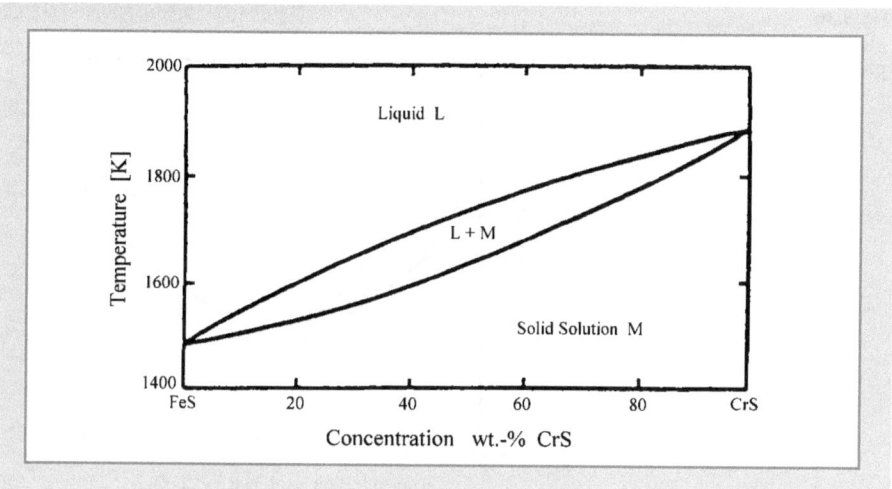

Fig. 3.34 Phase diagram of the FeS – CrS system [3]

Fig. 3.35
Phase diagram of the Pb – Tl system [2]

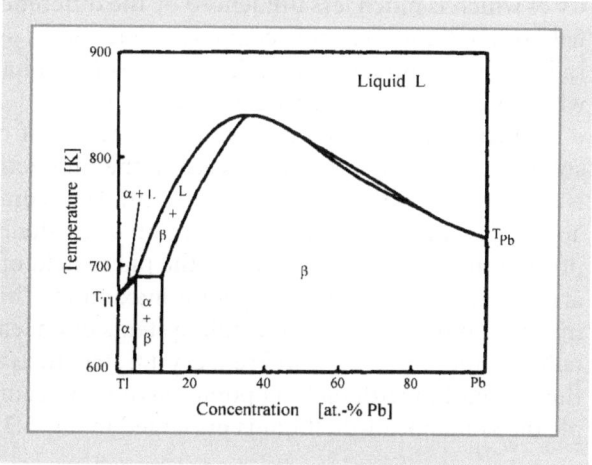

The frequency of the melting point minimum in metallic systems with complete solubility in solid state is due in the first place to the enthalpy of lattice distortion, which has to be spent during the formation of solid solutions from atoms of different size. Such a large enthalpy is not needed during the formation of a similar liquid solution. The effect of the enthalpy of lattice distortion on the phase equilibria, and the subsequent position of the melting point minima can be deduced quantitatively on a thermodynamic basis. It follows qualitatively, that when forming substitutional alloys from atoms with significantly different size, the large expenditure of the enthalpy of lattice distortion, displaces the stability difference between solid and liquid in favor of the liquid, the stabil-

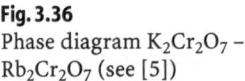

Fig. 3.36
Phase diagram $K_2Cr_2O_7$ –
$Rb_2Cr_2O_7$ (see [5])

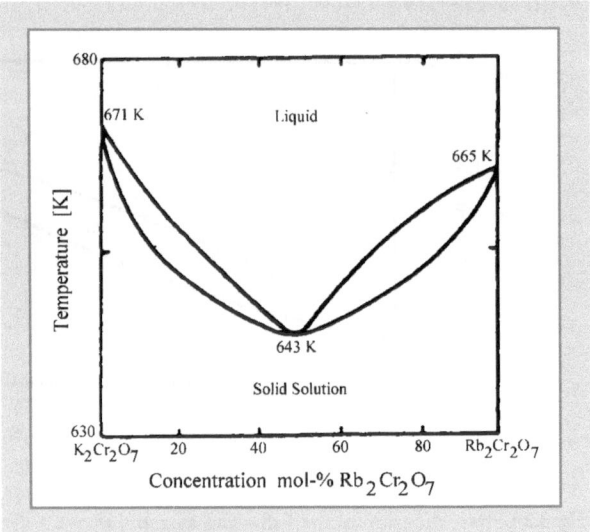

ity of which is much less influenced by the difference in atomic radii. The phase field of the liquid expands compared to the ideal system with identical atomic radii to lower temperature, at the expense of the phase field of the solid solution, which is the case at a melting point minimum.

A melting point maximum can be expected, if during the formation of the solid solution a significant increase in the interaction parameters occurs. This is not possible in the same degree in a similarly composed liquid, because of the absence of long range order. The region of existence of the liquid will be compressed through the expansion of the phase field of the solid to higher temperatures. The result is a melting point maximum. This significant increase in the interaction parameters in metallic systems often leads to formation of intermetallic compounds with a different crystal structure, thus, an extended solid solution formation with melting point maximum occurs only seldom. As an example the system lead-thallium is presented in Fig. 3.35.

3.18
Miscibility Gap in the Solid Phase

A substitutional solid solution, built up of atoms of different sizes, and where the lattice distortion enthalpy plays a significant role, must, at sufficiently low temperature, decompose into two phases, as can be easily shown using thermodynamic considerations. Below a certain critical temperature T_K, a solid-solid two phase region appears. The solubility lines c and c′ in Fig. 3.37 correspond exactly to the limiting solubility lines c and c′ in Fig. 3.6. At T_1 two solid phases are present in the middle of the composition range. The concentration of the two coexisting phases connected by the tie line K_1 are $(x_B^S)_1$ and $(x_B^S)_2$. With in-

Fig. 3.37
Phase diagram with a miscibility gap in the solid state

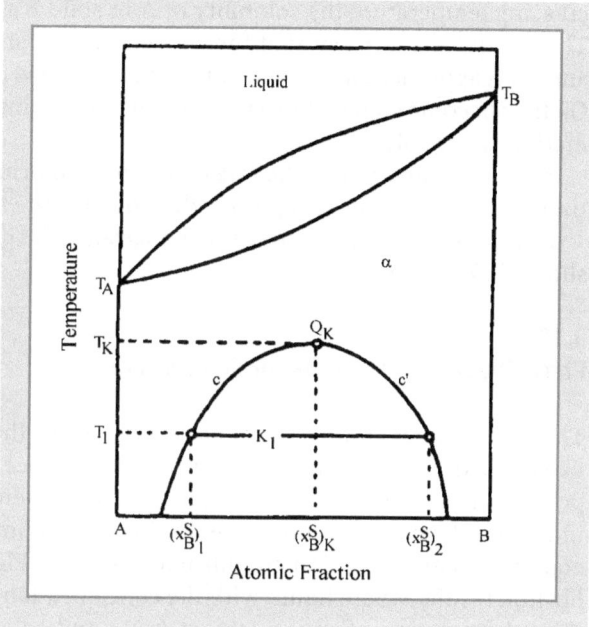

Fig. 3.38
Phase diagram Au – Ni [2]

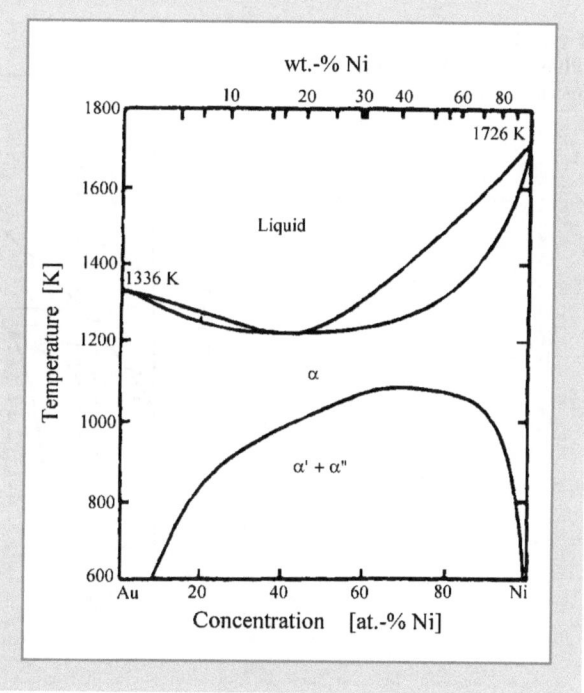

creasing temperature the solubility of A in solid B and of B in solid A increases. This is equivalent to say, that the concentrations of the coexisting phases $(x_B^S)_1$ and $(x_B^S)_2$ approach each other, until at the so-called critical temperature at point Q_K they become equal. Above the critical point uninterrupted solid solution formation is possible.

At lower temperatures the rate to attain equilibrium in the solid is low. Sometimes rather long tempering is needed to reach the true phase equilibria.

As an example of this type of real diagrams, the phase diagram gold-nickel is shown in Fig. 3.38.

3.19
Phase Diagram with Peritectic Equilibrium

The critical temperature of a miscibility gap in the solid state, caused by lattice distortion effects, increases with the difference in atomic radii of the components. T_K can also be higher than the solidus temperature of solid solutions in the middle composition range. In this case an intersection of the solid-solid equilibria with the fusion equilibria occurs (see Fig. 3.39). A three phase equilibrium results, where liquid with the concentration $(x_B^L)_P$ is in equilibrium with the solid solution of concentration $(x_B^S)_\alpha$ and solid solution of concentration $(x_B^S)_\beta$. According to the Gibbs phase rule, this is an invariant equilibrium. Even

Fig. 3.39
Phase diagram with a peritectic reaction

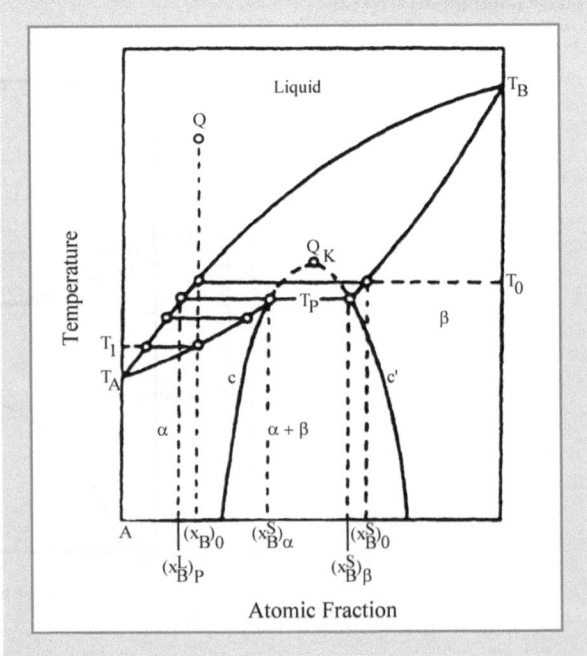

when the lattice structure of the A-rich and B-rich solid solution are the same, the stability ranges of the two structures are designated differently (α, β), to describe the phases differently, which can coexist with the liquid below T_K in an univariant equilibrium.

Cooling the liquid Q, crystallisation of β-solid solutions takes place, when the liquidus temperature T_0 is reached. Crystallization continues, until the temperature of the peritectic equilibrium is reached. The liquid now has the composition $(x_B^L)_P$, and the equilibrium solid solution (x_B^S). These two phases react under formation of solid phase α, with the concentration $(x_B^S)_\alpha$. Only when during this reaction all of the phase β is reacted, can the cooling continue. The concentration of liquid and coexisting solid solution follow the trace of the liquidus and solidus line below T_P, until, in the case of equilibrium, the solid α reaches the composition $(x_B)_0$, when the quantity of the liquid is zero. Because during solidification in the univariant solid-liquid equilibrium under rapid cooling conditions, total equilibrium is not reached, the leftover liquid must solidify during continued cooling.

The three phase equilibrium at T_P is an intermediate equilibrium (peritectic equilibrium). The first formed β-solid solution reacts with the liquid and forms α-solid solution. This reaction begins on the surface of the precipitated β-crystallite (Fig. 3.40).

During this process A-atoms from the liquid must penetrate through diffusion into the solid. For a complete reaction in the interior of the β-crystals more time is required, as is generally available in the normal cooling process of thermal analysis. After merely a local equilibrium is reached on the boundary liquid-β-solid solution, where only a layer close to the boundary of β-crystals reacted to α-crystals, the temperature falls below T_P. During this process, α-solid solution precipitates. The rest of the β-solid solution in the center of the primary crystallites remains untouched. No complete equilibrium of the whole system

Fig. 3.40
Reaction layer (α) between primary crystal (β) and liquid as a consequence of a peritectic reaction

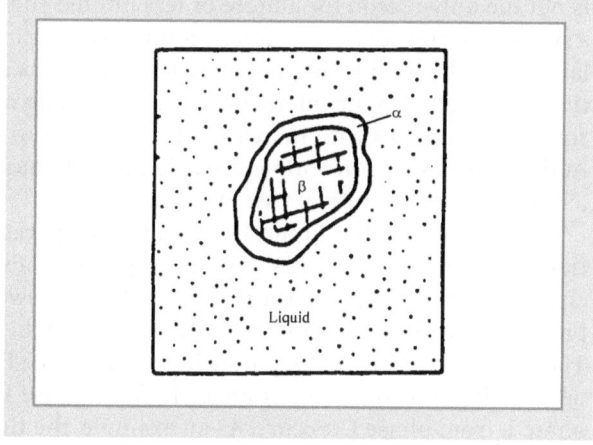

Fig. 3.41
Structure of a solidified Cu-Fe alloy containing 35 at.-% Fe (source: A. Klein). By freezing a peritectic reaction has been fast. Fe-dendrites enveloped by Cu-solid solutions. Light microscopy, magnification: 750-times

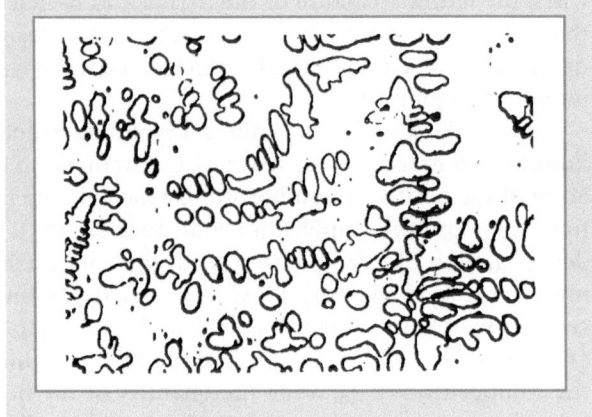

is reached. A texture results, with grains, in the center with β-solid solutions, at the surface with α-solid solutions. Figure 3.41 shows this fact using a Cu-Fe alloy as example. Sufficiently long tempering below the equilibrium solidus temperature of the alloy will finally yield the equilibrium situation. It is for an alloy with the concentration $(x_B)_o$, single phase as required, and shown in Fig. 3.38. Combination of fusion equilibria, where both, T_A and T_B, are decreased, with the solid-solid equilibria of a miscibility gap, results in a simple eutectic system, as represented in Fig. 3.6.

3.20
Miscibility Gap in the Liquid Phase

It is possible, that no complete miscibility exists in the liquid phase, if because of the large difference in atomic radii the structural and binding relationships are strongly disturbed. Then a miscibility gap in the liquid appears. This, however, is not the only reason for a more or less limited solubility of the components of a system. The reasons are generally found in the form of the atomic resp. molecular built of the liquid components, which often has as a consequence the reduction of the interatomic interactions during mixing of the solution partners. The solubility increases with increasing temperature, as is the case in solid systems. At the critical solution temperature T_K the two phases, which coexisted at lower temperature, are identical (see Fig. 3.42).

During cooling of a liquid L_0 when the temperature T_1 is reached, a separation into two liquids occurs (see also Fig. 3.43b). First the phase rich in A precipitates as fine droplets. Let this phase have a lower density than the starting phase from which it precipitates. The droplets of the A-rich phase rise to the top. They assemble there to a separate liquid layer (see Fig. 3.43c, d). During continued cooling, precipitation of phase L′ from phase L, as well as precipitation of phase L from phase L′ occurs. As an example, the tie line K_2 indicates the phase

Fig. 3.42
Phase diagram with a
monotectic reaction

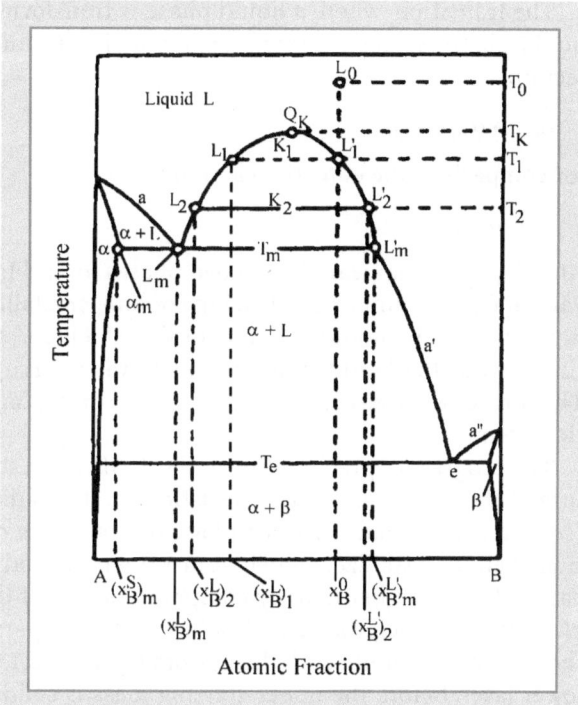

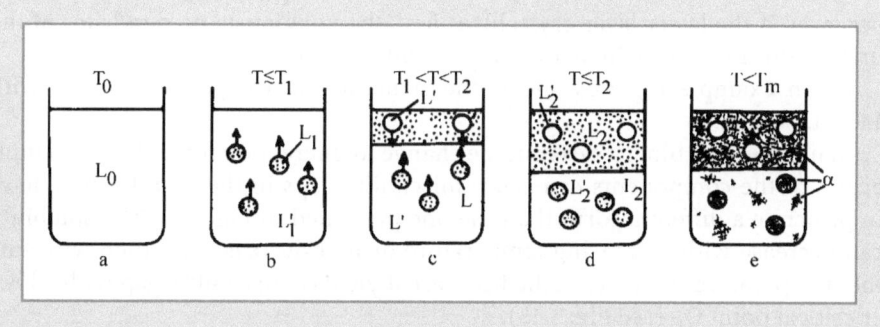

Fig. 3.43 Explanation of the reaction of the mixing and freezing of liquids passing a miscibility gap in the liquid state. L_0 liquid before starting (see Fig. 3.42); L liquid rich in A; L' liquid rich in B, α α-phase (solidified dendritically)

equilibrium at T_2. Finally when T_m is reached, the A-richer phase L_m participates in the two phase equilibrium L_m–L_m' and also the solidification equilibrium L_m–α_m. The phase L_m transforms at constant temperature T_m into the two phases α_m–L_m', where the liquid L_m' is naturally united with the phase L_m' formed during cooling through the liquid-liquid equilibria. The quantities of α_m and L_m', which are formed from L_m can be determined with the lever rule.

The transition, where a liquid phase is transformed into a solid and another liquid phase is called a monotectic reaction (monotectic equilibrium, monotectic):

$$L_m \rightarrow \alpha_m + L'_m \tag{3.35}$$

for comparison the eutectic reaction is

$$L \rightarrow \alpha + \beta \tag{3.36}$$

After L_m has disappeared, only two phases (α and L') are in equilibrium. A univariant equilibrium exists. During cooling, crystallization of α occurs, and the liquid passes the liquidus points of line a'. Line a' can be considered as a continuation of the liquidus line a, which was interrupted by the miscibility gap. The further progress of the solidification corresponds to that of a simple eutectic system.

The progress of precipitation and solidification show themselves in the structure of the sample in a characteristic way. If L is the upper layer of the sample (see Fig. 3.43c, d) and also the phase disappearing during the monotectic reaction, then α crystallizes, which occurs often as dendrites. The dendrites sink because of their higher density compared to that of the liquid, and get into layer L'. Liquid L', precipitated as droplets in the top part of the sample, sink also to the bottom, and unite with the bottom layer. A part, however, does not reach the lower layer, before the upper starting mass is solidified, and thus, remains enclosed in it. Droplets of L can also be found in the lower layer. In addition one can also find dendrites here, which sank down during crystallization of the upper layer. If the lower layer crystallizes first, then obviously no dendrites of the first solidifying layer will be found in the other layer.

As an example Fig. 3.44 shows the structure of a copper-lead alloy with 14.7 at.-% lead.

In nonmetallic binary systems, the change in interaction forces between molecules of the components can cause miscibility gaps in the liquid, which have in principle a different form than the one presented in Fig. 3.41. The solubility can decrease with increasing temperature or first decrease and then with further temperature increase. In the first case it yields a miscibility gap with a lower critical point Q_K (see Fig. 3.45).

The system triethylamine – water is an example. In the second case, a closed miscibility gap is present with a lower critical point Q_{K1} and an upper critical point Q_{K2} (see Fig. 3.46). Heating a single phase solution in the middle concentration range causes a demixing, which is eliminated with further raising of the temperature. A typical example is the system nicotine-water.

In aqueous or organic solutions, the start of the precipitation of a second liquid phase can be observed in clouding of solutions which are generally transparent for visible light. This sensitive proof of the appearance of a second phase can be used to accurately determine the solubility lines.

Fig. 3.44
Microstructure of a Cu-Pb
alloy with 14.7 at.-% Pb
(concentration of the monotectic, source: Ch. Staats),
bright: Cu; light microscopy;
magnification: 75-times

Fig. 3.45
Miscibility gap in the liquid
state with a lower critical
point Q_K

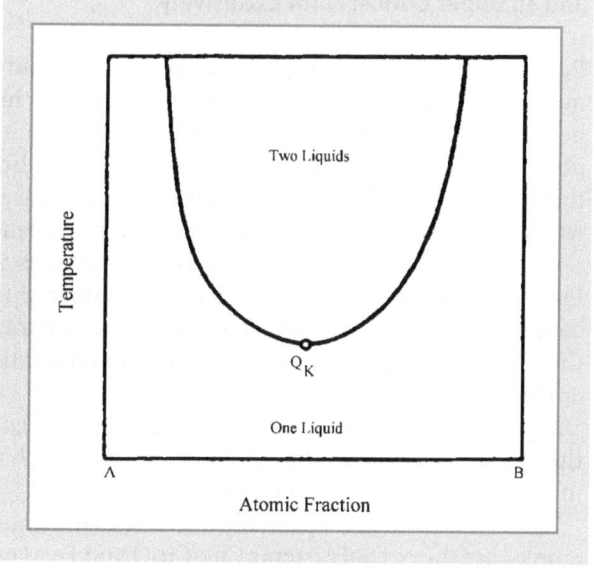

Fig. 3.46
Miscibility gap in the liquid
state with an upper and
lower critical point

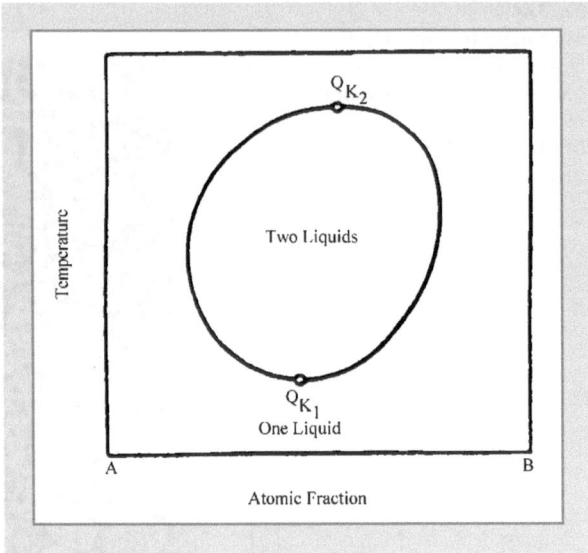

3.21
Real Phase Diagrams with a Miscibility Gap in the Liquid Phase
and an Upper Critical Point Exclusively

Equilibrium is reached rapidly between two liquid phases because of the high mobility of the atoms in the liquid. On the other hand, the thermal effect connected with the demixing is small, as no change in structure occurs. It is still possible to determine the solubility lines in the liquid state, using differential thermal analysis. Figure 3.47 represents the phase diagram gallium-bismuth, where the miscibility gap was determined by thermal analysis.

The partial miscibility of metals with sulfides is of industrial importance. Figure 3.48 shows the partial phase diagram copper-sulfur. In the winning of copper through the Bessemer process, the separation of blister copper from Cu_2S is done by layer formation in the molten mixture. "Copper" forms the lower layer, Cu_2S the upper.

As visible in the phase diagram (Fig. 3.48) the "copper"-layer contains even at the monotectic temperature 2.9 at.-% sulfur, which is removed with appropriate means, as it embrittles the copper.

Some metals are also partially miscible in the liquid state with their oxides. Examples are the partial systems $Cu - Cu_2O$ and $Fe - Fe_2O_3$. It is important, that metals and silicates practically do not mix in the liquid state, on which a large number of metal winning processes are based , when metal and slag are separated. On the other hand, a number of metals are miscible with their liquid halogenides. As Fig. 3.49 shows, liquid potassium and liquid potassium chloride are completely miscible above 1063 K. At lower temperatures a miscibility gap exists.

Fig 3.47
Phase diagram Ga – Bi [6]

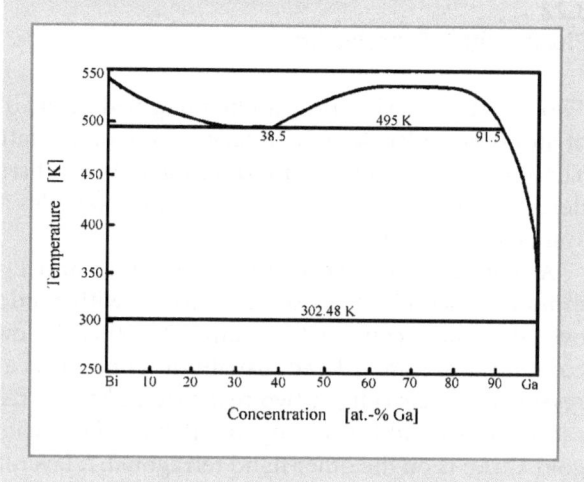

Fig. 3.48
Cu-rich portion of the Cu – S
phase diagram [2]

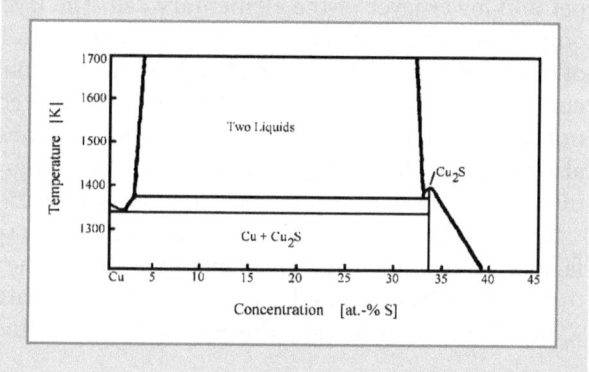

Fig. 3.49
Phase diagram of the K – Cl
system [7]

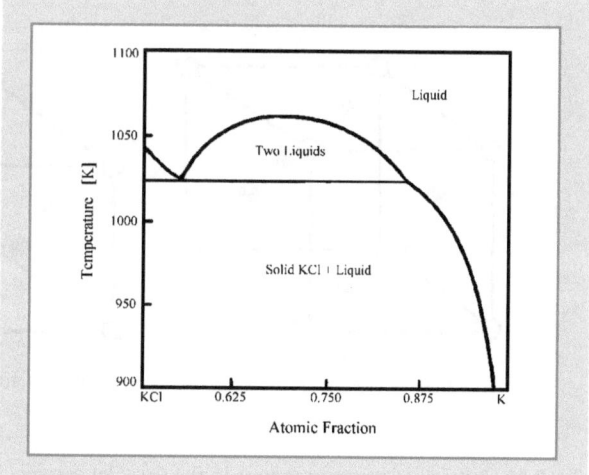

3.22
Phases with a Superlattice

Two extreme possibilities can be imagined in the distribution of the two types of atoms or molecules of a binary system on the lattice sites of a binary system. These are the completely random (statistical) distribution, and a strongly ordered distribution of the species. Simple examples are the solid solutions in the system copper-gold.

At temperatures between 683 K (410 °C) and 1162 K (889 °C, melting point minimum, see Fig. 3.36) solid solutions with random distribution are formed over the whole composition range 0–100 %. At lower temperatures at certain concentrations an ordered distribution of atoms can be stable. Figure 3.50 represents the basic cells of two real cases. They correspond to the stoichiometric ratios Cu_3Au and $CuAu$. Cu_3Au has the same basic cubic lattice as copper and gold. $CuAu$ is on the other hand tetragonal. A layer like structure exists. The two lattice parameters are fixed by the size of the gold atoms, which occupy the upper and lower layer of the elementary cell. The layer at the half height consists only of copper atoms. They are smaller, than the gold atoms, so that at this type of ordering a shortening of the height of the originally elementary cubic cell occurs. Thus, $c/a < 1$. This "solid solution" with ordered distribution of atoms does not belong anymore to the series of solid solutions with random distribution, also not in the limit, which can be attained by progressive increase of the ordering. $CuAu$ is a separate phase, a superlattice phase.

Cu_3Au is also a separate supperlattice phase, in spite of the fact, that its basic face centered cubic lattice is identical to that of the solid solution with random distribution. The ordered structure – the superlattice – can be recognized by X-ray diffraction through additional diffraction lines (superlattice lines).

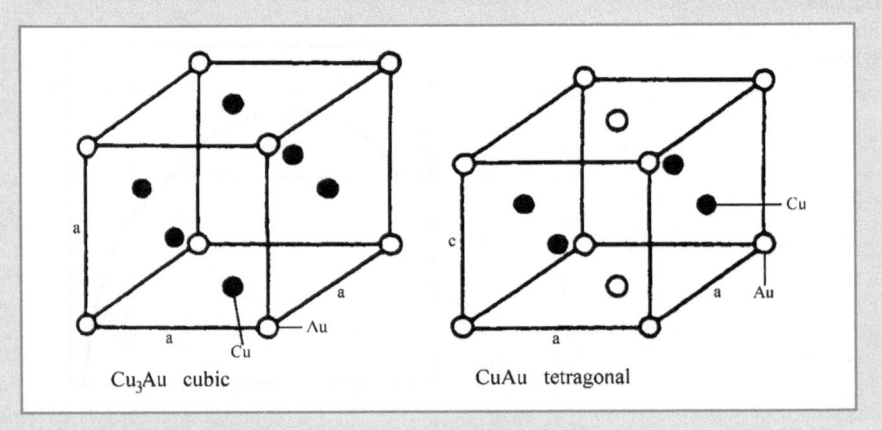

Cu₃Au cubic   CuAu tetragonal

Fig. 3.50 Unit cell of superstructures in the Cu – Au system

During heating the superlattice phase CuAu transforms at 683 K (410 °C) into a solid solution with random distribution. This can be compared to the melting of a pure substance, where the crystal with long range order transforms into a liquid with short range order. Though during the transition ordered distribution → disordered distribution the basic structure remains the same, the strictly ordered distribution of the atoms is removed. As can be shown based on thermodynamic considerations, with decreasing temperature increasingly ordered phases appear in every system (example: steam → water → ice).

During cooling of a solid solution with 50 % Au from above 683 K (410 °C) ordering only occurs, if the cooling rate is sufficiently slow. The kinetics of the order-disorder reaction is strongly controlled by the diffusion of the atoms, so that during rapid cooling, the sample is not long enough at a temperature with significant atomic mobility, to reach complete ordering. Quenching the solid solution with random distribution (for example throwing the sample at 973 K (700 °C) into water at room temperature) will practically prevent completely the order-disorder reaction. The solid solution with random distribution can be maintained at room temperature indefinitly as a metastable phase. With this procedure the random distribution in other systems can also be "frozen" and studied comfortably at room temperature.

The superlattice phases distinguish themselves significantly from the randomly distributed phases with respect to their chemical and physical properties. G. Tammann discovered the superlattice phases, because Cu_3Au and CuAu have greater resistance to attack by nitric acid than random solid solutions having the same composition. In addition, electric, magnetic and mechanical behavior show significant differences, of which industry takes advantage.

Superlattices appear in a solid solution, when, during the formation of the solid solution, the interatomic interaction energy is increased. Below a critical temperature T_K (critical ordering temperature) ordering takes place, which is such, that A–B pairs, whose interaction is stronger than the average of A–A and B–B pairs, appear in larger numbers, than corresponds to a random distribution. The Gibbs energy of the system is reduced. Differences in atomic radii can also promote ordering. The lattice distortion energy present in randomly distributed system can be significantly reduced through ordering. The maximum reduction of the free energy occurs obviously at the optimum stoichiometry. Deviations from this stoichiometry can, in the simplest case, decrease the stability of the superlattice phase.

At the correct stoichiometric composition the superlattice phase is stable to a higher temperature, than when it deviates from it. The observed phase diagrams correspond to these simple considerations.

Figure 3.51 shows the phase equilibria in the copper – gold system, conditioned by the order-disorder reactions. Around 50 at.-% gold, in the vicinity of the critical ordering point Q_K, similar phase equilibria exist as at that of a melting point maximum (see Fig. 3.32b). Below T_K (683 K for CuAu) two phase equilibria appear, where solid solutions with random atomic distribution and super-

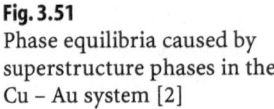

Fig. 3.51
Phase equilibria caused by
superstructure phases in the
Cu – Au system [2]

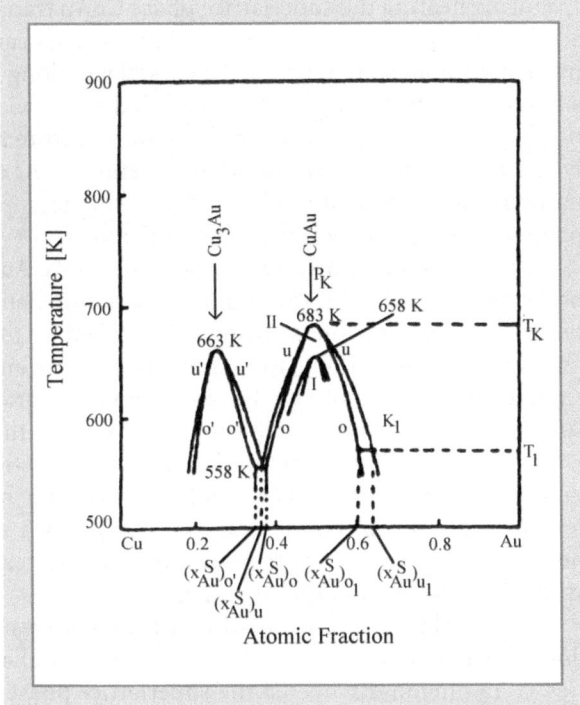

lattice phases are in equilibrium. The lines o (resp. o′) show the temperature-concentration values for the "ordered phase" which are in equilibrium with the "disordered phase". The lines u (resp. u′) represent the totality of the temperature-concentration points which correspond to the "disordered phase" in equilibrium in this two phase equilibrium. The tie line K_1 shows, that at T_1 for concentrations within the tie line, the superlattice phase participating in the two phase equilibrium has the concentration $(x_{Au}^S)_{o1}$ and the concentration of the corresponding solid solution is $(x_{Au}^S)_{u1}$.

In the system copper – gold, used as an example here, in addition to the superlattice phase CuAu the superstructure phases Cu_3Au and $CuAu_3$ appear. In Fig. 3.51 the phase equilibria conditioned by the superlattice phase Cu_3Au are shown. In principle, they are similar to that of CuAu. Meanwhile around 36 at.-% Au the two phase equilibria connected with Cu_3Au encounter the two phase equilibria involving CuAu. At the temperature of the intersection of lines u and u′ (at 558 K; 285 °C) a three phase equilibrium occurs, which is formally equivalent to the eutectic solidification equilibrium (eutectoid equilibrium). The disordered solid solution with the eutectoid concentration $(x_{Au}^S)_u$ decomposes during cooling into an ordered phase having the Cu_3Au structure with the composition $(x_{Au}^S)_{o'}$ and into an ordered phase having the CuAu structure with the composition $(x_{Au}^S)_o$.

The formula for the eutectoid transformation is in general

$$S_1 \rightarrow S_2 + S_3 \tag{3.37}$$

where S_2 and S_3 are two solids, which differ in composition and possibly in structure from solid S_1. According to the Gibbs phase rule, this is an invariant equilibrium.

The superlattice of a phase can change with decreasing temperature. Structural changes can occur at certain temperatures, where a certain ordered structure is transformed into another ordered structure. This is the case also with CuAu. Above 658 K (385 °C) a somewhat different atomic arrangement occurs as represented in Fig. 3.50 for the low temperature modification CuAu I. Above this temperature the stable modification is called CuAu II. It has an orthorhombic structure.

3.23
Systems with a Congruently Melting Compound

The critical temperature T_K of a phase with a superlattice can be higher than the melting points of their components. In this case fusion equilibria appear, in which the superlattice phase participates. An unlimited mutual solubility of the components is no longer possible. This applies not only to the superstructure phases, whose stability region reaches into the temperature range of the fusion equilibria, but also for any compound, which is formed from the components within certain composition ranges, and possesses a separate lattice with a specific melting point, and which cannot be deduced from the basic structure of the components through specific ordering of the atoms. As examples are given in the system Fe – SiO_2 the compound Fayalite Fe_2SiO_4 and in the system copper – sulfur the compound Cu_2S.

Figure 3.52 represents schematically the fusion equilibria for a system in which the compound V occurs, which has its own melting point T_V. The compound creates systems of two phase equilibria, similar to the ones created by a superlattice structure in a solid solution phase.

The tie lines K and K' indicate the solid-liquid equilibria occuring at this temperature in the different composition ranges. The liquidus lines a and a' encounter in their extension to lower temperatures at T_e resp. T_e' the liquidus lines a_1 and a_1'. Three phase eutectic equilibria occur at these temperatures:

$$L_e \rightarrow \alpha_e + V_e \tag{3.38}$$

and

$$L_{e'} \rightarrow \beta_{e'} + V_{e'} \tag{3.39}$$

The solubility lines d and d' (see Figs. 3.52 and 3.53) are temperature dependent. Generally with decreasing temperature the composition range, where the compound exists as the only phase, is compressed.

Fig. 3.52
Schematic phase diagram
including a congruently
melting compound

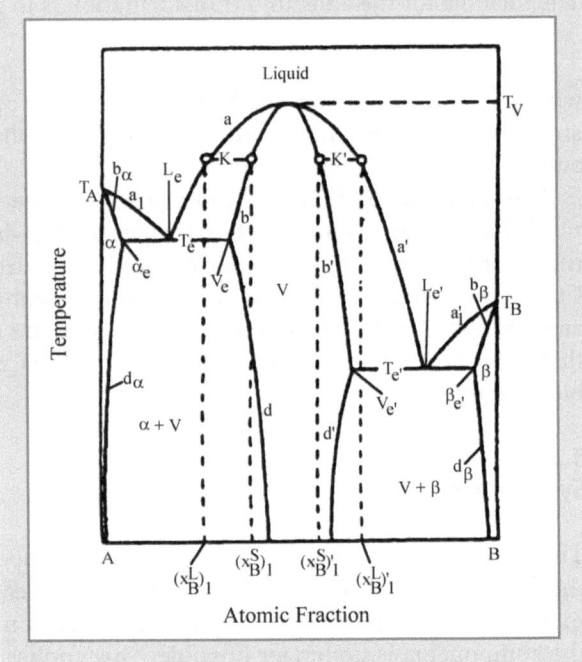

Some compounds are not stable below a temperature T_e' (see Fig. 3.53). The solubility curves d and d′ meet at this temperature. The compounds with the composition $(x_B^S)_{e'}$ decomposes when cooled to Te′ in an eutectoid three phase reaction into solid solutions with the composition $(x_B^S)_\alpha$ and $(x_B^S)_\beta$:

$$V \rightarrow \alpha + \beta \qquad\qquad (3.40)$$

On the other hand not all compounds are stable to such a high temperature that they can participate in the fusion equilibria of the system. During heating they can decompose into a mixture of two phases as shown in Fig. 3.54. Here decomposition products α and β with concentrations $(x_B^S)_\alpha$ and $(x_B^S)_\beta$ and in amounts determined by the lever rule are produced from the compound with the composition $(x_B^S)_{p'}$.

If such a mixture of α- and β-solid solutions is cooled from the temperature range $T > T_{p'}$ the compound V is formed by a peritectoidic reaction (analogous to the peritectic solidification equilibrium):

$$\alpha + \beta \rightarrow V \qquad\qquad (3.41)$$

The peritectoidic reaction is an inversion of the eutectoid reaction (see Eqs. (3.40) and (3.41))

The phase equilibria represented in Fig. 3.52 apply to a compound, in which the composition in one (temperature independent) range can be varied, with-

Fig. 3.53
Phase equilibria including the composition of a compound stable only at high temperatures

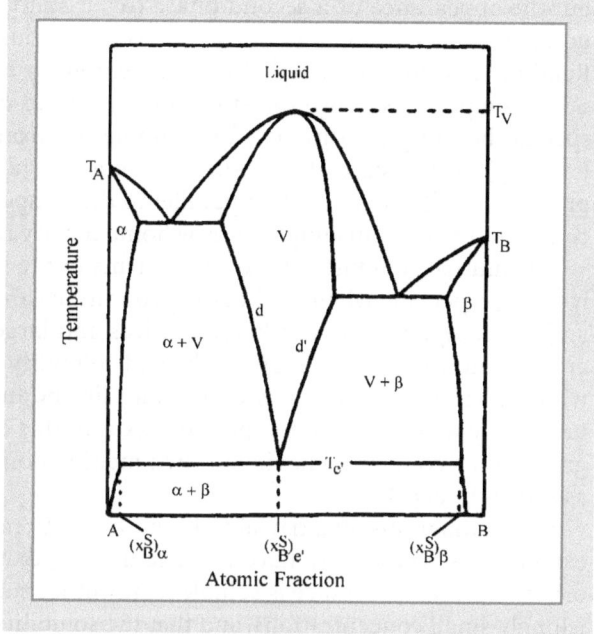

Fig. 3.54
Phase equilibria including the formation of a compound stable only at low temperatures

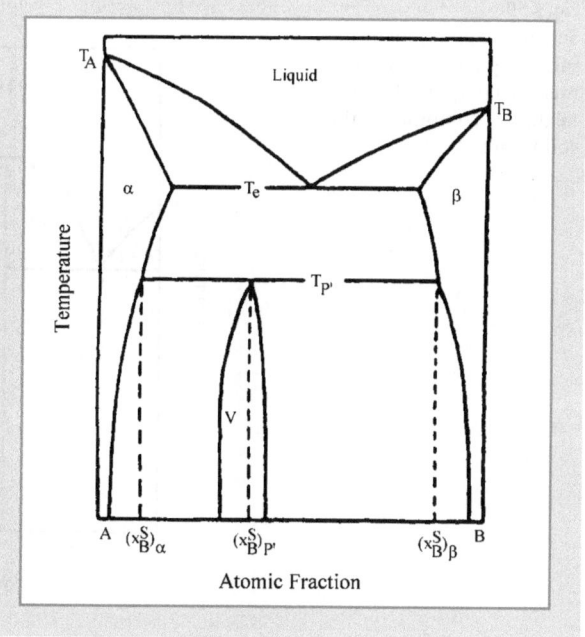

out the appearance of a second phase (α-, β-solid solution or liquid). A large number of chemical compounds, however, are bound to a fixed stoichiometry. Remind us of KCl, SiO_2 and H_2O. Only extremely small excesses of one or the other components above the stoichiometric ratio can be tolerated because of specific bonding conditions (for example electroneutrality with heteropolar bonding). In this case, the solidus lines b and b' and solubility lines d and d' practically collapse into a line parallel to the temperature axis. In this case the compound has a vanishingly narrow homogeneity range.

The mutual solubility of the components in the solid state is often extremely small, especially when the bonding conditions are different in the compound and the components and different lattices and large difference in atomic radii are present. Here the solidus line b_a and solubility line d_a coincide practically with the left temperature axis, and solidus line b_b and solubility line d_b coincide practically with the right temperature axis. In this case, a schematic phase diagram with a compound having its own melting point is simplified to a diagram as shown in Fig. 3.55.

In real diagrams, in a trivial fashion, solubility ranges are not drawn, if their extent is not known. We have to be reminded, that based on thermodynamic principles, each substance is soluble in any other substance, even if only in vanishingly small concentrations, and that the solubility is temperature dependent. If in principle this would not be so, than the Gibbs phase rule would be violated in the concentration range A – V below T_e (see Fig. 3.52). Based on the simple

Fig. 3.55
Schematic phase diagram including one congruently melting compound and negligible solubility in the solid state

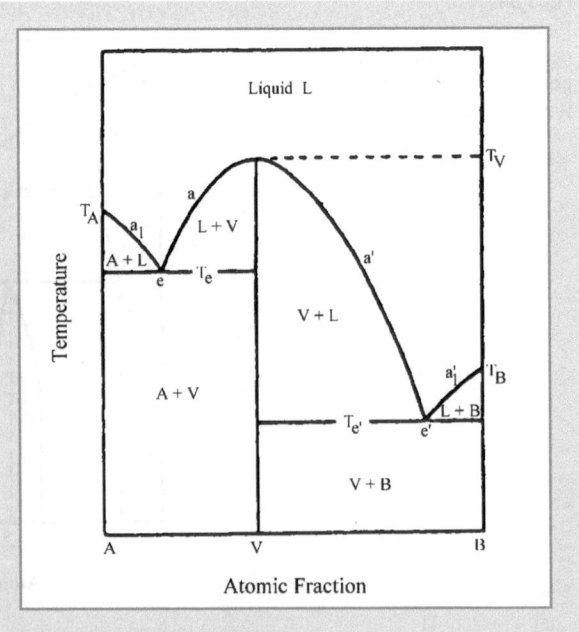

drawing in Fig. 3.55, we have to postulate that A is in a two-phase equilibrium with V. In a two-component system with a two phase equilibrium the composition of the phases must forcibly change. This would not be the case for the relations represented in Fig. 3.55, if the automatic assumption of a limited, temperature dependent solubility was not accepted, which due to its minimum extent could not be drawn.

The phase diagram in Fig. 3.55 can be regarded with a gross simplification as consisting of two partial diagrams, A – V and V – B. These are two simple eutectic systems, where one of the components V, is common to both systems. There is, however, a difference between a real eutectic system and the partial systems A – V and V – B. One can show, based on thermodynamic and kinetic considerations, that with low solubility in the solid, the liquidus line a_1 meets the T axis with an angle, that is different from $90°$. The melting point T_V of the compound on the other hand represent a real maximum, which connects without break e to e′ through liquidus line a – a′. The tangent at the maximum of a curve has always the slope zero. The perpendicular to the concentration axis A – V at V meets the liquidus line a – a′ with an angle of $90°$.

Based on the geometric situation, a compound with its own melting point is called an "open" melting compound, or a compound with a melting point maximum, or as a congruently melting compound.

3.24
Phase Diagram with a Non-Congruently Melting Compound

If V, the only compound appearing in a system, has its own melting point T_V or not, depends, as can be seen in Fig. 3.56, from its height as compared to T_A, the height of the pure component A. With fixed T_v and T_A, with $T_A = (T_A)_1$ one can expect a phase diagram with a congruently melting compound, also with $T_A = (T_A)_2$. $T_A = (T_A)_3$ represents a limiting situation. The liquidus line a_3 starting at $(T_A)_3$ meets the liquidus line a′ exactly at the melting point of the compound. If the melting point of component A is even higher, for example $T_A = (T_A)_4$, the liquidus line a_4 reaches the liquidus line a′ only at a temperature $T < T_V$, that is T_p. The compound does not have its own melting point. If the compound V is heated, it decomposes at T_p in a three phase reaction into a liquid of composition $(x_B)_p$ and solid A. The quantities are determined by the lever rule.

Correspondingly, the compound does not crystallize first from a liquid of composition of V. When the liquid L_V reaches the liquidus line a_4 during cooling, A crystallizes. With this an enrichment of the liquid in B takes place, which conditions a decrease in the liquidus temperature. When the temperature T_p is reached, liquid of composition $(x_B)_p$ reacts with the precipitated A and forms compound V:

$$L + A \rightarrow V \tag{3.42}$$

This is a peritectic reaction, analogous to the peritectic solidification of a solution (see Fig. 3.39). It begins during cooling when temperature T_p is reached

Fig. 3.56
Explanation of the influence
of the higher melting component on the character of
an intermediate compound

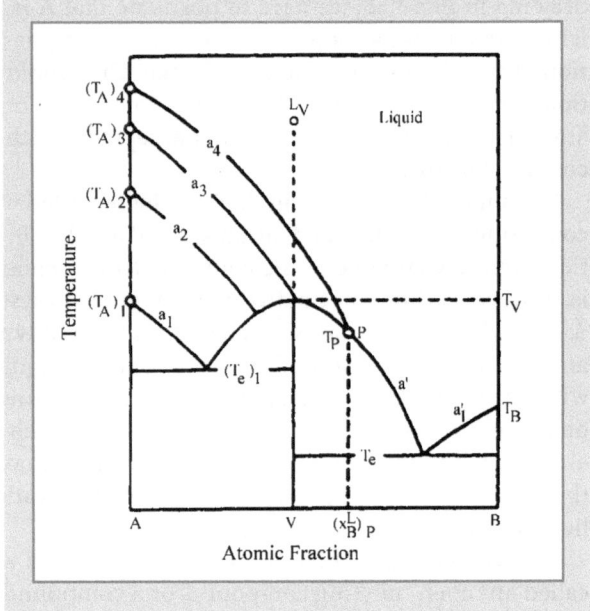

on the boundary liquid – solid A, and leads to completion only after long time. Under usual cooling conditions a structure results, where V is on the surfaces of the primary A crystallites (analogous to Fig. 3.40). As an example of such reaction zones Fig. 3.57 represents an Au-Al alloy solidified peritectically. After sufficiently long tempering below T_p the phase $AuAl_2$ can be transformed completely into AuAl.

In a cooling curve of a thermal analysis of a liquid L_V, an arrest occurs when the invariant peritectic equilibrium is reached, if equilibrium is present. The same applies for liquids with concentrations between $x_B^L = 0$ and $(x_B^L)_p$. The peritectic at T_p is indicated by the tie line for the three phase equilibrium (peritectic line), arising from the totality of the arrests of the cooling curves in the composition range $x_B < (x_B^L)_p$ where p is the peritectic point.

During cooling of liquids, with compositions $x_B^L > (x_B^L)_p$, crystallization starts when the liquidus line a′ is reached. Here compound V is the primary crystal.

Compounds, whose existence leads to fusion equilibria as shown in Fig. 3.56 for the case $T_A = (T_A)_4$, are called compounds with a "covered" maximum, or incongruently melting compounds.

Fig. 3.57
Microstructure including an incongruently melting compound which has been found by peritectic solidification (source: M. Kluge) Au-Al alloy containing 45 at.-% Al; Light microscopy, magnification: 150-times. Black: Primary formed $AuAl_2$ enveloped by white AuAl phase. Eutectic consists of $AuAl + Au_2Al$

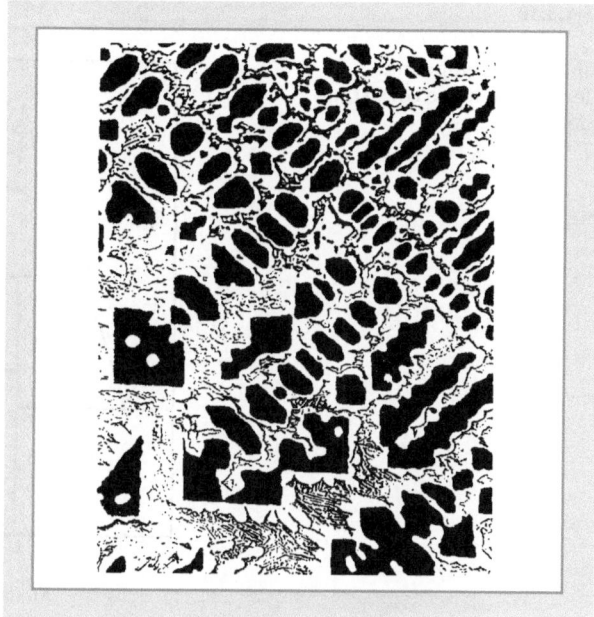

3.25
Phase Diagram with a Compound Forming from Two Melts

Figure 3.58 shows schematically the phase diagram of a system, where a compound V is formed from two coexisting liquid phases. During cooling of a single phase liquid L_V, the composition of which corresponds to that of V, decomposition into two liquid phases occurs when the miscibility gap is reached, which when the temperature T_t is reached, have the compositions $(x_B^l)_1$ and $(x_B^l)_2$. These phases form in an invariant three phase reaction the crystalline compound V:

$$L_1 + L_2 \rightarrow V \tag{3.43}$$

The transformation in Eq. (3.43) is called a syntectic reaction.

The crystallization of compound V occurs on the boundary between the two liquid phases. The formation of V separates the two reaction partners L_1 and L_2, so that a very long time is needed for a complete reaction, as the components of the two liquid phases have to diffuse through the solid phase formed.

In the case of only a local equilibrium at the boundary compound V – Liquid L_1 and compound V – Liquid L_2, after separation of the phases L_1 and L_2 from each other, a separate solidification of these liquids occurs. Here L_1 solidifies with V precipitating, until the temperature T_e is reached, and a final eutectic crystallization takes place. From L_2 also V crystallizes. In this case the solidification is terminated with the eutectic crystallization at $T_{e'}$.

Systems with compounds, which crystallize from two liquids, are relatively rare.

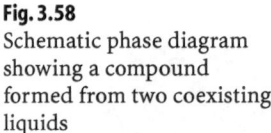

Fig. 3.58
Schematic phase diagram
showing a compound
formed from two coexisting
liquids

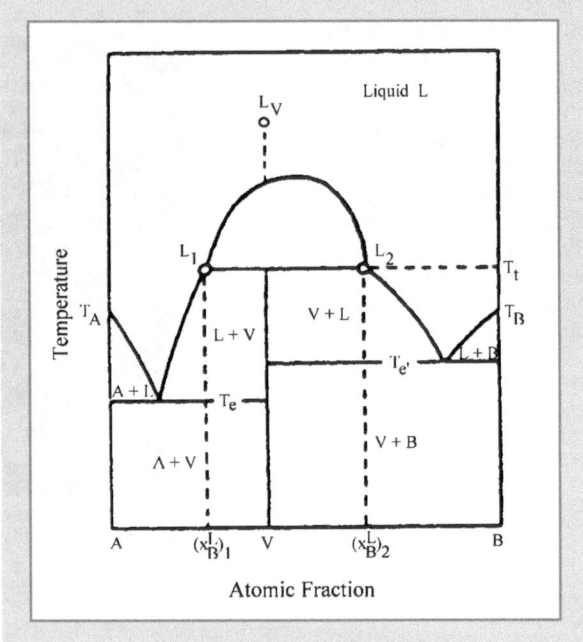

3.26
Real Diagrams with Compounds

A large number of real systems contain a congruently melting compound. As an example, the system FeO – Cr_2O_3 is represented in Fig. 3.59, where the compound $FeCr_2O_4$ appears. The compound $FeCr_2O_4$ belongs to the group of spinels, and is ferrimagnetic. Ferrimagnetic materials have great industrial importance. They are used for example to prepare permanent magnets.

Several compounds can appear in a system. In the simplest case, the fusion equilibria of the partial systems combine to an equivalent number of eutectic three phase equilibria. As an example for several congruently melting compound Fig. 3.60 shows the silver-strontium system.

Figure 3.61 shows a system with an incongruently melting compound, which at lower temperature decomposes through an eutectoid reaction.

Very often the situation exists, that in a system congruently melting as well as incongruently melting compounds are present. As example, Fig. 3.62 reproduces the system RbF – ThF_4. From the total seven compounds in this system, the congruently melting phase $3RbFThF_4$ decomposes by eutectoid reaction into two other compounds, while $5RbFThF_4$ forms from solid RbF and solid $7RbF2ThF_4$ during cooling only below the temperature range of the fusion equilibria.

Fig. 3.59
Phase diagram FeO – Cr_2O_3
[5]

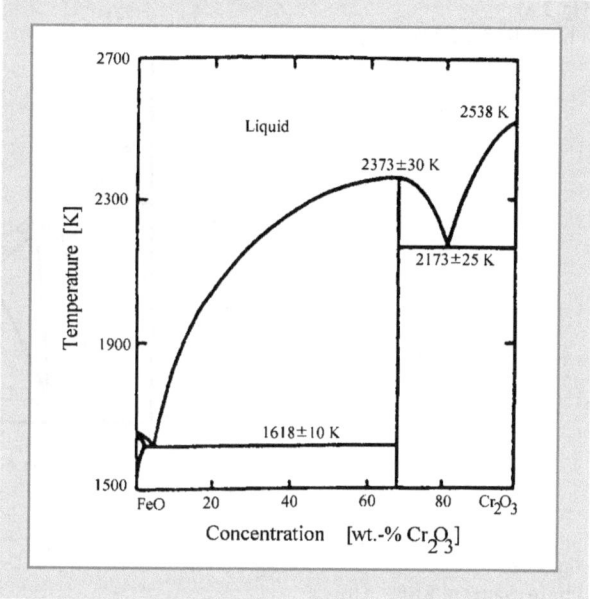

Fig. 3.60
Phase diagram Ag – Sr [2]

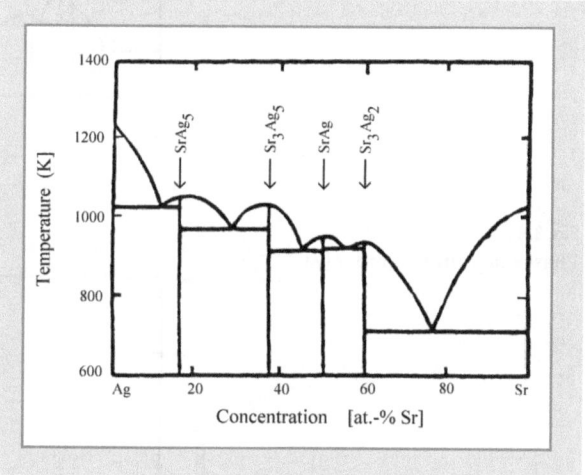

As an example for the formation of a compound through a reaction of two coexisting liquids Fig. 3.63 represents the phase diagram bismuth – bismuth-bromide. Above the critical demixing temperature, visible in the Figure, complete miscibility between liquid metallic bismuth and liquid salt $BiBr_3$ exists. At lower temperatures the subbromide BiBr is stable.

Fig. 3.61
Phase diagram Pb – Bi [8]

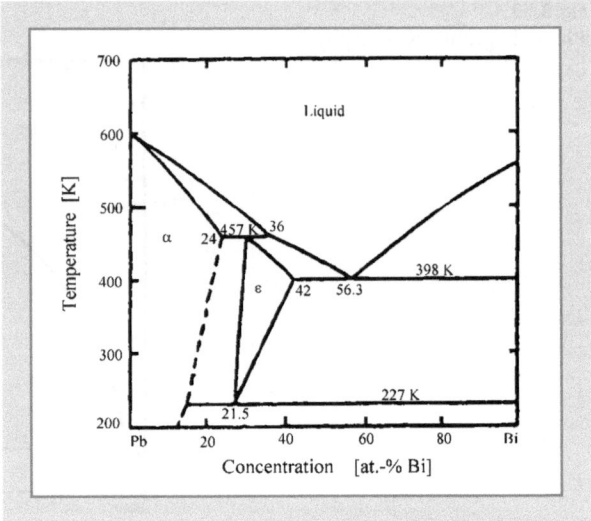

Fig. 3.62
Phase diagram RbF – ThF$_4$
[7]

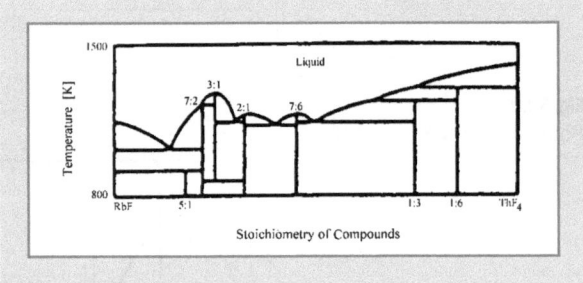

Fig. 3.63
Phase diagram Bi – BiBr$_3$ [7]

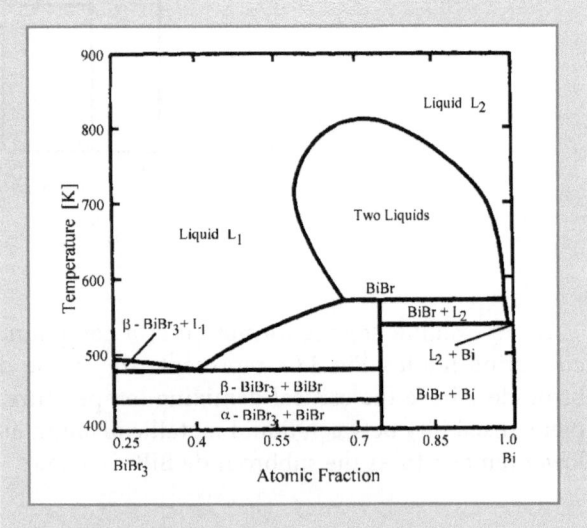

3.27
Transformation Equilibria

The polymorphic forms in many substances yield transformation equilibria in binary systems. The anisotropic transformation can be treated similarly to fusion from an energetic and kinetic point of view. The total process is the sum of two partial processes, the forward and backward reactions. For the rates of the partial reactions similar atomic factors as in the solid-liquid reaction are responsible. Therefore, in analogy to Eq. (3.11), we have for the transformation equilibrium, when forward and backward reactions occur at the same rate:

$$\frac{\Delta H^u}{RT^u} \approx \ln \frac{A_{\alpha \to \beta}}{A_{\beta \to \alpha}} \tag{3.44}$$

ΔH^u is the change in enthalpy during transformation of the substance from its α-form to its β-form, R is the gas constant, T^u the absolute transformation temperature. $A_{\alpha \to \beta}$ represents the accommodation coefficient for the partial reaction of the transition of the low temperature modification α to the high temperature modification β, and $A_{\beta \to \alpha}$ represents the accommodation coefficient for the partial reaction $\beta \to \alpha$.

If addition of a second component forms solid solutions, a displacement ΔT^u of the transformation point occurs, similar to the melting point displacement in solid-liquid equilibria. For a two-component system A – B, and solutions having a small B content, similarly to Eq. (3.19) we have:

$$\frac{\Delta H_A^u}{R(T_A^u)^2} = \frac{x_B^\beta}{\Delta T_A^u} - \frac{x_B^\alpha}{\Delta T_A^u} \tag{3.45}$$

ΔH_A^u is the transformation enthalpy of component A, T_A^u, the transformation temperature of component A, ΔT_A^u the displacement of the transformation temperature of A and x_B^β and x_B^α the mole fraction of component B in the coexisting solid solutions of the low temperature modification α and the high temperature modification β. Depending on the energetic situation, a transformation point lowering, or a transformation point increase is possible.

If both components have a transformation point, and the high temperature modifications form complete solid solutions, but the low temperature modifications do not, solid-solid equilibria occur in the simplest case, as shown schematically in Fig. 3.64. Figure 3.65 shows the phase diagram calcium-strontium. Both components appear in the represented region in three modifications, which pairwise form complete solid solutions. The similarity between the transformation equilibria and the fusion equilibria also shown is unmistakeable.

Iron has at low temperature a cubic body centered structure (α-Fe). It transforms during heating at 1184 K (911 °C) into a cubic face centered modification (γ-Fe). At 1665 K (1392 °C) γ-iron transforms into δ-iron, which has the same crystal structure as α-iron. This creates a pecularity, which consists in the fact, that by adding alloying elements which form solutions with iron, the transfor-

Fig. 3.64
Schematic representation
of simple equilibria of transformation for the case that
high temperature modifications of components are un-
interruptedly forming solid
solutions and on the other
hand the low temperature
modifications show limited
solubility in each other

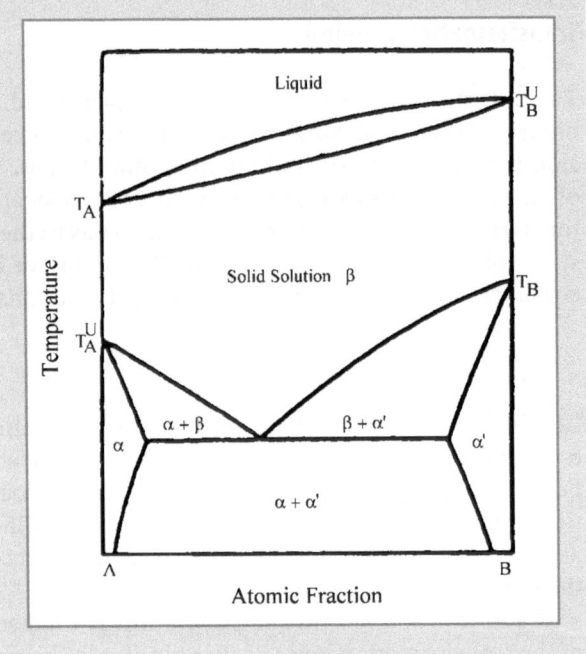

Fig. 3.65
Phase diagram Ca – Sr [6]

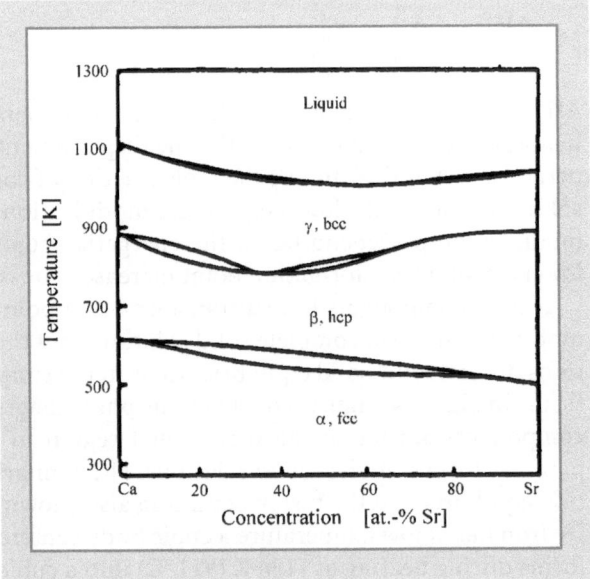

mation temperatures for the transformation $\alpha \rightarrow \gamma$ and $\gamma \rightarrow \delta$ can be displaced in the opposite direction. Elements, which due to energetic reasons, stabilize the γ-phase against the α-phase and δ-phase, expand the existence region of the γ-solid solutions, so that the α-γ transformation point decreases and the γ-δ transformation point increases (see Fig. 3.66a). On the other hand, if the addition element stabilizes the body centered cubic modification of iron, against the γ-modification the α-γ transformation temperature is increased, and the γ-δ transformation temperature decreased (see Fig. 3.66b). In this case, if the range of the body centered cubic solid solution extends to high enough concentration of the addition element, the α-γ and γ-δ equilibrium lines meet and join one another. A "closed" γ-solution field (γ-loop) exists. Both possibilities (Fig. 3.66a, b) are realized in industrially important systems with iron one of the components.

Fig. 3.66
Schematic representation of possible equilibria of transformation in binary systems with Fe as the main component

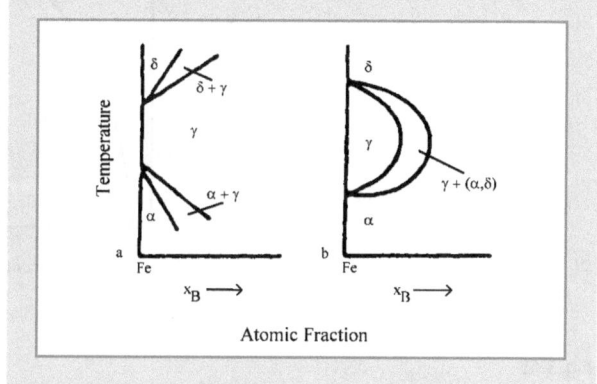

Fig. 3.67
Phase diagram Fe – Ni [2]

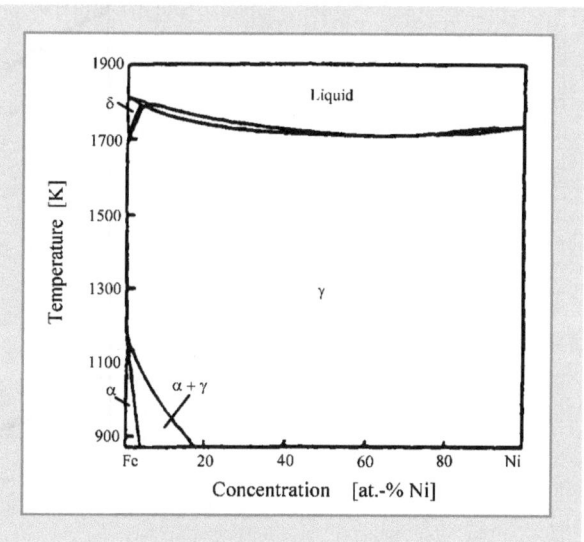

As an example the system iron – nickel is shown in Fig. 3.67 and iron-molybdenum in Fig. 3.68.

A simultaneous lowering of the transformation temperature of the $\alpha \rightarrow \gamma$ and $\gamma \rightarrow \delta$ transformation is possible, as happens with the addition of chromium to iron. Figure 3.69 shows a corresponding part from the iron-chromium phase diagram. The equilibrium lines closing the γ-field present here a minimum.

Fig. 3.68
Phase diagram Fe – Mo [2]

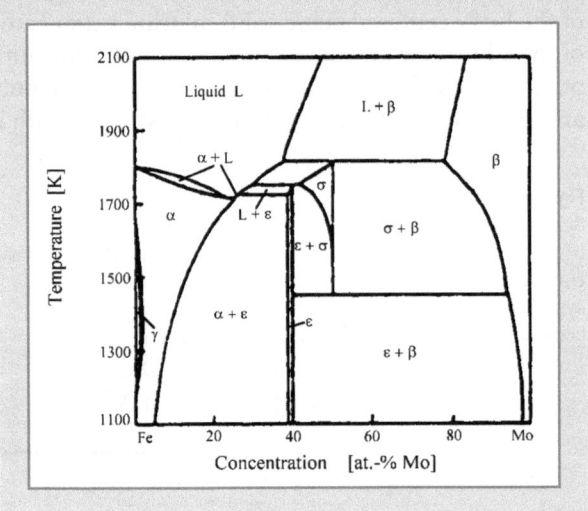

Fig. 3.69
Fe-rich part of the phase diagram Fe – Cr [2]

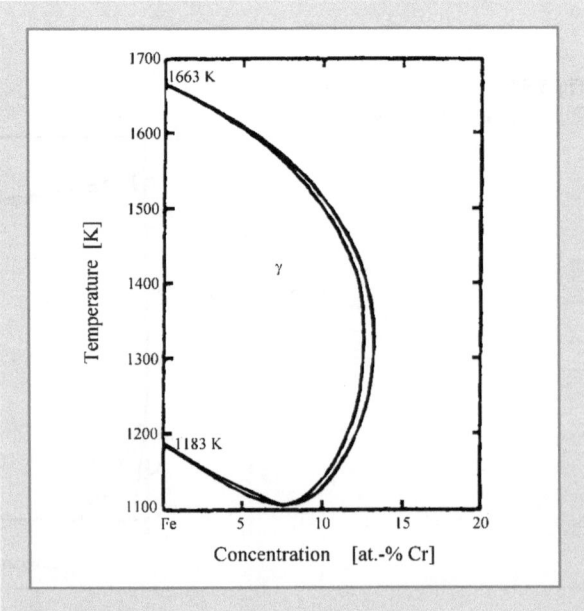

3.28
The Iron-Carbon Phase Diagram

Reduction of iron ore in the blast furnace yields pig iron with a carbon content of about 2 wt.-%. The preparation of steel from pig iron consists mainly in lowering the carbon content. Not less important is the heat treatment, through which the certain phases, correctly distributed, are produced in the structure, to obtain the typical properties of the steel. An overview about the most important possibilities to influence the structure can already be obtained from the iron-carbon phase diagram. It is shown in Fig. 3.70.

Addition of carbon lowers the melting point and the α-γ transformation point. The γ-δ transformation point is raised. The γ-solid solution is stabilized with respect to the body centered cubic solid solution, by the addition of carbon, the γ-field is expanded at the cost of the α- and δ-fields. This is due to the fact, that carbon, because of its small atomic radius does not substitute for iron atoms on its regular lattice sites, but is placed in interstitial sites (octahedral interstitial sites). These interstitial sites offer less place to the carbon atoms in the cubic body centered iron, than in the cubic face centered. Lattice deformation energy has to be expended in both cases, when creating an interstitial solid solution. It is, however, significantly higher for the formation of a body centered

Fig. 3.70
Phase diagram Fe – C [2].
----- Equilibrium lines for
stable equilibria with graphite as the C-richest phase,
-.-.-. magnetic transfor-
mation

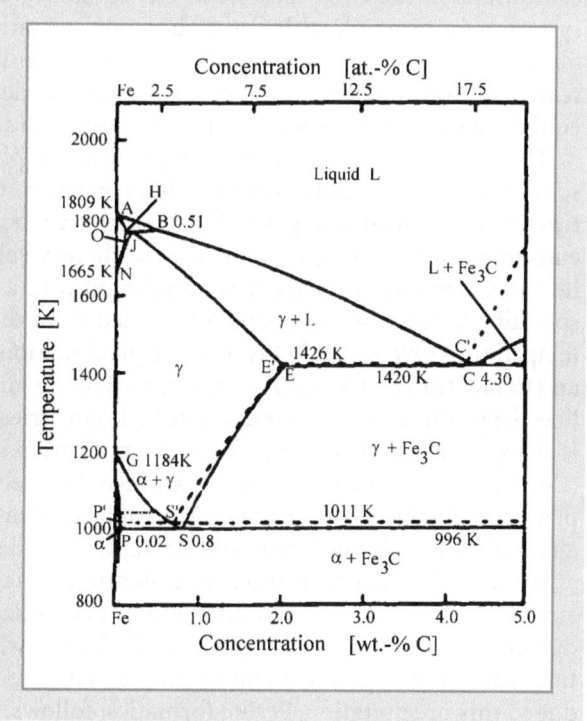

cubic iron solid solution than for the formation of a face centered cubic solid solution. Thus, carbon is enriched in the γ-phase in the α-γ and γ-δ equilibria, because it is connected with a much smaller lattice distortion energy. In each case, for the coexisting phases we have

$$x_C^\alpha < x_C^\gamma \text{ or } x_C^\delta < x_C^\gamma$$

The γ-δ transformation equilibria combine with the fusion equilibria, creating an invariant three phase equilibrium, where liquid, γ-solid solution and δ-solid solution coexist.

In the iron-carbon system only graphite appears as the stable form, besides the α-, γ-, and δ-solid solutions. An eutectic results with about 4.3 wt.-% C. From an iron rich liquid as used in industry, however, graphite generally does not crystallize under customary cooling conditions. Rather, during solidification of iron alloys with more than 4.3 wt.-% C, the compound Fe_3C forms. This compound, called Cementite, is metastable. Its structure has greater similarity to the atomic structure of iron rich liquids, than to graphite. Thus, nuclei of Cementite are formed easier than those of graphite. The liquidus curve of the Cementite also combines with the liquidus line which starts at the melting point of iron, to an eutectic.

The solubility limit of the γ-phase, called Austenite, is different, depending if the graphite or Cementite is the phase which appears at the higher carbon concentrations in the two phase field. The solubility limit belonging to the stable system is in principle at lower carbon concentrations. The stable phase equilibria involving graphite are only of importance if one waits long enough to reach equilibrium, until the Cementite present decomposes. Here only phase equilibria involving Cemetite as partner are considered.

The solubility line E–S, which limits the single phase field of the Austenite (γ-solid solution) against the two phase field Austenite-Cementite combines with the two phase equilibrium α-γ, which starts at the α-γ transformation of pure iron. An eutectoid three phase equilibrium arises, where α-solid solution (Ferrite), Austenite and Cementite coexist. If γ-solid solution with a carbon concentration corresponding to the eutectoid point S is cooled from the γ-field, when the eutectoid temperature (996 K; 723 °C) is reached, precipitation of Ferrite (α-solid solution) and Cementite (Fe_3C) begins simultanously. If the Austenite sample is polycrystalline, the precipitation begins at the grain boundaries, because here the nucleation is energetically favored compared to the nucleation in the undisturbed grain. The two phases Ferrite and Cementite arrange themselves in a lamellar form, completely analogous to the crystallization of a lamellar eutectic. This structure with alternating lamellae of Ferrite and Cementite is called Pearlite.

If the carbon concentration in Austenite is smaller than corresponds to the eutectic point, primary α-solid solution crystallizes on the grain boundaries, during cooling when reaching the line G–S. Only, when lowering of the temperature, when the carbon content of the Austenite reaches the concentration of S due to this precipitation, Perlite formation follows.

Similarly, primary Cementite precipitates at the grain boundaries, if starting with an Austenite with a carbon concentration above 0.8 wt.-% C. Microstructures for these three situations are shown in Fig. 3.71.

The microstructures shown in Fig. 3.71 are obtained through slow cooling of Austenite. With increasing cooling rate the time the sample stays a few degrees below 996 K, where the Perlite formation is rapid, decreases. With decreasing temperature the diffusion rate decreases. The lamellae become thinner, to permit rapid distribution of the carbon on the phases forming. With sufficiently

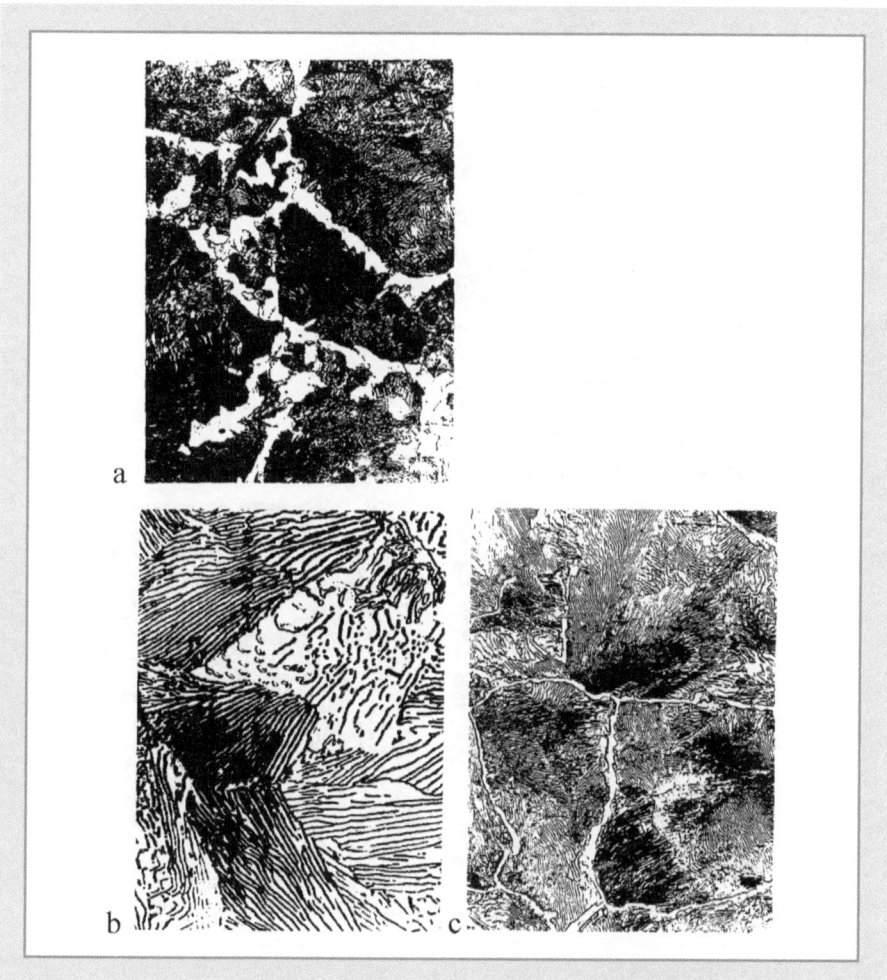

Fig. 3.71 Microstructure of Fe-C alloys; Light microscopy. **a** 0.5 wt.-% C, primary Ferrite (bright, on grain boundaries) and Perlite; magnification: 600-times (source: U. Albrecht), **b** 0.8 wt.-% C, Perlite; magnification: 375-times (source: A. Gerold), **c** 1.4 wt.-% C, primary Cementite (bright, on grain boundaries) and Perlite; magnification: 300-times (source: E. Scheil)

rapid cooling, the thickness of the lamellae becomes so thin, that they can not be resolved using light microscopy. Such a microstructure is called Sorbite (see Fig. 3.72a). The lamellar nature can be seen using an electron microscope.

Finally, if the cooling rate is so high, that the sample rapidly reaches temperature ranges where the diffusion controlled transformation stops before the Austenite is completely transformed, only sphere like microstructures are visible in a few places, where the Pearlite formation started, but then stopped. This microstructure is called Troostite (see Fig. 3.71b).

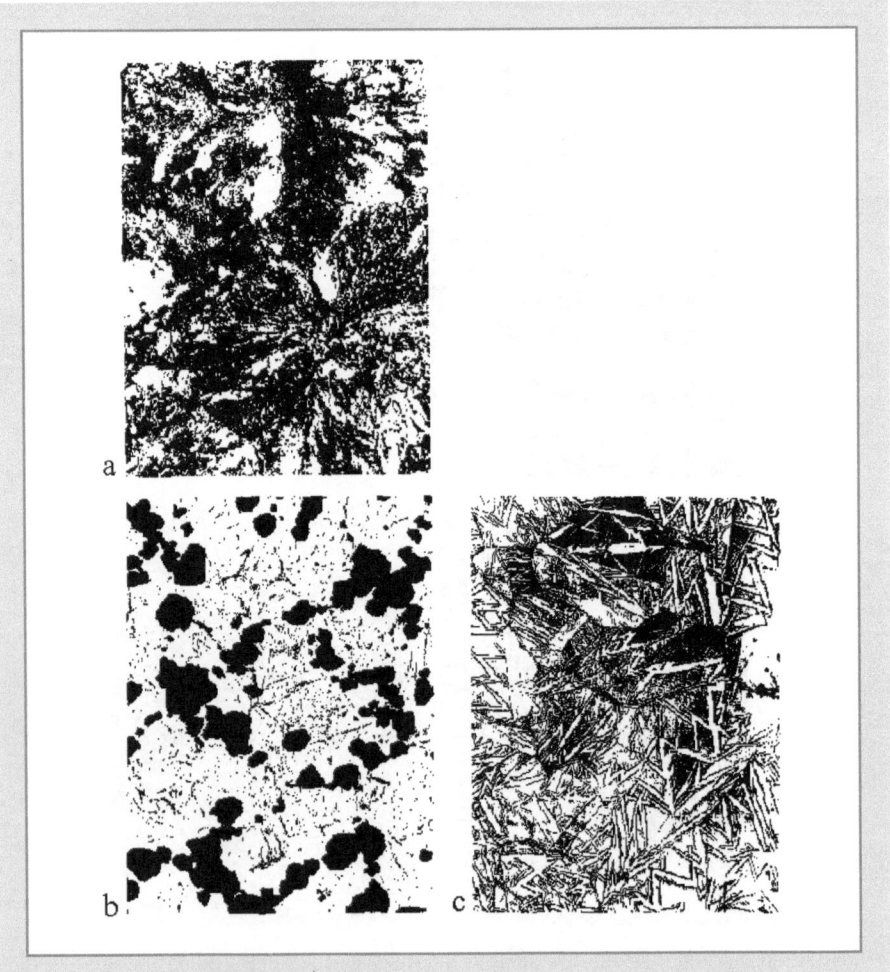

Fig. 3.72 Microstructure of rapidly cooled Fe-C alloy; Light microscopy. **a** 0.8 wt.-% C: Sorbite, magnification: 600-times (source: I. Atmer), **b** 0.8 wt.-% C: Troostite (black) and rest Austenite with Martensite, **c** 1.7 wt.-% C: Martensite (bright) in rest – Austenite; magnification: 150-times (source: A. Gerold)

At even higher cooling rates the formation of Pearlite is completely suppressed. Austenite can be undercooled as a metastable phase way below the equilibrium temperature of Pearlite formation. The difference in energy between the stable and metastable phases increases with increasing undercooling below an equilibrium point. The tendency to transform into the equilibrium phases increases. In the present case, this is primarily the tendency for transforming the γ-phase into the body centered cubic α-phase. At a sufficiently low temperature (approx. 470 K) a diffusionless transformation occurs, resulting in a distorted form of the α-Fe lattice, called Martensite. This diffusionless phase transition is called the martensitic transformation. Individual atoms are not moved by diffusion until they effect a transformation step. Rather all atoms in a large volume of Austenite undertake identical steps, similarly to the mechanical formation of crystallographic twins. This cooperative mechanism is not thermally activated. The transformation rate is practically temperature independent. It corresponds to the rate of propagation of an elastic deformation in the corresponding solid (speed of sound).

If a certain volume element is transformed into the new structure, it changes its form, and imposes an elastic distortion onto the surroundings in the original matrix, which did not transform yet. If a critical amount of distortion is reached, it hinders further transformations. With a preset end temperature of the cooling process the transformation process stops after the transformation of a specific amount of the starting matrix. Only further cooling can extract more Gibbs energy from the transformation process, which is available to increase the deformation on the Austenite matrix. The Austenite sample is not completely transformed. One part of the sample remains Austenite (leftover Austenite). The microstructure of a martensitically transformed Austenite sample is shown in Fig. 3.72c. Also in the microstructure of the "Troostite", in addition to spherolites one can recognize martensitic lines (see Fig. 3.72b).

The Martensite formed through a diffusionless transformation, has a body centered tetragonal structure. It is similar to the aspired α-Fe structure.

A simple representation of the geometrical connection between the structure of the Martensite with the starting phase (Austenite, cubic face centered) is shown in Fig. 3.73. One can see, that the face centered cubic lattice can also be represented as a body centered tetragonal lattice. To transform the tetragonal body centered structure shown into a face centered cubic structure, a compression in the c-direction and an expansion in the a- and b-directions is required. The carbon atoms (x) sitting in the lattice holes disturb such a hypothetical transition to an exactly cubic face centered lattice. Thus, the tetragonally distorted lattice of the Martensite results, the c/a ratio depending on the carbon content. The simple geometric description shows principally the starting and end positions of the atoms. The atomic mechanism, according to which the transition occurs, does not consists in a jolting or dilation. The process is complicated. Jolting and dilatation are not transition processes which can be carried out in an atomic cooperative process. The complex process of the transition is

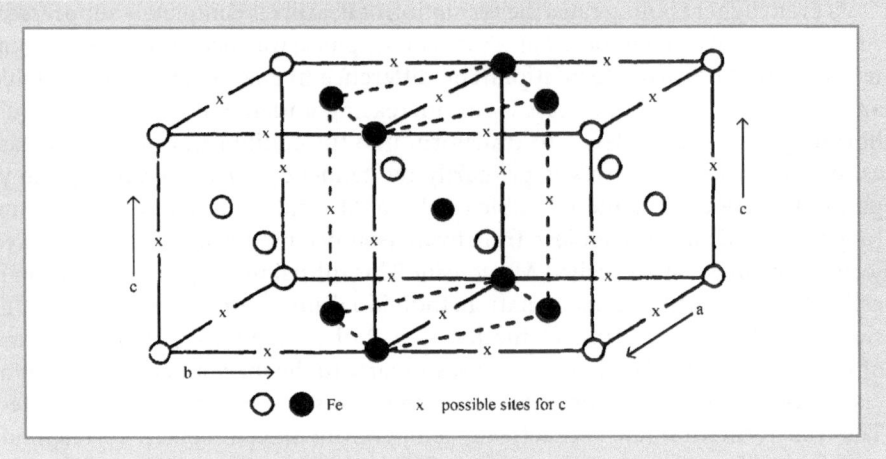

Fig. 3.73 Correspondence of bc cubic and bc tetragonal structure to explain the martensitic transformation

dependent in addition on the carbon content and the presence of other alloying components.

Martensite is hard and brittle. Pearlite is soft and tough. Through suitable heat treatment one can obtain the desired properties. If Martensite is heated to higher temperatures, where the diffusion of carbon is high enough, this metastable phase decomposes with the formation of Ferrite and Cementite. This tempering process permits also to regulate the properties of steel.

References

Citations
[1] B. Chalmers, "Principles of Solidification", J. Wiley, Sons, New York (1964)
[2] M. Hansen and K. Anderko, "Constitution of Binary Alloys", McGraw-Hill Book Comp., New York (1958)
[3] R. Vogel, "Einführung in die Metallurgie", Musterschmidt-Verlag, Göttingen (1955)
[4] T.B. Massalski, N. Okamoto, P.R. Sabramanian and Linda Kacprzak, "Binary Alloy Phase Diagrams", Second Edition, A.S.M. International (1992) Materials Information Society
[5] E.M. Levin and H.F. McMurdie (Editor: M.K. Reser), "Phase Diagrams for Ceramists, 1975 Supplement", The Amer. Ceramic Soc., Columbus, Ohio, USA (1975)
[6] R.P. Elliott, "Constitution of Binary Alloys", First Supplement, McGraw-Hill Book Comp., New York (1965)
[7] M. Blander (Editor), "Molten Salt Chemistry", Interscience Publishers, New York (1964)
[8] B. Predel and W. Schwermann, Z. Metallkde., _58_, 553 (1967)

General References
"Das Zustandsschaubild Eisen-Kohlenstoff und die Grundlagen der Eisen-Kohlenstoff Legierungen", 4. Edition, Verlag Stahleisen, Düsseldorf (1961)
P. Gordon, "Principles of Phase Diagrams in Materials Systems", McGraw-Hill Book Comp., New York (1968)

A.G. Guy and G. Petzow, "Metallkunde für Ingenieure", Akademische Verlagsgesellschaft, Frankfurt/M. (1970)

R. Haase und H. Schonert, "Solid-Liquid Equilibrium", in "The International Encyclopedia of Physical Chemistry and Chemical Physics", Topic 13, Mixtures, Solutions, Chemical and Phase Equilibria, Vol. I, Editor: M.L. McGlashan, Pergamon Press, Oxford (1969)

W.G. Moffatt, "The Handbook of Binary Phase Diagrams", General Electric Company. Schenectady (1978)

J. Nyvlt, "Solid-Liquid Phase Equilibria", Elsevier Scientific Publ. Comp., Amsterdam (1977)

G Petzow, "Metallographic Etching", Amer. Soc. for Metals, Metals Park, Ohio, USA (1978)

G Petzow, "Metallographisches Ätzen", 5. Edition, Gebr. Bornträger, Stuttgart (1976)

Prince, "Alloy Phase Equilibria", Elsevier Publ. Comp., Amsterdam (1966)

F.A. Shunk, "Constitution of Binary Alloys", Second Supplement, McGraw-Hill Book Comp., New York (1969)

H. Schumann, "Metallographie", VEB Deutscher Verlag für Grundstoffindustrie, Leipzig (1975)

R Vogel, "Die heterogenen Gleichgewichte", 2. Edition, Akademische Verlagsgesellschaft Geest u. Portig, Leipzig (1959)

CHAPTER 4

Phase Equilibria in Three-Component Systems and Four-Component Systems with Exclusion of the Gas Phase

4.1
The Composition Triangle

In a three-component system (A, B, C) the composition of a mixture is fixed by the concentration of two components. For the mole fractions we have

$$x_A + x_B + x_C = 1 \qquad\qquad (3.2a):$$

The graphical representation of the composition of a ternary mixture is done generally in the composition triangle (see Fig. 4.1a).

In an equilateral triangle the sum of the length of all verticals from a point Q to the sides of the triangle equals the height h of the triangle. If one sets $h = 1$, than for point Q:

$$\overline{Qa} + \overline{Qb} + \overline{Qc} = 1 \qquad\qquad (4.1)$$

For the representation of the concentration (see Eq. (4.2a)) the distances Qa, Qb and Qc can be set equal to the mole fractions x_A, x_B and x_C. This, however, is not the optimal method of representation. It is evident from Fig. 4.1a that the following relations hold:

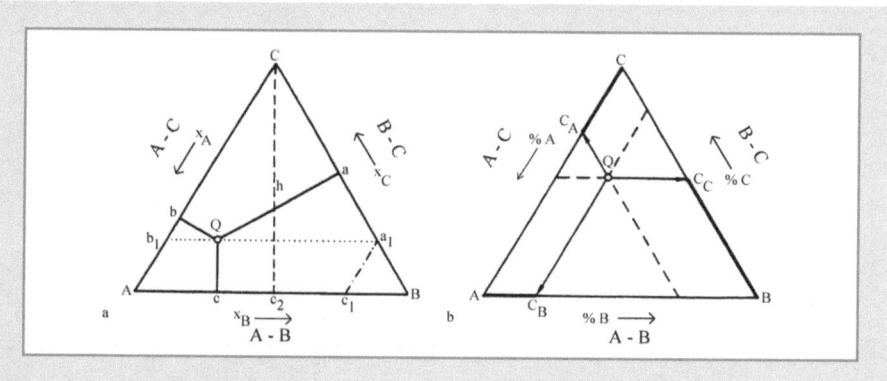

Fig. 4.1 Representation of concentration of ternary mixtures in triangle coordinates

$$\overline{Qc}:\overline{Qc_2}=\overline{AB_1}:\overline{AC}=\overline{Ba_1}:\overline{BC} \tag{4.2}$$

Thus, instead of $Cc_2 = h = 1$, one can set

$$\overline{AB}=\overline{AC}=\overline{CB}=1 \tag{4.3}$$

Now the corners of the equilateral triangle correspond to the pure components. On the sides the concentration of the bounding binary systems are marked. The interior surface of the triangle represents the totality of the composition points of the ternary mixture.

A special case of a set of compositions exists, when the ternary mixtures lay on a line parallel to one side of the triangle. The distance of the line $c_1 - a_1$ from the corner B, or equivalently the distance of all the points on the line $c_1 - a_1$

Fig. 4.2
Analysis of a tie-line in a ternary system

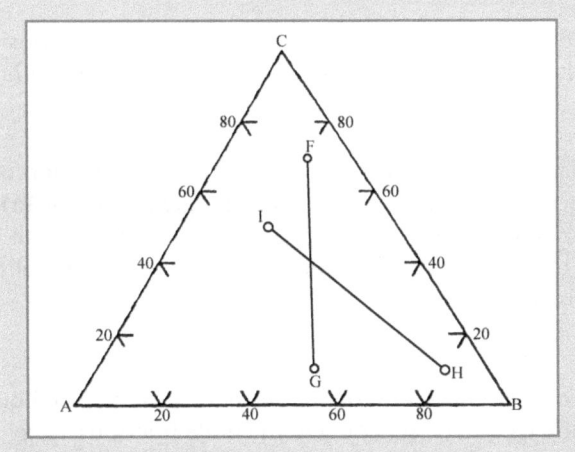

Fig. 4.3
Analysis of the tie triangle

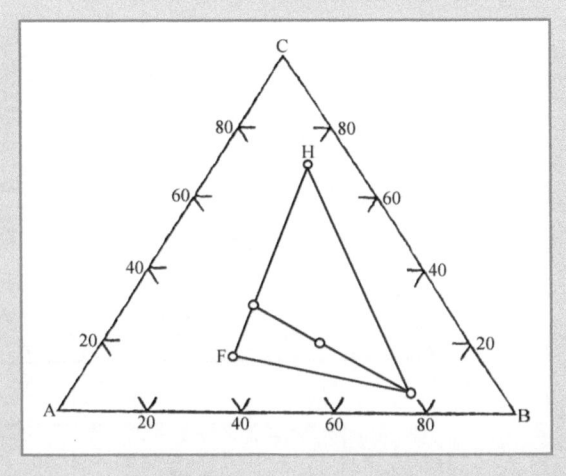

from the line A – C is constant. All mixtures, whose compositions correspond to points on the line $c_1 - a_1$ have the same content in B (Figs. 4.2 and 4.3).

Another special case arises for a line in the composition triangle, which passes through a corner. Mixtures, whose composition lay on such a line are characterized by a constant ratio of the concentrations of the other two components, through whose corner the line does not pass.

Instead of atom fractions the concentration is often given in atom-% or weight-%. The distances AB = AC = CB correspond then to 100 at.-% or 100 wt.-%. To determine the concentration of the components which correspond to a given point Q in the composition triangle, lines parallel to the sides of the triangle are drawn through this point (see Fig. 4.1b). The share of A (concentration c_A) is given by the distance Cc_A. The distance Ac_B represents the concentration of B and the distance Bc_c the concentration of C.

4.2
Lever Rule in Ternary Systems

If a ternary mixture Q decomposes into two phases α and β, the concentrations Q, α and β lay on a line, as in a binary system, a tie line.

The ratio of the amounts of the phases α and β is given by the lever rule.

Let M be the totality of the atoms in alloy Q. Is m_α the fraction of A atoms contained in the α-phase, and m_β the fraction of atoms present in the β-phase, then one has (see Fig. 4.4)

$$\frac{m_\alpha}{m\beta} = \frac{\overline{Q\beta}}{\overline{\alpha Q}} = \frac{x_B^\beta - x_1}{x_B - x_B^\alpha} \quad \text{or} \quad \frac{m_\alpha}{m\beta} = \frac{x_C - x_C^\beta}{x_C^\alpha - x_C} \tag{4.4}$$

Fig. 4.4
Explanation of the lever-rule in a ternary system

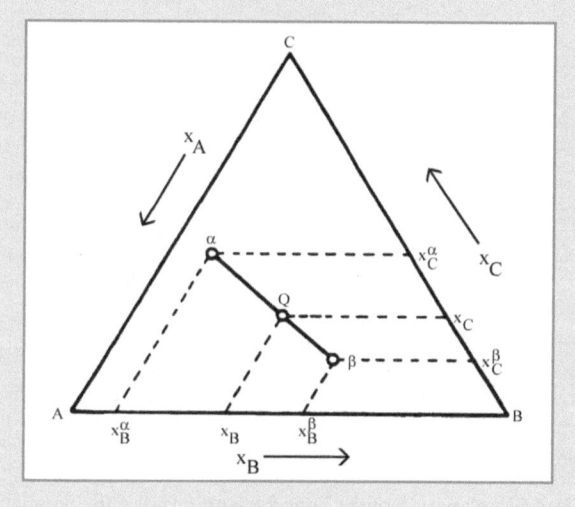

4.3
Compatibility Triangle

If a ternary solution Q decomposes into three phases α, β and γ, the points representing their compositions form a triangle. The compositions of the starting phase Q lay in the interior of this triangle $\alpha\beta\gamma$ (see Fig. 4.5a). The compositions are represented by atom fractions and the total number of atoms in the three phase alloy is M. The fractions of M, which are present in the various phases are m_α, m_β and m_γ. The composition of Q corresponds to the center of gravity of these three quantities of atoms, which are suspended at the corner of the triangle $\alpha\beta\gamma$ (see Fig. 4.5b). This triangle, connecting the three coexisting phases is called a compatibility triangle. If the composition is expressed in mass fractions or wt.-%, then Q is the center of gravity of the masses of the three phases α, β and γ.

The atom fractions x_A, x_B and x_C of the center of gravity of a compatibility triangle are given by

$$x_A = m_\alpha \cdot x_A^\alpha + m_\beta \cdot x_A^\beta + m_\gamma \cdot x_A^\gamma$$

$$x_B = m_\alpha \cdot x_B^\alpha + m_\beta \cdot x_B^\beta + m_\gamma \cdot x_B^\gamma$$

$$x_C = m_\alpha \cdot x_C^\alpha + m_\beta \cdot x_C^\beta + m_\gamma \cdot x_C^\gamma \qquad (4.5)$$

The mass fraction of a phase (for example phase α) of the total mass M of the phase mixture is, as can be seen using the lever rule from Fig. 4.5a:

$$\frac{m_\alpha}{M} = \frac{\overline{Q\alpha}}{\alpha a} \qquad (4.6)$$

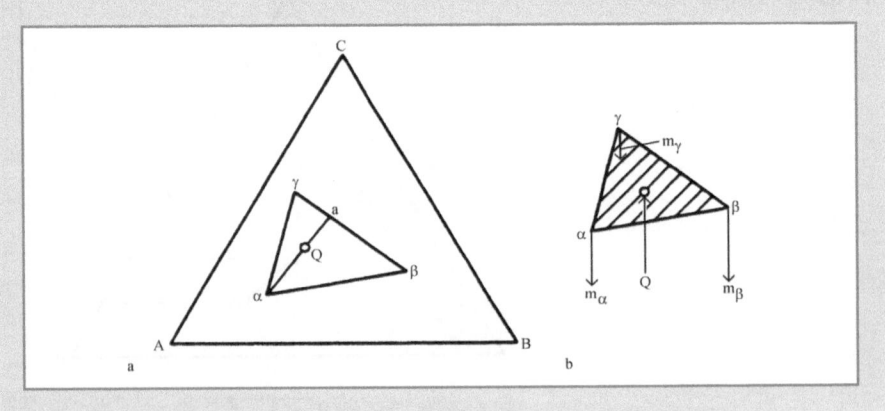

Fig. 4.5 Explanation of the konod triangle and of the compatibility triangle

4.4
Four Phase Equilibria

The rule of the center of gravity is naturally valid, even if a material consists of a mixture of more than three phases. In analogy to the compatibility triangle, the center of gravity of the multiple compatibility shape must be determined taking into consideration the quantities of each phase present.

Three cases have to be distinguished with a mixture consisting of four phases (solid phases α, β, γ and liquid L):

a) The composition of the liquid (L) is inside the triangle formed by the other three components (see Fig. 4.6a).

b) The composition of the four phases form a tetrahedron. The composition of each phase lays outside of the triangle, formed by the other three phases (see Fig. 4.6b).

c) In a ternary system, according to the Gibbs phase rule, in an invariant equilibrium at constant pressure four phases are present. The composition points of coexistent phases in such equilibria can have the arrangements shown in Fig. 4.6a, b. The coexistence of the individual phases is shown by single tie lines. The proportion of the quantities of the phases in this invariant equilibrium are changed by the addition or removal of heat. L is a liquid, and α, β and γ are solid solutions which are formed during heat removal from L, thus, we have a ternary eutectic equilibrium:

$$L \leftrightarrow \alpha + \beta + \gamma \tag{4.7}$$

The composition of the liquid L in an eutectic equilibrium must necessarily lay inside the triangle, which can be formed from the compositions of the solid phases formed.

The case represented in Fig. 4.6b corresponds to another invariant equilibrium. Point a represents a mixture of the two phases α and β as well as a mixture of phases β and L. Also, according to the lever rule, the correct ratios of the

Fig. 4.6
Four phase equilibria in ternary systems. **a** Eutectic equilibrium, **b** transition equilibrium, **c** peritectic equilibrium

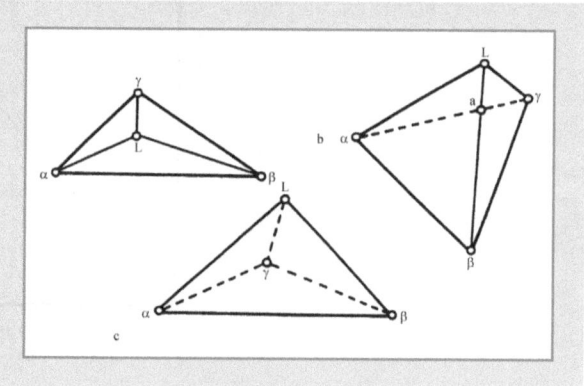

quantities m_α/m_γ and m_β/m_L must be present. The only equilibrium which satisfies these conditions is:

$$L + \beta \leftrightarrow \alpha + \gamma \tag{4.8}$$

Equation (4.8) represents a reaction, where, during cooling the first formed β-solid solution reacts in equilibrium with the liquid L and forms the two solid solutions α and γ. This reaction is called an intermediate equilibrium.

The conditions of the ternary peritectic reaction are shown in Fig. 4.6c. In this case the liquid reacts during cooling with the solid solutions α and β formed earlier, and creates the solid solution γ:

$$L + \alpha + \beta \leftrightarrow \gamma \tag{4.9}$$

The point L represents the support point of the triangle in Figure 4.6a if at the corner α the mass μ_α at the corner β the mass μ_β and at the corner γ the mass μ_γ are suspended. The same is valid for the point a in Fig. 4.6b and point γ in Fig. 4.6c.

4.5
Representation of Ternary Phase Diagrams

For a complete representation of phase equilibria in a ternary system three state variables are necessary while the pressure is kept constant: There are two compositions and the temperature. The temperature is plotted perpendicularly to the composition triangle (see Fig. 4.7). A three edged prism results. The edges

Fig. 4.7
System of coordinates for representation of ternary phase diagrams at constant pressure

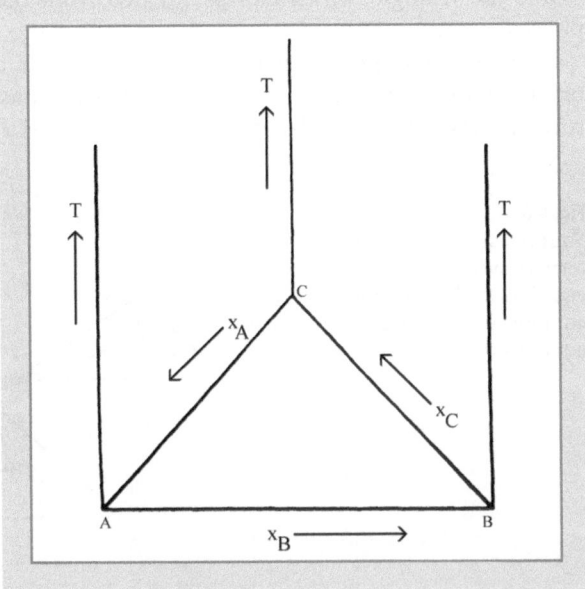

of the prism show the state of the pure components A, B and C as a function of the temperature. The three lateral prism surfaces represent the phase diagrams of the bounding binary systems A–B, A–C and B–C. Finally the inside of the prism indicates the changes in composition of the ternary mixture.

4.6
A Simple System with a Ternary Eutectic

Let us consider a system, for which the following conditions apply (see Fig. 4.8):
a) The components are practically immiscible in the solid state, and completely miscible in the liquid state. The bounding binary systems are therefore simple eutectic systems.
b) A ternary eutectic is present.
c) No intermetallic compounds are present.

The liquidus lines u and v in Fig. 4.8, which for example, start at the melting point of component C, T_C, and belong to the two boundary systems A–C and B–C, are in the ternary the limiting lines of an arched liquidus surface L_C. Similar considerations apply for the liquidus lines and liquidus surfaces L_A and L_B, which originate at T_A and T_B. The liquidus surfaces L_C and L_A meet in an eutectic channel, which starts at the eutectic point e_1 of the bounding system A–C.

Fig. 4.8
Three-dimensional representation of the liquidus spheres of a ternary system with total miscibility of the components in the liquid and with total immiscibility in the solid state and no occurrence of an intermediate phase

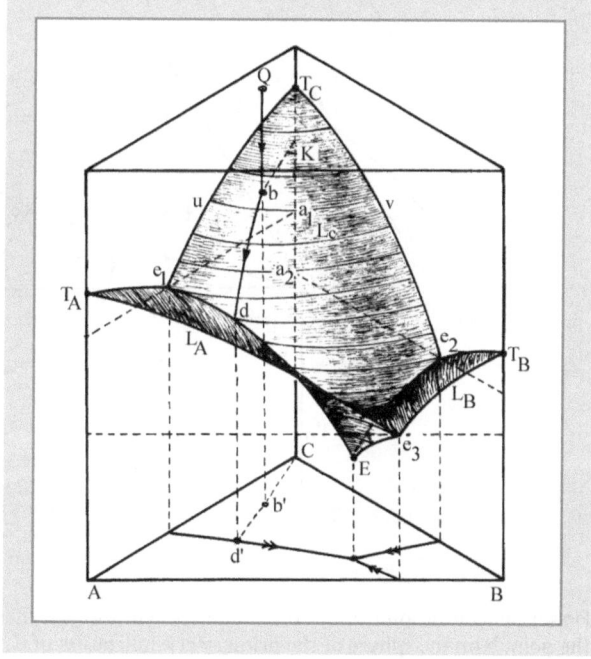

Correspondingly the same applies to the other liquidus surfaces and eutectic channels. The three eutectic channels meet in the ternary eutectic point E.

As a rule it is more sensible, than a drawing in perspective, to represent the important parts of a three-dimensional phase diagram in the form of a two-dimensional projection as shown as an example in Fig. 4.9.

The binary bounding systems are turned by 90° into the plane of the drawing. The eutectic channels start at the binary eutectic points. They are indicated by double arrows. The arrows show the direction, in which the channel runs to lower temperatures.

Under the simplified assumption that no significant mutual solubility of the components in the solid phase exists, the solidification of a liquid Q occurs in the following manner (see Figs. 4.8 and 4.9). During cooling the liquidus surface L_C is met at b. Practically pure C crystallizes. K is the tie line, which at the beginning of the solidification connects the coexisting phases (Liquid b and solid C).

The liquid becomes poorer in C by precipitation of C. The melting point decreases. The change in concentration of the liquid follows the straight line extension of the tie line K in the direction of the drawn arrows, assuming negligible solubility of A and B in solid C (see Fig. 4.9).

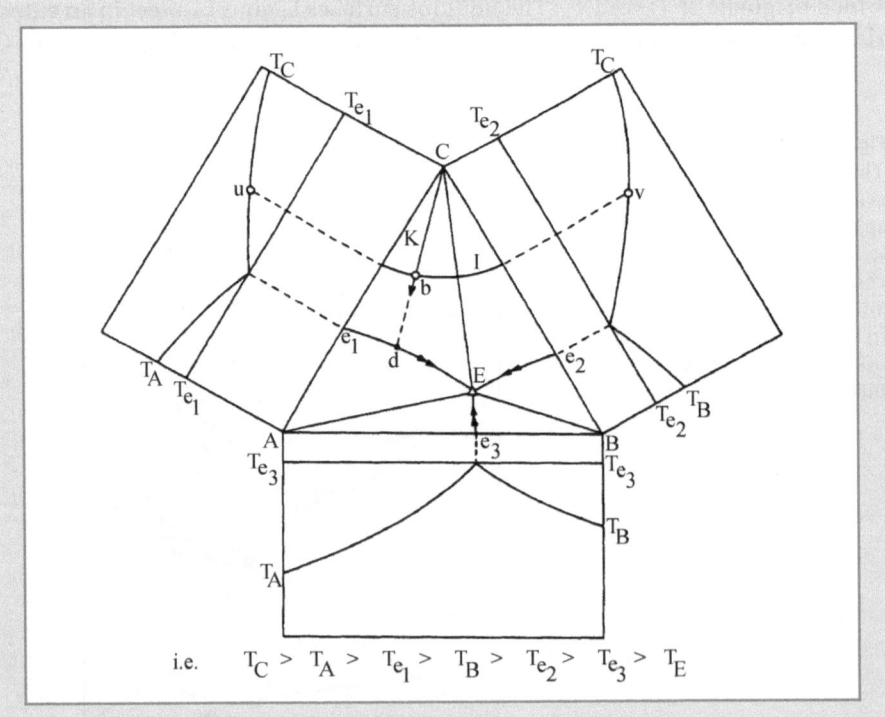

i.e. $T_C > T_A > T_{e_1} > T_B > T_{e_2} > T_{e_3} > T_E$

Fig 4.9 Concentration triangle of a simple system with a ternary eutectic. I = Isotherm through the point b on the sphere of the primary crystallization of C

During this primary solidification two phases are in equilibrium. According to the Gibbs phase rule, two degrees of freedom are present. The crystallization can occur on a surface, the liquidus surface. In a binary system for the primary crystallization $F = 1$, and the precipitation of the primary crystals occurs along a line (liquidus line).

If point d is reached during continued cooling, the liquid is not only saturated in C, but also in A. Now C and A crystallize simultaneously. This represents a binary eutectic crystallization of a ternary alloy, where one degree of freedom is still available. The solidification occurs along a line, the channel of the binary eutectic crystallization, $e_1 - E$ (see Figs. 4.8 and 4.9). The temperature can further decrease. It must be remembered, that the eutectic solidification in a binary system is an invariant reaction, and thus, occurs at constant temperature.

The liquid's content on B increases during precipitation of A and C while the eutectic channel $e_1 - E$ is followed, until the ternary eutectic E is reached. Here three phases A, B and C crystallize simultaneously. An invariant equilibrium occurs (four phase equilibrium). It is visible in Fig. 4.9 that in the composition range of the triangle AEB, after primary crystallization, the binary eutectic solidification of A and B follows. In the range AEC the binary eutectic crystallization of A and C, and in the triangle BEC the binary eutectic precipitation of B and C which finally leads to the ternary eutectic.

4.7
Phase Fields in a Ternary Eutectic System

In a phase diagram there are fields, in which various number of phases exist and which limit each other (see Fig. 4.8). These fields are – depending on the number of the phases in the field – two-dimensional or three-dimensional. Only in the field in which only one phase is present, does the temperature and composition indicate the state immediately. In all other cases, where the state variables describe a point in a multiphase field, the compositions of the coexisting phases are on the boundaries of this field. They are naturally in a certain relationship with the total composition.

The primary crystallization of C is assigned a field in space, which is limited to high temperatures by the liquidus surface L_C (see Fig. 4.8). The flat surfaces $T_C e_2 a_2$ and $T_C e_1 a_1$ act as further boundaries. Towards lower temperatures the primary crystallization field of C is limited by the curved surfaces $a_2 E e_2$ and $a_1 E e_1$, on which all temperature-composition points lay, where the binary eutectic crystallization starts. The curvature is defined by the course of the eutectic channels $e_1 E$ and $e_2 E$. The surfaces $a_2 E e_2$ and $a_1 E e_1$ meet at the temperature of the ternary eutectic E and form a "rib". The same is valid for the fields of crystallization of A and of B.

The fields of binary eutectic crystallizations, which follow towards lower temperatures the fields of primary crystallization, are also three-dimensional. These field are bound towards high temperatures by the fields of primary crystallization. For example these are in the bounding system C – B the fields of primary crystallization of C and B, which meet in the channel of the binary eutectic crystallization. The thus formed two limiting surfaces and the third limiting surface, represented by the side surface of the binary system B – C, form a three sided "tube". Towards lower temperatures it is limited by the surface of the ternary eutectic crystallization. The base of the three sided "tube" is shown in the partial triangle BEC. Towards higher temperatures the sides BE and EC approach the line BC of the triangle BEC, until at the temperature of the eutectic e_2 it degenerates completely into the line BC. In the three sided tube of the binary eutectic crystallization just described, the phases liquid, solid B and solid C are present side by side. The situation is identical for the other two fields of binary eutectic crystallization, where liquid, solid A and solid C resp. liquid, solid A and solid B are in equilibrium.

During cooling, after the binary eutectic precipitation, the ternary eutectic crystallization occurs. It takes place at a constant temperature and thus, is represented by an isothermal plane.

After ternary eutectic crystallization the almost pure solid components A, B and C are present. According to the Gibbs phase rule $F = 1$.

Thus, the stability field of these solid phases has an extension in the T direction: it is three-dimensional. With mutual solubility practically absent among the solid phases the space corresponds to a prism, formed by the concentration triangle and the temperature axis as edges.

In a simple system with a ternary eutectic alltogether eight stability fields are present:

1 phase field of the single phase liquid,
3 fields of primary crystallization,
3 fields of binary eutectic crystallization,
1 field exclusively of the three solid phases.

It must be emphasized, that this applies only to the ideal system with insignificant mutual solubility of the components in the solid state. In real systems, as was already mentioned, there is always a solubility in the solid state, even if it is difficult to prove. Thus, in addition to the fields already mentioned, there are three fields of solid solutions around the pure components, and three fields of partial solubility, where the third, solid component is in equilibrium with the binary eutectic of the other two components. Thus, in a real ternary eutectic system 14 phase fields occur.

4.8
Cuts at Constant Temperature

Isothermal cuts can provide a rapid overview about phase equilibria.

Above the melting point of the highest melting compound the composition triangle represents trivially the existence field of the liquid (see Fig. 4.10a).

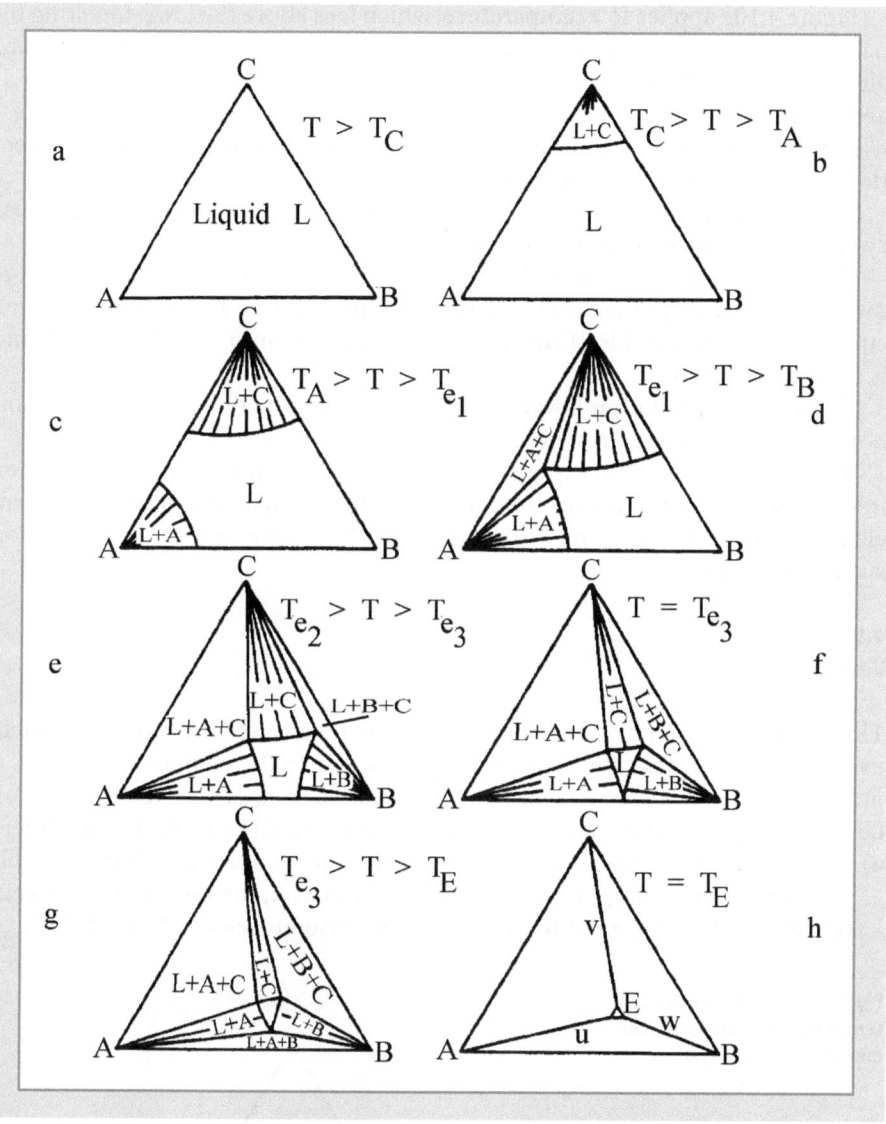

Fig. 4.10 Isothermal cuts in a simple ternary phase diagram including a ternary eutectic

At a temperature T, below T_C but above T_A and T_B, a cut through the two-phase field of the primary crystallization of C occurs (see Fig. 4.10b). In this region the tie lines of the two coexisting phases (liquid and solid C) are shown. If T is above T_B, but below T_A and T_C, then Fig. 4.10c can result.

Figure 4.10d represents a cut for a temperature which intersects the channel of the binary eutectic crystallization A + C.

Figure 4.10e applies to a temperature, which lays above the lowest melting binary eutectic e_3, but below the binary eutectics e_1 and e_2. In addition to the single phase field of the liquid and three two phase fields also two three phase fields are present.

Figure 4.10f shows the isothermal cut at the temperature of the binary eutectic e_3 of the bounding system A – B.

Figure 4.10f shows the isothermal cut at the temperature between the ternary eutectic reaction and the lowest melting binary eutectic e_3.

The three phase fields of the binary eutectics touch each other at the temperature of the ternary eutectic (Fig. 4.10h). The lines u, v and w are the saturation curves of the phase fields of the primary crystallization. The field of the liquid is reduced to a point E.

No liquid exists at even lower temperatures. The practically pure solid components A, B and C are present.

The isothermal cuts in Fig. 4.10 differ significantly from the representation in Fig. 4.9. Figure 4.9 is not an isothermal cut, but a projection of the significant elements of the ternary phase diagram onto the concentration plane (fusion surface projection).

4.9
Vertical Cuts

The experimental study of ternary systems by thermal analysis does not yield primarily isothermal cuts, but cuts perpendicular to the composition plane. It makes sense to pick cuts with straight traces in the composition triangle. As can be seen in Fig. 4.11, three types of temperature-composition cuts are possible:
a) The trace passes through a corner of the composition triangle. If for example the temperature-composition cut passes through A, the ratio of the concentrations of the mixture of this cut, x_B/x_C is constant (see Fig. 4.11, cut (a)).

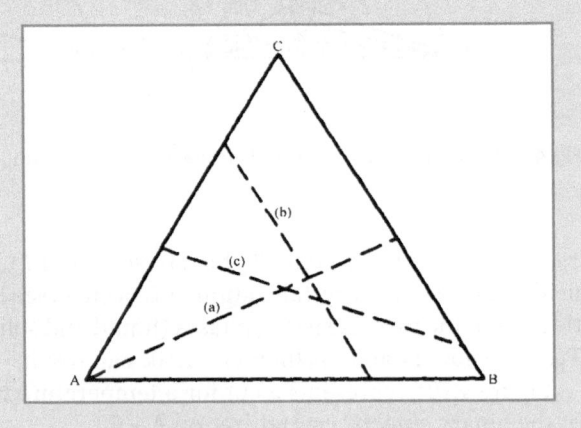

Fig. 4.11 Possible types of temperature-concentration-cuts in a ternary system

b) The temperature-composition cut is parallel to a side of the composition triangle. The concentration of component A, which here does not belong to the boundary system, to which the cut is parallel, is the same for all mixtures (see Fig. 4.11, cut (b)).
c) Less often, cuts of any position are considered (see Fig. 4.11, cut (c)). They are for example useful, if in the presence of binary compounds the cut is drawn from the composition of the compound in one binary system to the composition of the compound in the other binary system.

4.10
Temperature-Composition Cut through a Corner of the Composition Triangle

In the following a system with a ternary eutectic and negligible mutual solubility of the components in the solid state is considered again.

A temperature-composition cut along the trace b – B in Fig. 4.12a has the shape shown in Fig. 4.12b. The point T_b in Fig. 4.12b corresponds to the liquidus temperature of the binary bounding system A – C at the concentration b. This liquidus temperature decreases according to the line T_b – d in Fig. 4.12b by the addition of the third component B. Between b and d (see Fig. 4.12a) primary crystallization of component A occurs. At d it meets the saturation curve of the binary eutectic crystallization e_3 – E. Cooling the liquid of this composition precipitates immediately two solid phases, A and B. During cooling between d and B first primary crystallization of a phase occurs, in this case B. Through primary precipitation of B, the liquid becomes poorer in B, until it reaches finally the composition d and the binary eutectic solidification of A and B begins. Accordingly with all liquids in the composition range between d and B, binary eutectic crystallization of A and B follows the primary precipitation. T_d is the temperature of the saturation curve at the composition d.

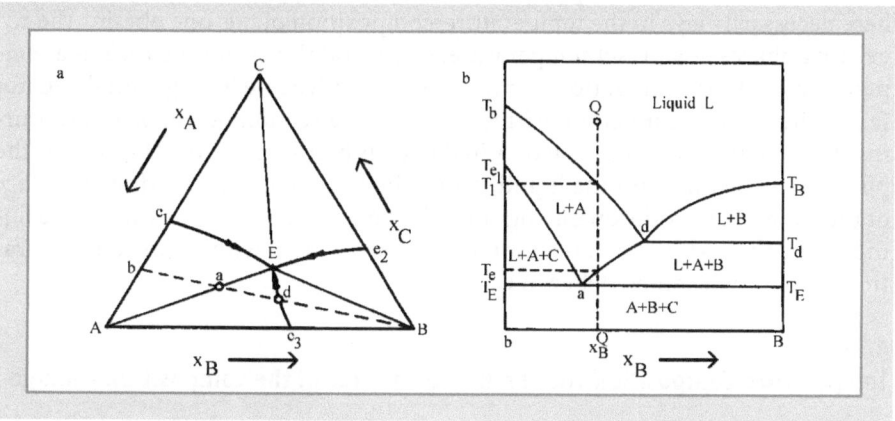

Fig. 4.12 Temperature-concentration-cut through a corner of a concentration triangle for the case of a simple system with a ternary eutectic

A binary eutectic reaction also follows the primary crystallization of A in the region b – d. Starting from the binary mixture with the composition b, the liquid is enriched in C due to the crystallization of A during cooling until the eutectic point e_1 is reached (see Fig. 4.12a).

The temperature at which the three phase reaction e_1 takes place is T_{e1} (see Fig. 4.12b). All liquids with the composition between b and a meet the saturation curve $e_1 - E$, because of the decrease in A content during primary crystallization, and the closer to E, the larger their content in B is. The temperature, at which this saturation curve is reached, decreases correspondingly with increasing content in B.

During the binary eutectic crystallization of A and C the content in B of the liquid increases. The composition follows the saturation curve $e_1 - E$ in the direction of the arrow, until the ternary eutectic is reached. Here the leftover liquid solidifies at T_E in a four phase reaction. If the starting liquid has the concentration of a, the binary eutectic crystallization of A and C is left out. The primary crystallization of A is followed immediately by the ternary eutectic reaction, when the temperature T_E is reached. The relationship between composition and temperature for the eutectic solidification processes in the range between b and a is given in Fig. 4.12b by the line T_{E1} and a, and by the straight line drawn through E parallel to the base line b – B.

Liquids whose composition lay between a and d encounter, after primary precipitation of A, the saturation curve $e_3 - E$ of the binary eutectic crystallization of A and B. The temperature where this saturation curve is reached is all the closer to T_E, the lower the content in B of the starting liquid is. The corresponding relationship between temperature and composition is give in Fig. 4.12b by the line a – d.

The individual fields of the temperature-composition cut in Fig. 4.12a, obtained in a simple fashion with little basic knowledge, are for a clearer overview, customarily marked by the phase or phases present. By drawing a tie line, which here necessarily lays in the temperature-composition plane, one obtains the coexisting phases at a preset temperature. In general this is not the case in a temperature-composition section in a ternary system. The tie lines generally do not fall in the temperature-composition plane of the section. As an example, during the primary crystallization of A in the fields between b and d (Fig. 4.12a) the phase A, coexisting with the liquid, cannot be represented in the drawn surface of the cut b – B. In addition during solidification of A the composition of the liquid leaves the series of composition in the section b – B with a constant A : C ratio. The liquid becomes richer in C.

4.11
Temperature-Composition Cut Parallel to One Side of the Composition Triangle

The construction of a temperature-concentration section a-b with a constant content of component C is shown in Fig. 4.13.

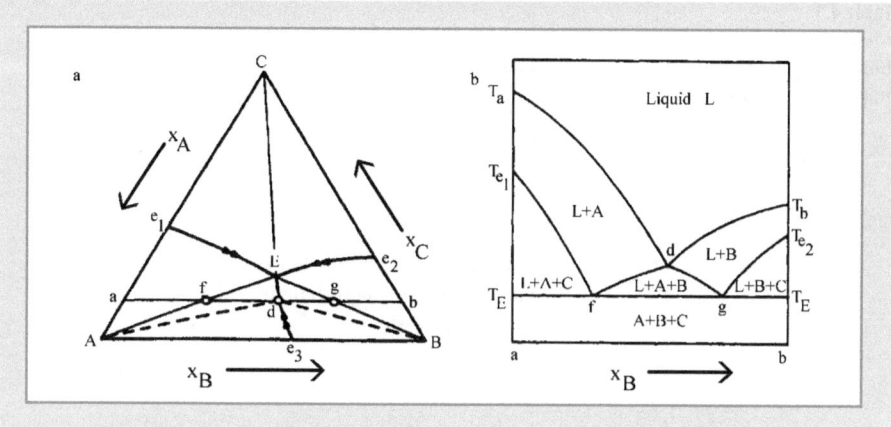

Fig. 4.13 Temperature-Concentration-Cut parallel to a side of the concentration triangle for the case of a simple system with a ternary eutectic

During cooling of liquids with the composition between a and d primary crystallization of A occurs first. The temperatures, at which these two phase reactions begin is fixed by the line T_a – d (see Fig. 4.13b). Similarly in the region between d and b (see Fig. 4.13a) primary crystallization of B takes place along the line d – T_b, shown in Fig. 4.13b.

When melting mixtures with the composition between a and f the binary eutectic reaction along the line e_1 – E (Fig. 4.13a), respectively T_{e1} – f (Fig. 4.13b) follow the primary crystallization. Correspondingly in liquids with the composition between g and b the binary eutectic crystallization of B and C along the lines e_2-E (Fig. 4.13a) resp. T_{e2}-g (Fig. 4.13b) follow the primary precipitation of B.

Liquids with compositions between f and d reach, after primary crystallization of A, the saturation curve e_3 – E of the binary eutectic crystallization of A + B. The same applies for mixtures with compositions between d and g after primary precipitation of B. The temperatures, at which the saturation curve e_3 – E is reached, are shown in Fig. 4.13b by the lines f – d and d – g.

All mixtures finally reach the ternary eutectic; those with the starting compositions of f and g immediately after the primary crystallization, all others after the primary crystallization and the binary eutectic crystallization.

4.12
Simple Real Diagrams with a Ternary Eutectic

In a simple ternary eutectic system the temperature of the ternary eutectic can be significantly lower than that of the lowest melting eutectic of a bounding binary system. As an example Table 4.1 contains the melting points of the components, of the binary eutectics and of the ternary eutectic in the system benzol-diphenylamine-naphtalene.

Table 4.1
Melting points in the system
benzol-diphenylamin-
naphthalene (after [1])

Material	Temperature [K]
Benzol	278
Diphenylamin	326
Naphthalene	353
Eutectic diphenylamin-naphthalene	305
Eutectic diphenylamin-benzol	269
Eutectic benzol-naphthalene	270
Ternary eutectic	260

Fig. 4.14
Phase diagram Pb – Cd – Sn
(after [2])

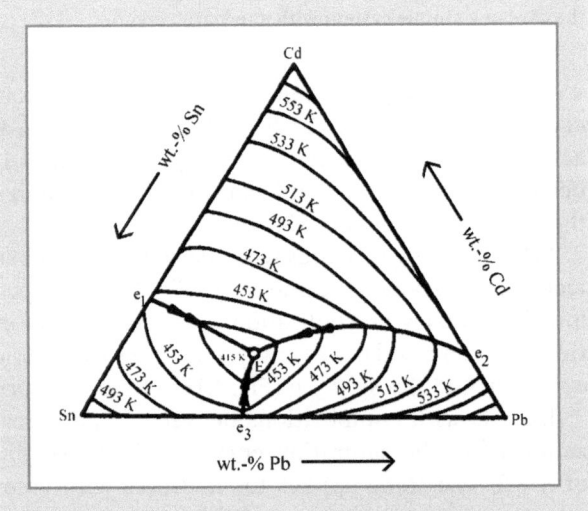

Simple binary eutectic systems are often formed by low melting metals.

There binary systems can in many cases be combined into simple ternary eutectic systems. Figure 4.14 represents the liquidus projection of the system Pb – Cd – Sn. To give an idea about the steepness of the primary crystallization surfaces, in addition to the saturation curves of the binary solidifications, isotherms of the liquidus surfaces are drawn, as "elevation indicators". To be sure, the mutual solubility of the components in the solid state, in the vicinity of the temperature T_E is no longer negligibly small. Furthermore, an intermetallic compound appears in the system Cd – Sn at 95 at.-% Sn, which also affects the fusion equilibria in the ternary system. They are being researched now (T. Gödecke) and are not yet considered in Fig. 4.14.

The so-called Rose alloy is an eutectic mixture of lead, tin and bismuth. The melting point lays below the boiling point of water, namely at 369 K (96 °C). This system is likewise not a simple eutectic system, because in the bounding binary system Pb – Bi an intermetallic compound exists.

Similarly to cutting total binary systems with a congruently melting compounds into eutectic subsystems, a complex ternary system can also be separated into easily understandable subsystems. In the silicate system $CaO - Al_2O_3 - SiO_2$ a subsystem $CaO - CaO \cdot SiO_2 - CaO \cdot Al_2O_3 \cdot 2SiO_2$ appears, which has a relatively low melting ($T_E = 1438$ K, 1165 °C) eutectic. The components of the subsystem melt above 1813 K. Low melting silicate mixtures are of importance in the formations of slags in metallurgical processes and in the preparation of silicate containing glasses. The viscosity of molten silicates is very high, due to structural effects. It increases rapidly with decreasing temperature. The crystallization of viscous liquids can be absent, because of difficulties in nucleation, so that at low temperatures an amorphous, metastable solid is obtained. For example, this fact is taken advantage on industrially in the system $SiO_2 - CaO - Na_2O$ for the manufacture of glasses.

4.13
Thermal Analysis and Structure of Simple Ternary Eutectic Systems

Similarly as for binary systems, thermal analysis is used for the study of ternary fusion diagrams.

The cooling curve of a liquid Q of composition x_B^Q (see Fig. 4.12b) is shown in Fig. 4.15. After a first period u, which corresponds to the simple cooling of

Fig. 4.15
Cooling curve of thermal analysis of a mixture Q in Fig. 4.12

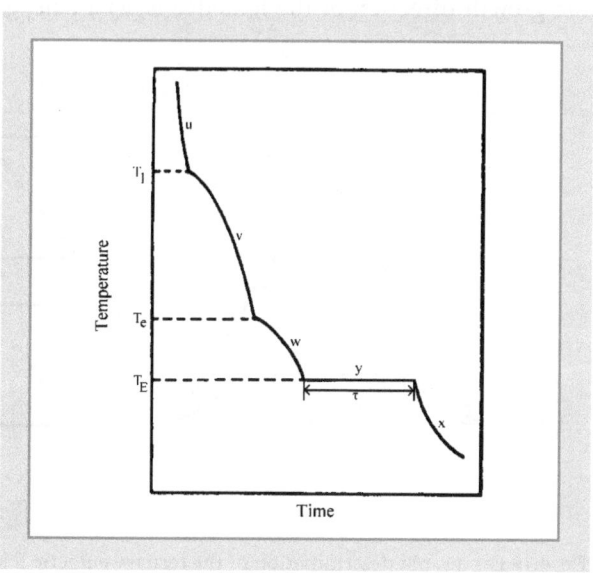

the liquid, the primary crystallization of A (bivariant reaction) takes place along the curve v. It is followed in the range of w by the univariant binary eutectic reaction of A and B, and then the invariant ternary eutectic solidification along y. The curve z shows the undisturbed cooling of the mixtures of the three solid phases A, B and C.

Often, in the case of a simple eutectic system, the position of the eutectic point is of primary interest, and not a complete knowledge of the fusion equilibria. This can – as in binary system – be elucidated by a few thermal analysis curves. Plotting of the arrest time t of the ternary eutectic occurence (see Fig. 4.15, y) perpendicular to the composition triangle yields a three sided pyramidal with the top at the composition of the lowest melting mixture. In Fig. 4.16 l, m, and n are the projections of the pyramidal edges onto the composition plane. The plotting of the arrest time t against the concentration of one component (for example B), yields, depending on the position of the section, a triangular or four sided form. The position of the ternary eutectic can be placed from a few sections of this nature.

An important aid in the construction of a ternary phase diagram is the microscopic investigation of the solidified structure, as in the case of binary systems. Primary precipitations appear as a rule in isolated polyhedral or dendritic regions. If it is followed by a binary eutectic reaction, the primary crystals are surrounded by the two phase structure formed, because on the primary crystals at least one of the components of the binary eutectic does not have to nucleate. Also during normal cooling rates, at the boundary liquid solid, the concentration of the point on the saturation curve is reached sooner, than farther away from the primary crystals.

If the primary precipitation occurs in the form of idiomorphic crystals, the growth direction of the lamellae is, as a rule, given by the primary crystal

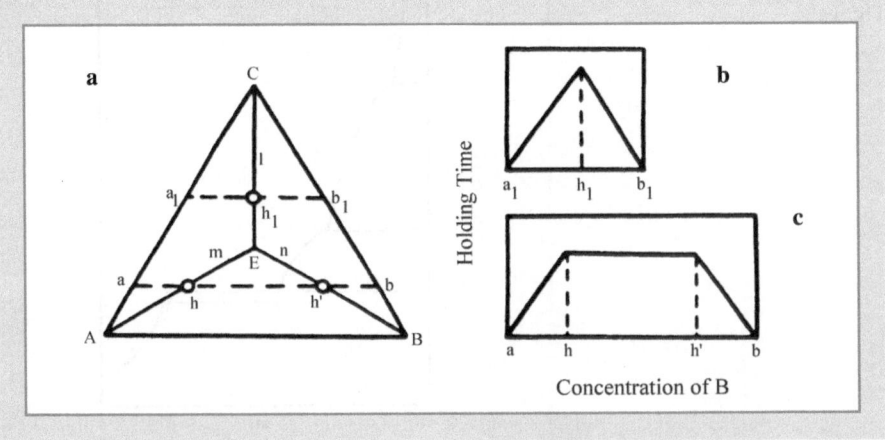

Fig. 4.16 For a rapid determination of the ternary eutectic E by thermal analysis

present, serving as a nucleus (leading phase). In this case the binary eutectic layer grow uniformly onto the primary crystal. So called "eutectic polyhedrons" form. Their presence proves, that during solidification secondarily a binary eutectic crystallization occurred. As an example Fig. 4.17 shows a microstructure with "eutectic polyhedrons".

The solidification of the ternary eutectic occurs seldom in a lamellar shape of the three phases formed. As a rule, two phases crystallize in a lamellar structure,

Fig. 4.17
Microstructure of ternary alloy of the Bi-Sn-Pb system (source: T. Goedecke) 30 wt.-% Sn, 7 wt.-% Pb; primary crystallization of white Bi-crystals, coarse binary eutectic (Bi + Sn), "eutectic polyhedrons", fine ternary eutectic (Bi + Sn + ε-phase), light microscopy, magnification: 75-times

Fig. 4.18
Microstructure of ternary eutectic Cu-Ag-Cd (source: G. Petzow) 25 wt.-% Cu, 22.5 wt.-% Ag, 52.5 wt.-% Cd. White: Cu-solid solution, grey: β- Hume-Rothery-phase, Black: γ-Hume-Rothery-phase, light microscopy, magnification: 200-times

the third interposed in an irregular shape. It is possible, that no lamellar structure appears. In the latter case, the three structural components are distributed irregularly. Figure 4.18 represent the structure of a ternary eutectic.

4.14
Properties of Neighboring Phase Fields

A glance at Figs. 4.10b and 4.11b indicates that the number of coexisting phases, while going from a phase field into a neighboring one, changes in a regular fashion. During cooling of the liquid L in Fig. 4.10b a transition occurs from a phase field with one phase (L) into one with two phases (L + A), and finally into the phase field with three phases (L + A + B). Masing called this generally applicable regularity as the "rule of the touching phase fields". Accordingly, two phase fields have only then a common contact surface, when the number of phases in them differs by 1. The Gibbs phase rule has to be observed as well. Thus, there is no four-phase field in a three-component system, because at constant pressure four phases can only coexist at a single temperature. The "four phase field of the ternary eutectic solidification" degenerated into a four-phase surface.

While continuing the cooling of liquid L in Fig. 4.10b, the sample leaves the phase field of the binary eutectic solidification (L + A + B), enters the "four-phase field of the ternary eutectic solidification" (L + A + B + C), which has degenerated into a surface, leaves the latter during further cooling, to reach the phase field of the three solid phases (A + B + C). Masing's rule, generally valid based on thermodynamic foundations, is independent of the number of components in the system, keeping in mind the limitations imposed by the Gibbs phase rule. It can be very useful during construction and control of complicated four-component systems.

L.S. Palatnik and A.I Landau [3]summarized the rule about the contiguous phase fields into a simple formula:

$$r_1 \quad = r - d^- - d^+ \geq 0 \tag{4.10}$$

Where r is the dimension of the phase diagram or the section through the phase diagram and r_1 the dimension of the boundary between the neighboring phases d^- and d^+ the number of phases, which during transition from a neighboring phase field into another neighboring phase field are omitted or added. When applying Eq. (4.10) the limitations given by the Gibbs phase rule must be considered.

Table 4.2 summarizes some statements obtained using Eq. (4.10), when applied to the section in Fig. 4.19, with a constant concentration of one component.

In a two-dimensional phase diagram or a two-dimensional section, the boundaries between phase fields have the dimension 1 or 0. $r_1 > 1$ is impossible. If, for example $r_1 = 2$, then from Eq. (4.10) $d^- = d^+ = 0$ would follow. If no change

Table 4.2
Correlation after Palatnik and Landau on the temperature-concentration cuts in Fig. 4.19

Transition	d⁻	d⁺	r₁
$L \rightarrow L + A$	0	1	1; line
$L + A + B \rightarrow L + B$	1	0	1; line
$L \rightarrow L + A + B$	0	2	0; point

Fig 4.19
Cut through a ternary system to demonstrate the law found by Palatnik and Landau [3] concerning neighboring phase regions

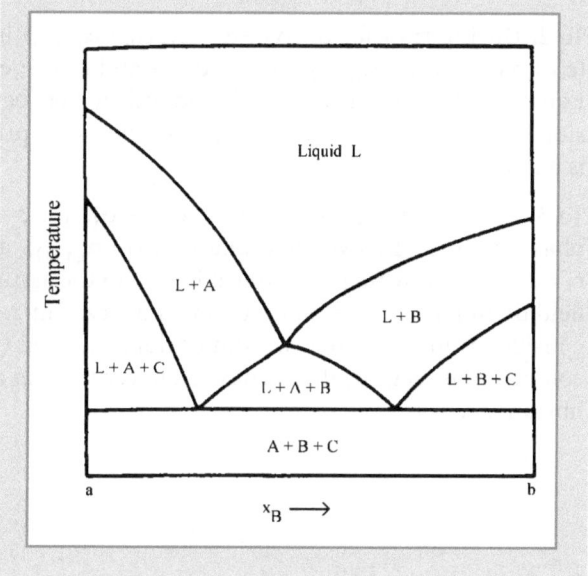

in the number of phases occur, then a phase transition does not take place. The two phases would be identical. Thus, always:

$$r_1 \leq r - 1 \tag{4.11}$$

The relations, formulated by Palatnik and Landau [3] from topological considerations, give the possibility for a fundamental determination of the number of lines in a phase diagram which can meet in a point. In Fig. 4.20a Eq. (4.10) for $r = 2$ is satisfied. The boundary between the phase fields I and II is one-dimensional (line). The number of phases in I ($\alpha_1 + \alpha_2 + \dots \alpha_n$) differs by 1 from the number of phases in II ($\alpha_1 + \alpha_2 + \dots \alpha_n + \alpha_{(n+1)}$).

Figure 4.20b permits to check, if three lines, which separate phase fields, can meet in one point. Based on the transition I $\rightarrow$ III the following combinations of phases for the phase field III are possible.

1. $\alpha_1 + \alpha_2 + \ldots \alpha_n + \alpha_{(n+1)}$
2. $\alpha_1 + \alpha_2 + \ldots \alpha_n + \alpha_{(n+2)}$
3. $\alpha_1 + \alpha_2 + \ldots \alpha_{(i-1)} + \alpha_{(i+1)} \ldots + \alpha_n$

In the last case, phase i present in phase field I, is missing in phase field III.

These three at first realistic looking phase mixtures of III in reality are the same like the phase mixture of II. One checks, if Eq. (4.10) is satisfied.

To 1: This first possible phase mixture III is identical to that of II. This is impossible.

To 2: During transfer of $(\alpha_1 + \alpha_2 + \ldots \alpha_n + \alpha_{(n+2)})$ into the phase mixture of II $(\alpha_1 + \alpha_2 + \ldots \alpha_n + \alpha_{(n+1)})$ there is $d^- = 1$ and $d^+ = 1$; consequently $r_1 = 2 - 1 - 1 = 0$. Equation (4.10) requires that the boundary between II and III has the dimension zero. This is in contradiction with the assumption, that the boundary II/III is a line.

To 3: If the mixture $(\alpha_1 + \alpha_2 + \ldots \alpha_{(i-1)} + \alpha_{(i+1)} \ldots + \alpha_n)$ is transferred into the phase mixture II, two new phases $(\alpha_1 + \alpha_{(n+1)})$ are added. Using Eq. (4.10) $r_1 = 2 - 0 - 2 = 0$, which is in contradiction to the assumed meeting of three-phase field boundary lines in a point, as was the case in 2. One can conclude from this considerations, that in a two-dimensional section (isothermal or temperature-composition section) three lines can never meet in a point, if no invariant equilibrium occurs.

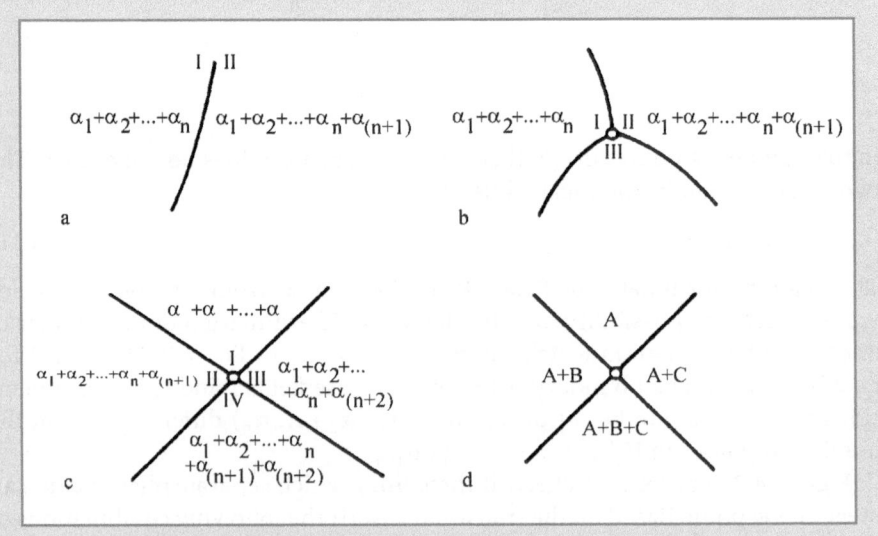

Fig. 4.20 Demonstration of the cross-rule

Also five lines cannot meet in a point in a two-dimensional phase diagram, which can be proven in a similar way, as was done for three lines.

Only four lines can have a common point. The phase mixtures of the four-phase fields meeting in one point must have compositions, which can be deduced from Fig. 4.20c, d.

One of the phase fields has the lowest number of phases (A). This field encounters at the point the phase field with the largest number of phases (A + B + C). The phase fields neighboring A each contain one phase more (A + B; A + C), than the phase field with the least number of phases. This cross rule is of use in constructing phase diagrams of multi-component systems. If the composition of one-phase field, close to the cross point, is known, only two possibilities exist for the composition of the phase mixtures in the three other phase fields. A few experiments can decide between these two possibilities.

4.15
Non Regular Sections

Sectioning reduces the available state variables. At an isothermal section the temperature is fixed, at a temperature-composition-section one of the concentration variables is constant. The dimension of a section is

$$r = B + 1 - \varphi \tag{4.12}$$

B is the number of components of the system and φ the number of fixed state variables. The three-dimensional diagram in Fig. 4.8 is an isobaric section across the four-dimensional temperature-pressure-composition phase diagram, with the pressure $p = 1$ atm. Considering a constant temperature section through this three-dimensional diagram, this section has, as experience taught us, the dimension

$$r = B + 1 + \varphi = 3 + 1 - 2 = 2$$

In an analogous fashion, the dimension of a boundary of two-phase fields is smaller in a section, as is the starting diagram. If in the section, φ is reduced by 1, the dimension of a boundary between two-phase fields is also reduced by 1. A surface between two-phase fields in a three-dimensional phase diagram becomes a line in a two-dimensional section. This applies to regular sections. Exceptions occur at non regular sections. An example is shown in Fig. 4.21.

The dimension of all boundaries between touching phase fields is, in the temperature-composition section $a_E - b_E$ of Fig. 4.21, smaller by 1 than in the three-dimensional starting diagram. However, the ternary eutectic point is an exception. Point E, which was already a point in the starting three-dimensional phase diagram, remains also a point in the section $a_E - b_E$.

The section in Fig. 4.21b is a limiting case for the transition of the general sections of the character shown in Fig. 4.21c to the large number of sections shown in Fig. 4.21d. Non regular sections are in general sections for a transition between two regular sections.

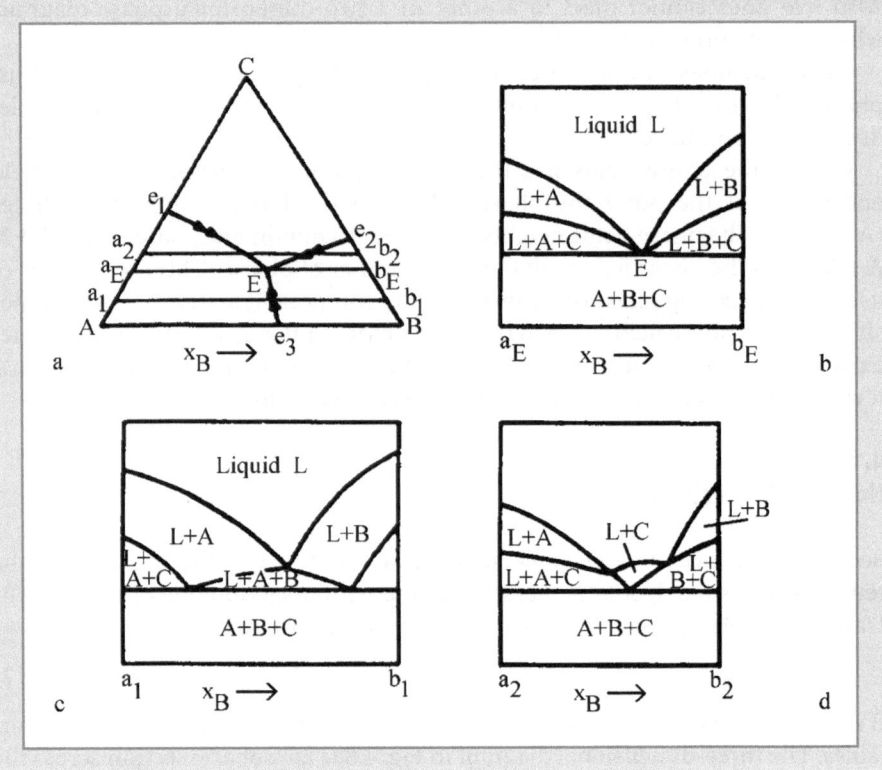

Fig. 4.21 Explanation of a non regular cut

4.16
Critical Point

The rules about neighboring phase fields do not apply at a critical point, such as the ones at miscibility gaps in the solid and liquid state. Rather, Palatnik and Landau [3] found the following relationship between the dimension r of the phase diagram (or section), the dimension r_1 of the boundary between the touching phase fields, and the number d_K of the phases, which unite to one phase at the critical point

$$r_1 = r - d_K \geq 0 \qquad (4.13)$$

For example, in the case of a solid miscibility gap in a binary system

$$r_1 = r - d_K = 2 - 2 = 0$$

A dimensionless critical element occurs – the critical point Q_K (see for example Fig. 3.42).

4.17
Schreinemakers' Rule

Two saturation curves, which for example in an isothermal section of a three-component system represent the saturation concentration on one hand of A and on the other hand of B in the liquid can meet at composition x_0. At x_0 the liquid is saturated in A as well as in B. The two saturation curves cannot cross randomly, as was shown by F.A.H. Schreinemakers, based on thermodynamic considerations.

There are in total four possibilities for two saturation curves to cross. They are represented in Fig. 4.22. Let us consider case a) first. At the composition a a mixture of liquid L and solid B is present. The quantities of the phases are given by the lever rule. If substance A is added to the mixture of composition a, the fraction of solid B in the two-phase mixture (L + B) decreases, until point b is reached, when all solid B disappears. The same applies, when adding A, to the saturation curve along which liquid and A are in equilibrium.

The case **b** shows reverse behavior. The composition a corresponds again to a two-phase mixture of liquid L and solid B. If A is added, no dissolution occurs, but precipitation of solid B. The fraction of B in the two-phase mixture increases with increasing content in A. If the overall composition of the phase mixture

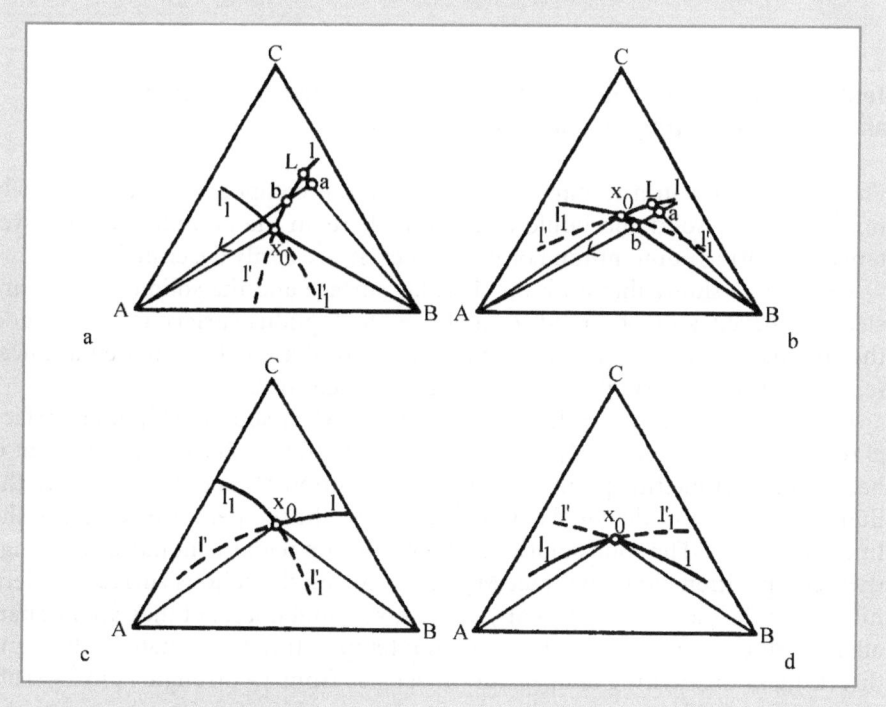

Fig. 4.22 Explanation of Schreinemakers' rule

reaches the point B, the liquid has the composition x_0. The liquid is now saturated in both A and B.

Cases **a** and **b** are the thermodynamically permitted possibilities for the intersection of two saturation curves. The first case (**a**), where the addition of A increases the solubility of B in a two-phase mixture (L + B), is characterized by the fact, that the extensions l' and l_1' of the saturation curves l and l_1 beyond the intersection x_0 lay inside the compatibility triangle Ax_0B. In the case **b**, where the solubility of B in a two-phase mixture (L + B) is decreased, the extensions l' and l_1' of the saturation curves l and l_1 beyond x_0 both lay outside the compatibility triangle Ax_0B.

In case **c** the extension of one of the saturation curves (l_1') lay inside the compatibility triangle Ax_0B, that of the other (l') outside. In the case **d** the extension of the saturation curves l' and l_1' enter the single phase region. The cases **c** and **d** are not possible, based on energetic considerations. Thus, the Schreinemakers' rule says:

"In an isothermal section two saturation curves intersect in such a fashion, that their extension beyond the intersection either both lay inside the compatibility triangle, or both lay outside the compatibility triangle. Also in the latter case, the point, created by the intersecting saturation curves, must point towards the compatibility triangle."

4.18
Ternary Systems with Unlimited Solubility in the Solid and Liquid State, and without a Melting Point Minimum or Maximum

Fusion equilibria in a ternary system, where the components are completely miscible in the solid and liquid state, and where, in the bounding binary systems no melting point minima or maxima exist, are easily described.

Figure 4.23 shows that both the liquidus surface and the solidus surface are steadily curved, and, observed from above, the liquidus surface is convex, and the solidus surface concave. They enclose a two-phase field, bounded at lower temperatures by the single phase field of solid solutions.

Crystallization starts, while cooling the liquid Q, as soon the liquidus surface is reached. Solid solution M_1 crystallizes from liquid L_1. Because component C has the highest melting point (T_C), the solid solution M_1 is richer in C than the liquid L_1. The liquid, depleted in C, follows, while the temperature decreases, the line $L_1, L_2, \ldots L_5$. The solid solutions, in equilibrium with the liquid, also change their composition along the trace $M_1, M_2, \ldots M_5$ on the solidus surface. By definition, the tie lines $L_1 - M_1, L_2 - M_2, \ldots$ are horizontal lines. But they are not parallel for all temperatures. With decreasing temperature they rotate in the same direction, as the melting temperatures decrease from T_C through T_A to T_B (rule of Konovalov). The rotation direction and projection of the tie lines onto the base plane is shown in Fig. 4.23.

Fig. 4.23
Three-dimensional presentation of a ternary system with complete solubility in liquid and solid state

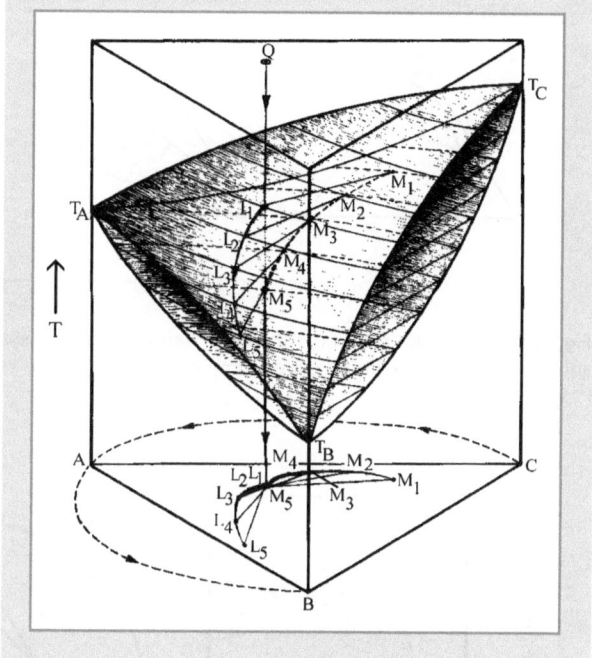

Naturally all tie lines pass through the line parallel to the T axis and pass through the point Q. All is solidified, when the solidus composition has reached the starting composition $(x_B^L)_1 = (x_B^S)_5$.

Under normal cooling rates, layered crystal formation can be expected during solidification in ternary systems with complete solubility, as is the case in a binary system with total miscibility.

4.19
Isothermal Section through a Ternary System with Unlimited Solution Formation

Figure 4.24 represents three examples of isothermal sections through the ternary phase diagram.

The tie lines of a two-phase region shown, do not pass through a point, the corner of the composition triangle, as is the case in the absence of solid solubility. Thus, it is not possible to determine the position of the tie line in a $(L+\alpha)$ equilibrium only from the composition of the liquid. To position the tie line, the composition of the coexisting α-solid solution is required. A tie line cannot cross a two-phase field, based on the Gibbs phase rule.

The rule of Konovalov about the rotation of the tie lines is based on the fact, that in a solid-liquid equilibrium, the component with the higher melting point

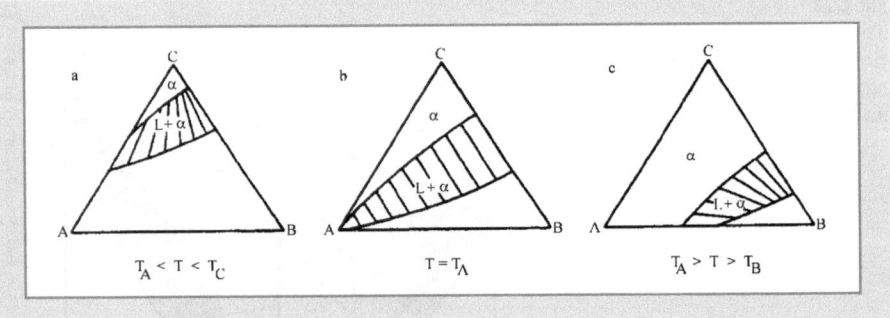

Fig. 4.24 Some isothermal cuts through a simple ternary system with total solubility in the solid as well as in the liquid state (compare with Fig. 4.23)

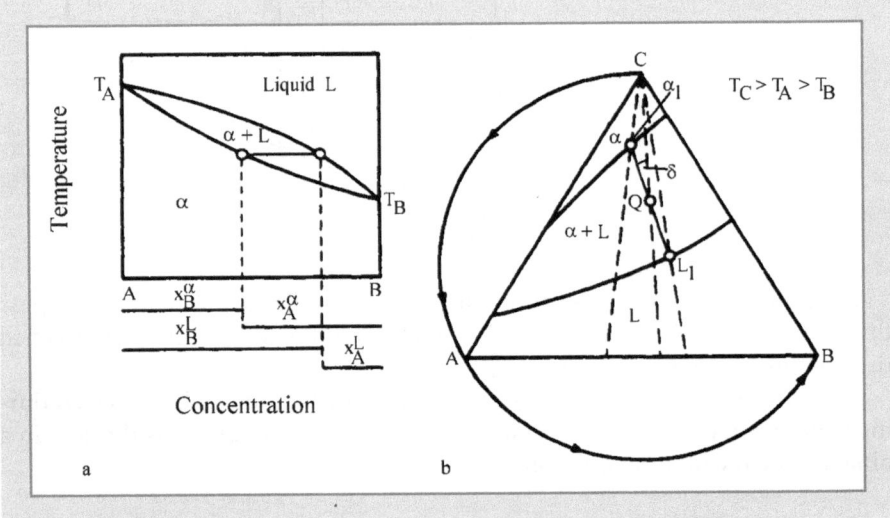

Fig. 4.25 Explanation of the rule of Konovalov concerning the turning of a knod

occurs in higher concentration in the solid solution. This is immediately visible for a corresponding two-component system in Fig. 4.25a. We have:

$$\frac{x_A^\alpha}{x_B^\alpha} > \frac{x_A^L}{x_B^L} \tag{4.14}$$

In Fig. 4.25b a ternary system is considered, whose components have the melting temperatures $T_C > T_A > T_B$. For a mixture Q, in the absence of solid solution formation, the tie line is given through the line C – Q. The direction of the tie line $a_1 - L_1$, as it occurs in the presence of solid solution formation, deviates from the above one by the angle δ. The rotation direction of the tie lines is the same, as the decrease in melting temperatures of the components in the concentration triangle (Konovalov rule). One can deduce from Fig. 4.25b:

$$\frac{x_A^{\alpha_1}}{x_B^{\alpha_1}} > \frac{x_A^{L_1}}{x_B^{L_1}} \tag{4.15}$$

This agrees with Eq. (4.14). Based on the assumptions the Konovalov rule is satisfied. The violation of the Konovalov rule, by assuming a rotation direction opposite to C–Q, can be proven in a similar fashion. Not only

$$\frac{x_A^{\alpha_1}}{x_B^{\alpha_1}} < \frac{x_A^{L_1}}{x_B^{L_1}} \tag{4.16}$$

but also

$$\frac{x_A^{\alpha_1}}{x_B^{\alpha_1}} = \frac{x_A^{L_1}}{x_B^{L_1}} \tag{4.17}$$

would contradict the Konovalov rule, expressed in Eqs. (4.15) and (4.14). This means, that the tie line of a two-phase field $\alpha + L$ cannot pass through the corner of the composition triangle, from where the α-field extends into the ternary.

4.20
Temperature-Composition Sections through a Ternary System with Unlimited Solid Solution Formation

Figure 4.26 represents two characteristic sections through the phase diagram shown in Fig. 4.23. In the section with constant A:B ratio, T_{C1}' is the liquidus temperature and T_{C1}'' is the solidus temperature of the binary A-B mixture at c_1. Similarly T_a' and T_b' are the liquidus temperatures in the binary base system A–C at a, and in the binary base system A–B at b. T_a'' and T_b'' are the corresponding solidus temperatures.

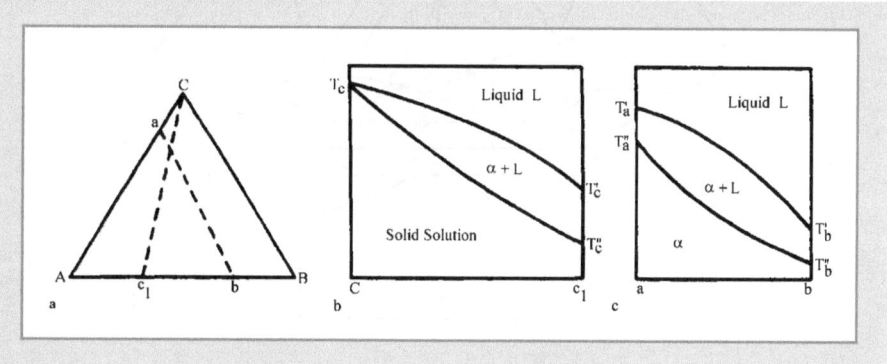

Fig. 4.26 Temperature-concentration cut through a simple ternary system with complete formation of solid solutions without a melting point extremum

A system, which does not show limited solubility in the solid and liquid phases, and does not have a melting point maximum or minimum, is realized for example by the combination of three Laves phases $UFe_2 - UCo_2 - UMn_2$ [9, 10]. The interval between liquidus and solidus temperatures is relatively small. The system $UFe_2 - UCo_2 - UMn_2$ is a ternary section through the quaternary system $U - Fe - Co - Mn$.

4.21
System with a Ternary Eutectic and Limited Solid Solution

As already mentioned substances which are completely soluble in each other in the liquid state, are also soluble at least to a limited extend in the solid state.

In the simplest case this leads to eutectic fusion equilibria, where, however, solid solubility has to be taken into consideration. Figure 4.27 represents the projections of a few curves of a ternary eutectic phase diagram with significant solid solution formation.

Around A, B, and C fields of solid solutions α, β and γ appear. In the invariant equilibrium of the ternary eutectic the liquid is now in equilibrium with the corresponding solid solutions α_E, β_E and γ_E.

Figure 4.28 shows an isothermal section for a temperature below T_E, the temperature of the ternary eutectic. Seven phase fields are present: three single phase fields of the solid solutions (α, β, γ), three two phase fields ($\alpha + \beta$), ($\alpha + \gamma$)

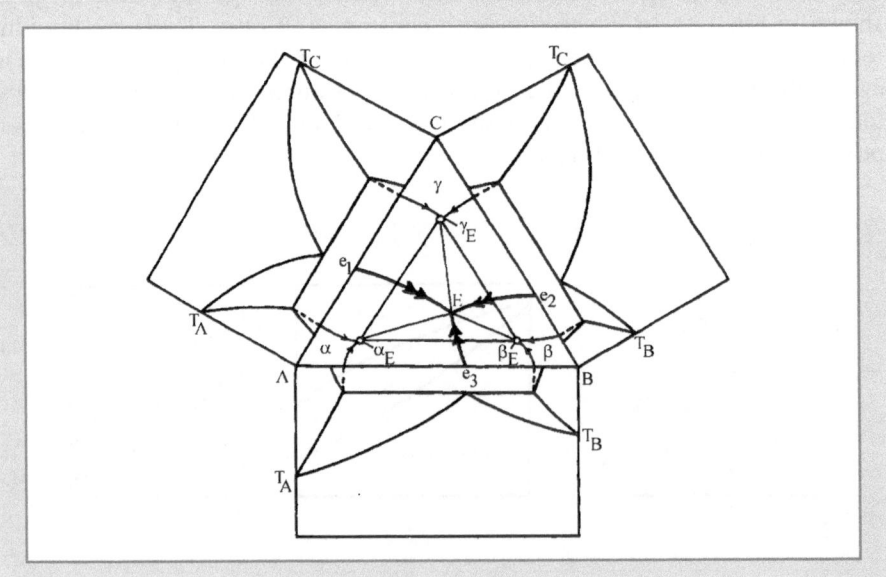

Fig. 4.27 Projection of outlines of a ternary eutectic phase diagram with limited solubility of the components in the solid state

Fig. 4.28
Isothermal cut at $T < T_E$
through a ternary eutectic
system with limited formation of solid solutions

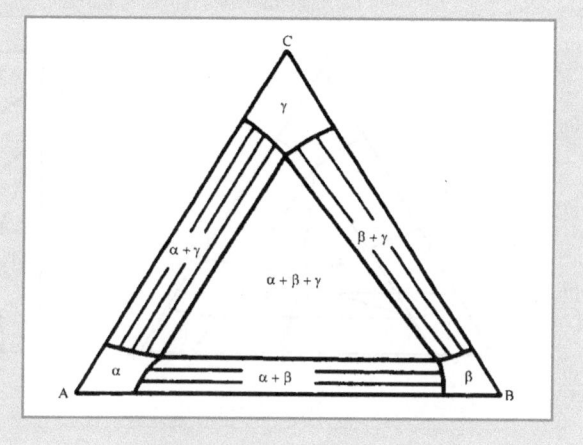

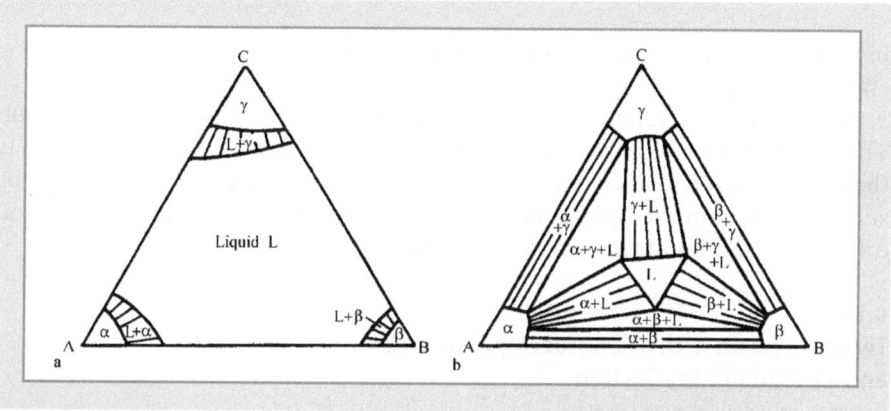

Fig. 4.29 Isothermal cuts at $T > T_E$ through a ternary eutectic system with limited formation of solid solutions

and $(\beta + \gamma)$, and a further area, where all three solid solutions are in equilibrium. The solubility equilibria in the solid are naturally temperature dependent. As a rule the size of the single phase fields decreases with decreasing temperature. Generally with decreasing temperature, point α_E approaches point A (similarly $\beta_E \rightarrow B$, and $\gamma_E \rightarrow C$).

Figure 4.29 represent two isothermal sections at temperatures above T_E. Figure 4.29a shows the phase equilibria at a temperature T, which is below that of the lowest melting component and above the temperature of the highest melting binary eutectic. Only two-phase equilibria are present.

Figure 4.29a shows schematically the phase equilibria occuring slightly above T_E and below the temperature of the lowest melting binary eutectic. In addition to solid-liquid and solid-solid two-phase equilibria, three regions appear,

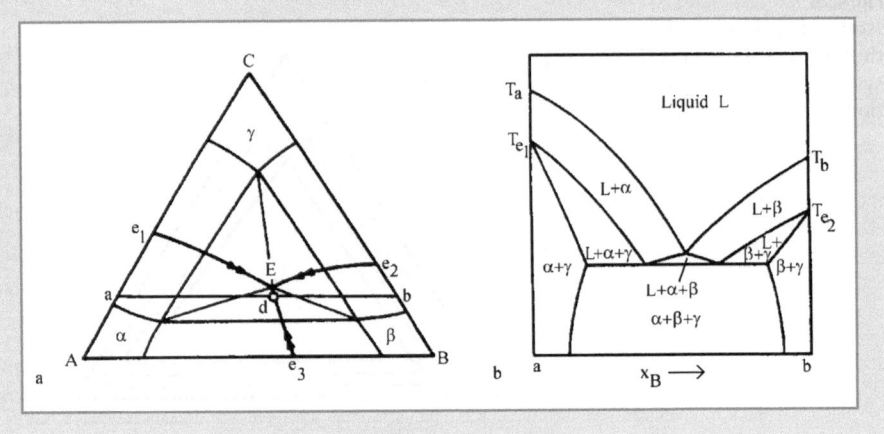

Fig. 4.30 Temperature-concentration cut through a ternary eutectic system with limited solid solutions

in which the binary eutectic crystallizations take place $(L \rightarrow \alpha + \gamma)$, $(L \rightarrow \beta + \gamma)$ and $(L \rightarrow \alpha + \beta)$.

An example of a temperature-composition section is presented in Fig. 4.30b. The section a-b for constant concentration of C corresponds in essential parts to the case of complete insolubility in the solid, as shown in Fig. 4.13. Figure 4.30b shows in addition the solid-solid two phase equilibria, $(\alpha + \gamma)$ and $(\beta + \gamma)$, conditioned by the significant solubility in the solid state.

4.22
Ternary System with a Congruently Melting Binary Compound and a Pseudobinary Section

In a system of three components, in addition to binary compounds in the bounding systems, ternary compounds can also be present. Easily understandable phase equilibria exist if, in one bounding system a congruently melting compound is present, and in the other two bounding systems, simple eutectic fusion equilibria occur. The basics are explained below, under the simplifying assumption of unnoticeable small solubilities in the solid state. In earlier chapters the steps necessary to consider solid solubility were already given.

Figure 4.31 shows the characteristic features of the possible equilibria in such a system. The binary A – B system, which in a simplified fashion can be decomposed into two eutectic subsystems A – V and V – B, the ternary can be handled in a similar fashion. The two ternary eutectic subsystems are A – V – C and V – C – B.

First the situation is discussed, where the saturation curves starting at the binary eutectics e_1 and e_3, meet in point E of the subtriangle ACV. Here a ternary eutectic occurs, because A, V and C crystallize simultaneously from the liquid,

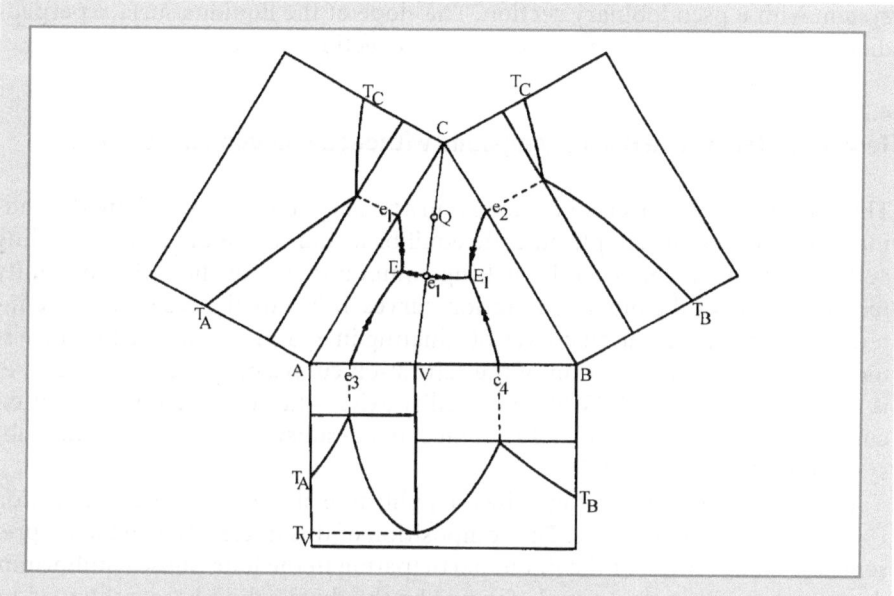

Fig. 4.31 Explanation of phase equilibria in a ternary system with a congruently melting binary compound V and with a quasi-binary cut

E is within the triangle, and can be formed in this invariant four phase equilibrium from the solid phases A, V, and C through heat removal. At the point E another saturation curve of a binary eutectic must arrive, that of the simultaneous crystallization of V and C. This can only be the case, if in the section V – C a binary eutectic solidification takes place. The solidification of the mixture in the V – C section occurs as in a binary eutectic system. A mixture of concentration Q solidifies under precipitation of C. The liquid, without deviating from the section V – C, arrives at the binary eutectic point e_5, where it solidifies completely in an invariant reaction under simultaneous precipitation of C and V. Though Q is a ternary mixture, its crystallization occurs as in a binary system. Section with this property is called pseudobinary section.

In the partial ternary system V – B – C a ternary eutectic (E_1) can occur, in a fashion explained above. The point e_5 of the binary eutectic in the pseudobinary section C – V has the highest temperature of the saturation curves of the binary eutectic crystallization e_5 – E and e_5 – E_1.

If a ternary system has a pseudobinary section, a separation into two subsystems is possible in principle. In the case of the phase diagram represented in Fig. 4.29, two subsystems result, the construction of which corresponds to a simple ternary eutectic system.

It must be mentioned, that the liquidus line, in the vicinity of a pure component has a finite slope, whereas the slope at the melting point of a congruently melting compound in a binary system is zero. The same applies for a ternary system with a pseudobinary section. The slope of the liquidus surface perpendicular to the direction of the pseudobinary section is zero.

4.23
Ternary System with a Binary Compound without a Pseudobinary Section

The appearance of two ternary eutectics in a ternary system, in which two bounding binaries presents simple eutectic equilibria, and where the third bounding system contains a congruently melting compound, is not the only possibility for the joining of the binary saturation curves. It is only the case, when the intersection of the saturation curves originating in e_1 and e_3 is inside the subsystem ACV, and the intersection of the saturation curves originating in e_2 and e_4 is inside the subsystem BCV. The possibility exists – however, that the saturation curves originating in e_1 and e_3 do not meet in the subsystem ACV, but in the subsystem BCV (see Fig. 4.32).

In this case an invariant four-phase equilibrium also exists at the intersection U of these saturation curves. The composition of this intersection, which represents the composition of the liquid participating in the four-phase equilibrium, does not lay within the triangle formed by the three solid phases, obtained by heat removal from the liquid L. Rather, a quadrangle AVUC can now be formed from the phases coexisting at the four-phase equilibrium, which indicates an intermediate equilibrium according to Fig. 4.6b.

In this case no pseudobinary section exists. Mixtures of the section C-V change their composition during solidification, so that finally they do not correspond to the points of the line C – V. During solidification of a liquid of composition Q, first C crystallizes, until the leftover liquid reachs the composition c. Now A and C precipitate simultaneously, B is enriched in the liquid, and thus – the composition leaves section C –V.

Fig. 4.32
Explanation of phase equilibria in a ternary system with a congruently melting binary compound V without a quasi-binary cut

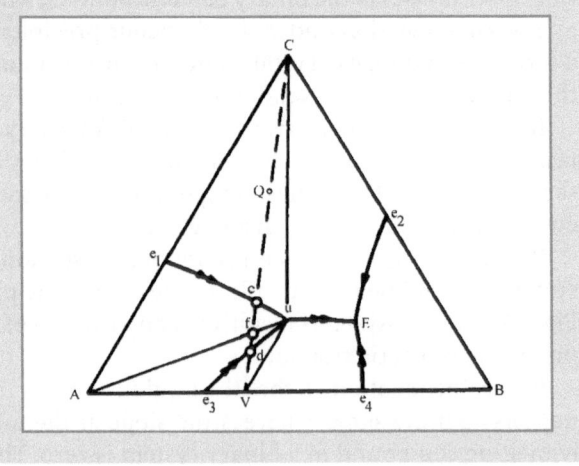

With compositions of the section C – V between V and d, primary crystalliza-tion of V occurs, until the saturation curve e_3 – U of the binary eutectic crystalli-zation is reached at d. During further cooling V and A precipitate together. The liquid is enriched in B, and deviates from the section C – V.

Liquids, which lay in the region c-d on the section C – V, deviate immediately from the section C – V because of primary crystallization of A. Liquids of the section C – V do not behave in any case as pseudobinaries.

Along the saturation curve e_1 – U solid phases A and C are in equilibrium with the liquid, and along the saturation curve e_3 – U solid phases A and V. Thus, at point U a four-phase equilibrium is present, with the phases L, A, C and V. An-other saturation curve must originate at this point U, along which a three-phase equilibrium between L, C and V exists. This is, as can be seen in Fig. 4.32, also the binary eutectic saturation curve U – E, which together with the binary eu-tectic solubility curves e_2 – E and e_4 – E leads to the ternary eutectic E. The tem-perature must thus, decrease in the direction U $\rightarrow$ E along the saturation curve U – E, as it decreases starting in e_2 and e_4 along the saturation curves e_2 – U and e_4 – U. From the direction of the saturation curve U – E the intermediate char-acter of the four-phase equilibrium in U is immediately visible. All liquids with compositions laying in the quadrangle AVUC, change their composition dur-ing solidification in such a manner that it finally corresponds to that of U. At the temperature which is assigned to the liquid at point U, the four-phase equilib-rium occurs

$$L_U + A \leftrightarrow C + V \qquad (4.18)$$

where L_U is the liquid with the composition of point U. Also the overall compo-sition of L_U and A corresponds to that of point f. At this equilibrium the partial structure of C + V also has the overall composition of f. During heat removal L_U reacts with A and forms C + V. L_U disappears first with starting compositions within the triangle ACV. The solidification is completed at the temperature T_U, which corresponds to the four-phase equilibrium at U. As a result of the transi-tional reaction the solid phases A, C and V appear.

With starting compositions within the triangle BCV the intermediate reac-tion, however, does not end at U. In this case, the solid phase A disappears first. A mixture of L, C and V is left. This corresponds to an univariant equilibrium. The temperature can decrease further along the saturation curve U – E of the binary eutectic crystallization of C + V. Finally the ternary eutectic E is reached, where the crystallization ends, through the simultaneous precipitation of B, C and V. Corresponding to the definition of a ternary eutectic equilibrium, the composi-tion of the liquid L_E, coexisting with the solid phases, lays in the triangle which can be constructed from the three phases B, C and V which form during simul-taneous crystallization.

The intermediate reaction cannot proceed undisturbed using usual cooling rates. As in peritectic reactions in two-component systems, local equilibrium is established only at the phase boundaries.

4.24
Isothermal Section and Temperature-Composition Section through a Ternary System with a Binary Compound with no Pseudobinary Section

Figure 4.33a represents schematically an isothermal section at a temperature which lays below the temperatures of the binary eutectics e_1, e_3 and e_4 (see Fig. 4.32) and above the binary eutectic e_2. In addition to the single phase field of the liquid L, four two-phase fields ($L + A, L + B, L + C$ and $L + V$) are present. Also, the three regions with univariant equilibria ($L + A + V$, $L + B + V$ and $L + A + C$) can be recognized.

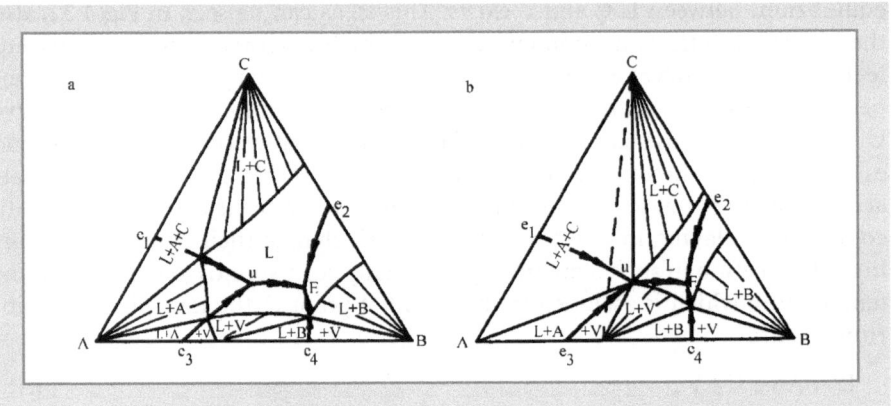

Fig. 4.33 Isothermal cut through a ternary system including a binary compound V without a quasi-binary cut. **a** For a temperature above T_{E2} and beneath temperatures of the other eutectica of the binary bounding system; **b** for the temperature of the peritectic point U

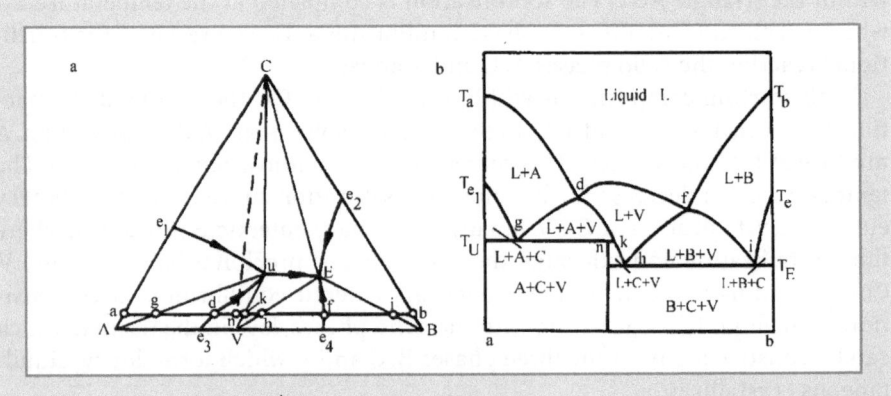

Fig. 4.34 Temperature-concentration cut through a ternary system with a binary compound V and without quasi-binary cut. **a** Projection of the melting sphere and auxilliary lines to determine the cut a-b, **b** Temperature-concentration cut a – b

The equilibrium conditions at temperature T_U, which corresponds to the intermediate reaction Eq. (4.18), which takes place when the liquid reachs the composition U, is shown in Fig. 4.33b. The phase field L + A present in Fig. 4.33a is contracted to the straight line A – U.

Figure 4.34b represents a temperature-composition section parallel to the bounding system A – B. In the composition range between a and d primary A, between d and f the compound V, and between f and b the component B crystallizes. Binary eutectic reactions, the invariant intermediate reaction and the ternary eutectic crystallization can follow. At starting compositions between a and k the result of the solidification is a mixture of A, C and V, at starting compositions between k and b after complete solidification a mixture of B, C and V is present.

4.25
Ternary System with Two Binary Compounds

A ternary system, where in each of two binary systems a compound is present, and the third has simple eutectic equilibria, can be divided into three ternary subsystems. The division, however, cannot be carried out immediately unequivocal, as shown in Fig. 4.35. The division can yield either the subtriangles V_1CV_2, V_1BV_2 and V_1BA, or V_1CV_2, V_1AV_2 and ABV_2. In addition, one must assume that the sections V_1V_2 on one side, and the sections AV_2 resp. BV_1 are pseudo-binary sections.

To obtain a rapid decision about the division into three subsystems, the "clearing cross" method of W. Guertler is used. A composition of the intersection point Q of the sections AV_2 and BV_1 is investigated. If in the solid state in addition to B also the compound V_1 is present, the division into the triangles V_1CV_2, V_1BV_2 and V_1BA is correct. One must investigate other compositions in the section BV_1, to ascertain that it is really a pseudobinary section. If the inves-

Fig. 4.35
The clearing cross method
(source: W. Guertler)

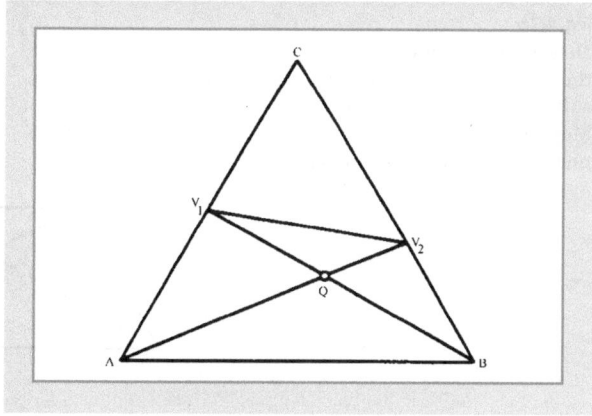

tigation of point Q shows, that in the solid state in addition to A V_2 is present, the division into the triangles V_1CV_2, V_1AV_2 and AV_2B is correct.

The clearing cross method permits obtaining a rapid overview about the possible phase equilibria in simple systems with two binary compounds. However, it is not applicable if in the composition range ABV_2V_1 in addition to eutectic also intermediate equilibria are present. In this case, the sections BV_1 and AV_2 are not pseudobinary sections. In addition intermediate equilibria practically never go to completion, and thus, do not give clear answers in structure investigations.

Based on the method of W. Guertler, the question, important in metallurgy, can be answered often, which sulfide of two metals Me_1 and Me_2 at preset exterior conditions has the larger Gibbs energy of formation, which of the two sulfides $(Me_1)S$ and $(Me_2)S$ thus, is less stable, and therefore is reduced by the metal, which forms the more stable sulfide.

Figure 4.36 represents schematically the ternary system sulphur – metal M_1 – metal M_2. If the investigation of the mixture Q shows, that $Me_2 - (Me_1)S$ is a pseudobinary section, thus, in the mixture Q, $(Me_1)S$ and Me_1 are not in equilibrium, then the thermodynamic stability of $(Me_1)S$ is greater than that of $(Me_2)S$. A mixture of $(Me_2)S$ and Me_1 with the overall composition of Q, reacts to $(Me_1)S + Me_2$ of the same overall composition Q. This is the method of "matte smelting". In earlier times lead was obtained on this basis through reaction of lead matte with iron:

$$PbS + Fe \rightarrow Pb + FeS \tag{4.19}$$

Also the elimination of sulfur from raw iron by addition of manganese is based on this principle:

$$FeS + Mn \rightarrow Fe + MnS \tag{4.20}$$

Fig. 4.36
Determination of the relative thermodynamic stability of sulfides of metals Me_1 and Me_2 due to the clearing cross method (source: W. Guertler)

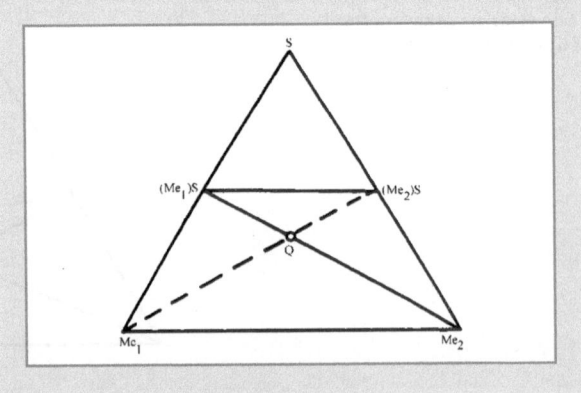

4.26
Ternary Compounds with Melting Point Maxima

If a ternary compound with a melting point maximum appears in a ternary system, fusion equilibria whose basic outlines are shown in Fig. 4.37 are present, assuming negligibly small solubilities in the solid state.

The liquidus surface of primary precipitation of the compound arching above V like a dome, is cut by the liquidus surfaces originating from the pure components A, B and C. The intersections are $E_1a_1E_3$, $E_3a_3E_2$ and $E_2a_2E_1$. It is E_1 the binary eutectic in the bounding system A–C, e_2 in B–C and e_3 in A–B. Three ternary eutectics appear: E_1 E_2 and E_3. Figure 4.37 represent the case, when A–V, B–V and C–V are pseudobinary sections. The composition triangle ABC can thus, be divided into the three subtriangles AVB, BVC and CVA. A ternary eutectic point is present in each of the subtriangles. The point a_1 corresponds to a maximum in temperature (saddle point) of the eutectic line E_1–E_3. The same applies for a_2 and a_3.

Fig. 4.37
Projection of melting sphere of a ternary system with the congruently melting ternary compound V

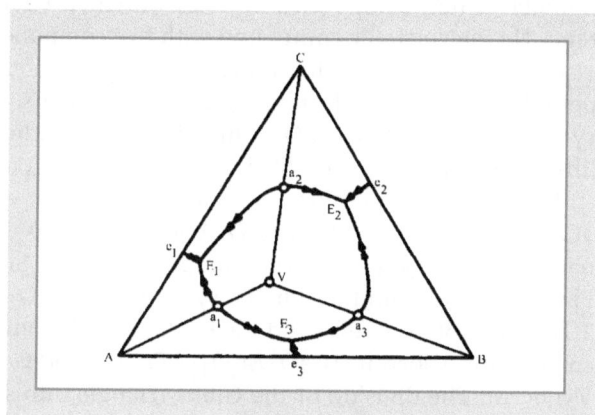

Fig. 4.38
Projection of the melting sphere for a ternary system with two congruently melting ternary compounds V_1 and V_2

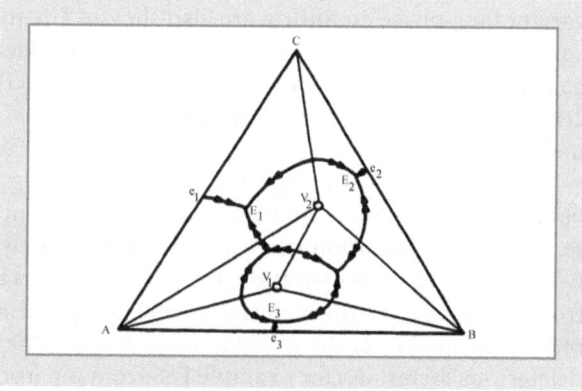

If more ternary compounds with melting point maxima are present, a corresponding division of the composition triangle into subtriangles is possible. Figure 4.38 shows that in the case of two ternary compounds.

4.27
Real Ternary Systems with Binary and Ternary Compounds

Three-component systems, which contain ternary compounds, and where the bounding binary systems present only simple eutectic fusion equilibria, practically do not occur. If a three-component system shows such a strong tendency for compound formation that ternary, congruently melting compounds form, then a strong tendency for compound formation also occurs at least in one of the binary bounding systems.

In the case of metallic systems experience shows that the addition of a third component to a two-component system forms only in very rare cases ternary intermetallic compounds, which are completely independent and do not represent solid solutions of binary intermetallic compounds of the bounding systems. The number of new compounds decreases further when going to quaternary systems. No intermetallic compound with five components is known.

As a rule in non-metallic systems ternary compounds also only occur if compounds are present in the binary bounding systems. This is the case in the ternary system $CaO - Al_2O_3 - SiO_2$, which is of great industrial importance. The outlines of the important equilibrium lines are shown in Fig. 4.39.

The binary bounding system $CaO - Al_2O_3$ exhibits four compounds, $3CaO \cdot Al_2O_3$, $5CaO \cdot 3Al_2O_3$, $CaO \cdot Al_2O_3$ and $3CaO \cdot 5Al_2O_3$. In the bounding binary system $Al_2O_3 - SiO_2$ the compound $Al_2O_3 \cdot SiO_2$ [$3Al_2O_3 \cdot 2SiO_2$] occurs. Finally in the bounding binary system $CaO - SiO_2$ three compounds appear: $2CaO \cdot SiO_2$, $3CaO \cdot 2SiO_2$ and $CaO \cdot SiO_2$ (Wollastonite). In addition the ternary compounds Anorthite ($CaO \cdot Al_2O_3 \cdot SiO_2$, V_1) and Gehlenite ($2CaO \cdot Al_2O_3 \cdot SiO_2$, V_2) occur. The division of the Gibbs triangle into subtriangles is immediately visible in Fig. 4.39.

The lines of the univariant three-phase equilibria and the points of the invariant four-phase equilibria are also shown. The melting point lowering of the components is, both in the binary bounding systems and the ternary system considerable. Above each of the points V_1 and V_2 a dome of the liquidus surface arises, representing primary crystallization of the corresponding congruently melting compound.

Slags for many metallurgical processes consist of mixtures in this system. To obtain low fusion temperatures, slags in the system $CaO - Al_2O_3 - SiO_2$ are chosen, whose composition correspond approximately to the eutectics E_1, E_2 and E_3. In a metallurgical operation, the job of the slags is to protect the liquid metal from air and to absorb materials, which are present in the ore used and which reduce the quality of the metal to be prepared. The capacity of a slag to absorb deleterious materials (for example FeS from pig iron) depends mainly on its ba-

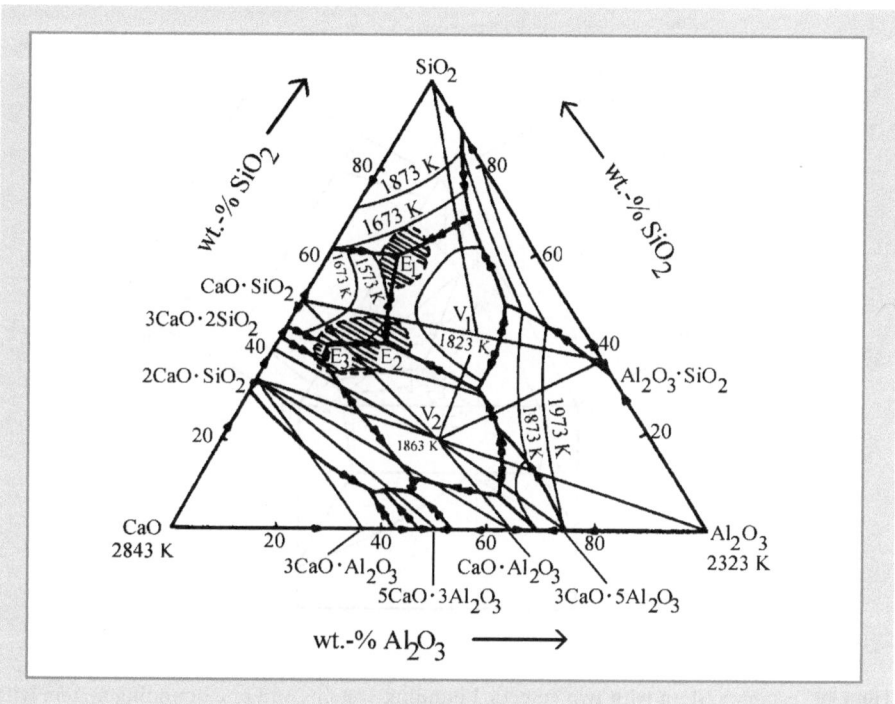

Fig. 4.39 The system $CaO - SiO_2 - Al_2O_3$ (after [4])

sicity. For acid slags, SiO_2 rich mixtures in the region of the ternary eutectic E_1 are used, whereas mixtures around E_2 and E_3 represent basic slags. The temperature of the eutectic point E_1 is $T = 1443$ K, and that of point E_2 is $T = 1538$ K. The slags used during iron winning, have compositions on one hand around E_1, on the other hand around E_2 and E_3, marked by the hatched regions.

4.28
Ternary System with Two Eutectic Bounding Binary Systems with Limited Solubility in the Solid and Complete Miscibility in the Third Bounding Binary System

Figure 4.40 shows the outlines of simple phase equilibria, where in the solid state, in two bounding binary systems miscibility gaps appear, and in the third bounding system complete solubility is present. A ternary eutectic reaction is not possible in this case. At the maximum two solid phases (α and β) can be present. In a ternary eutectic equilibrium three solid phases coexist with the liquid. Only a binary eutectic reaction can occur, along the eutectic line $e_1 - e_2$. A melting point maximum or a melting point minimum can appear in the ternary system.

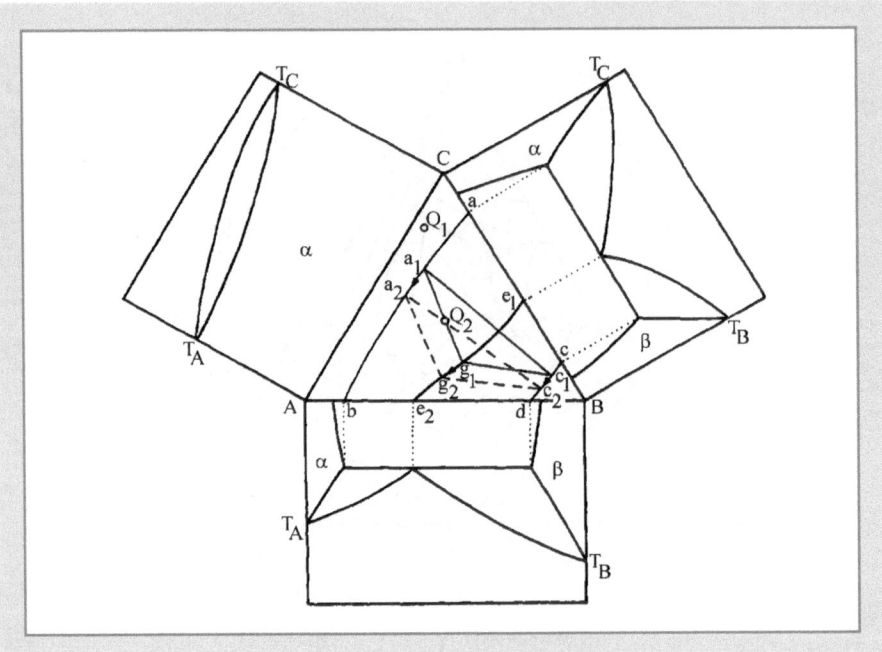

Fig. 4.40 Ternary system with two eutectic bounding systems and one bounding system with uninterrupted solubility in solid state

A liquid in the composition ranges ACab (for example corresponding to point Q_1) and cdB solidifies after passing the two-phase region solid-liquid into a solid solution. If during very slow cooling segregation can be avoided, the solid solution present at the end naturally has the same composition as the starting liquid. However, during crystallization of liquids in the region abdc, after reaching the liquidus surface, and precipitation of solid solution α (if the starting composition lay in the region bae_1e_2) or of solid solution β (if the starting composition lay in the region e_1e_2dc), a compositional change takes place in the liquid, so that finally the eutectic line $e_1 - e_2$ of binary crystallization is reached. If one neglects, for simplicity reason, the temperature dependence of the expansion of the α and β solid solutions, only α-phase of composition α_1, β-phase of composition c_1 and liquid of composition ε_1 are in equilibrium.

For an alloy of composition corresponding to point Q_2, the starting univariant reaction produces simultaneously α- and β-solid solutions. The composition of the liquid follows the eutectic line e_1-e_2, the composition of the α- resp. β-solid solution the lines $a - b$ resp. $c - d$. If finally the situation, represented by the triangle $a_2e_2c_2$ is reached, when the tie line $a_2 - c_2$ passes through the starting composition of the liquid (point Q_2) all the liquid is used up. The solidification is finished. As final products of the solidification process, phases α of composition a_2, and β of composition c_2 are present.

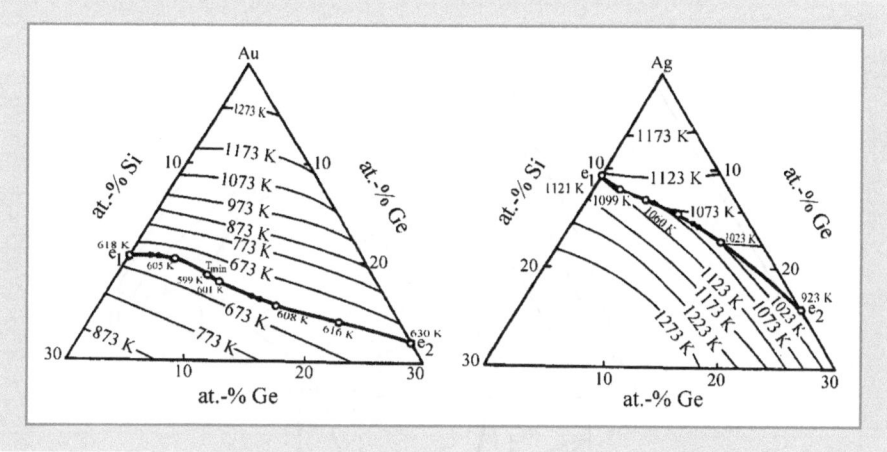

Fig. 4.41 Projection of the gold-rich, respectively Ag-rich corner of the system Si – Ge – Au, and Si – Ge – Ag, respectively [5]

As an example of a real diagram, the liquidus projection of the system silicon – germanium – gold, of great importance to semiconductor technology, is presented in Fig. 4.41. The bounding binary systems silicon – gold and germanium – gold have each an eutectic, the system silicon – germanium exhibits complete miscibility. The line of the binary eutectic solidification, which reaches from one binary eutectic e_1 to the other e_2, passes through a temperature minimum. Thermodynamic investigations indicate, that in the composition range of this minimum, increased interaction between dissimilar atoms, compared to the neighboring composition ranges, occurs. As this effect is not so pronounced in the solid state, a melting point minimum results. This is completely analogous to the melting point minimum in a binary system with unlimited solid solution formation.

In the homologous system silicon-germanium-silver the interaction between dissimilar atoms is not as pronounced. The line of the binary eutectic crystallization descends uniformly from e_1 (bounding system Ag – Si) to e_2 (bounding system Ag – Ge).

4.29
Ternary System with Two Peritectic Bounding Systems and Complete Solubility in the Third Bounding System

Figure 4.42 sketches the equilibrium conditions in a ternary system, where in two bounding systems limited solid solubility is present, and one of the solid solutions is formed by a binary peritectic reaction, and the third bounding system shows complete miscibility of the two solid solutions formed by the peritectic reactions. They are similar to the situation in Fig. 4.40. Maximum two solid

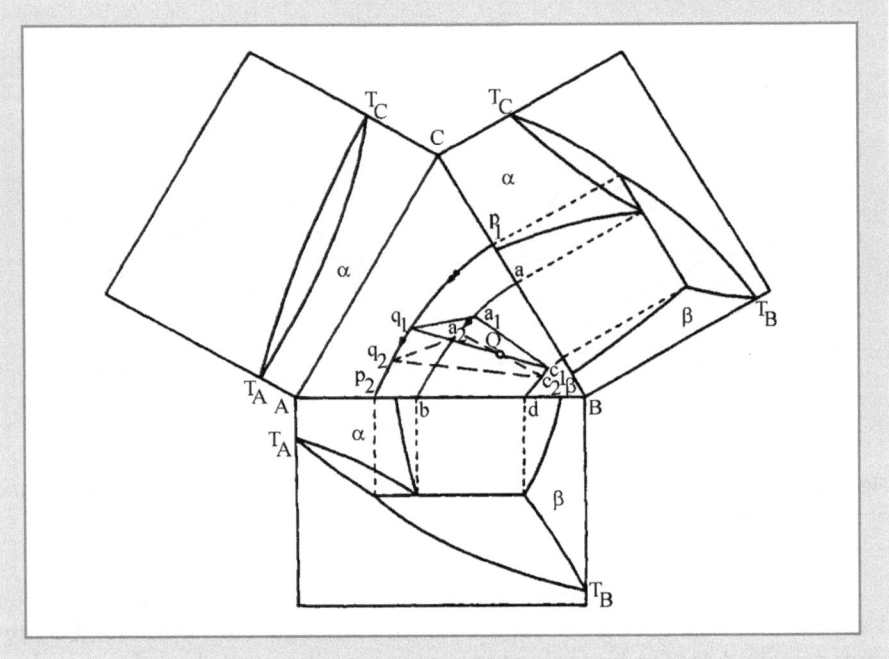

Fig. 4.42 Ternary system with two peritectical bounding systems and one bounding system with formation of uninterrupted solid solution

phases are present. The miscibility gaps in the bounding systems A – B and B – C are joined. Their composition limits during the peritectic solidification are indicated by the lines a – b and c – d. The line p_1-p_2 indicates the trace of the break in the liquidus surface, caused by the peritectic reaction. As at the maximum only two solid phases can be present, there is no possibility for a ternary eutectic or ternary peritectic reaction.

The solidification of liquids in the range ACp_1p_2 and Bcd yields single phase solids, if the equilibrium conditions are not disturbed by the formation of layer crystals (segregation).

The solidification of liquids in the region p_1cdp_2 (for example Q) yields first the primary crystallization of β-solid solutions, which causes depletion of B in the liquid, until a point on the line p_1 – p_2 is reached. Now α appears as a third phase. The compatibility triangle for the onset of the binary univariant peritectic reaction is $a_1c_1q_1$. During the peritectic reaction and with decreasing temperature, the compositions of the three coexisting phases are displaced along the arrows. The liquid reacts with the β-solid solution present and forms α-solid solutions. The whole liquid is used up when, after composition changes, the tie line between the two coexisting solid phases α and β passes through Q, the starting composition of the liquid. The univariant equilibrium, when the liquid just disappears, is given by $a_2c_2q_2$.

4.30
Transition between an Univariant Peritectic and an Univariant Eutectic Reaction

Ternary systems, in which, besides bivariant precipitation reactions, only univariant eutectic equilibria are present, are quite similar to ternary systems, where besides bivariant reactions only univariant peritectic phase equilibria occur. The possibility of a transition from one of the univariant reactions to the other exists. This is the case in a ternary system, where one bounding binary system has simple eutectic equilibria with limited solid solubility, the other is characterized by a peritectic, and the third shows complete miscibility. The equilibrium conditions are sketched in Fig. 4.43.

On the line $p_1 - e_1$ the binary peritectic undergoes a continuous change to a binary eutectic.

Figure 4.44 represents this in space. The notations are the same as in Fig. 4.43. In Fig. 4.43 line $p_1 - e_1$ crosses line a - b. The crossover point is not an intersection, as indicated in Fig. 4.44. In the region of the apparent intersection, the line a-b lays at a lower temperature than the line $p_1 - e_1$.

For two cases of an univariant equilibrium the tie triangles are indicated in Fig. 4.43. It is characteristic for the univariant eutectic equilibrium, that the

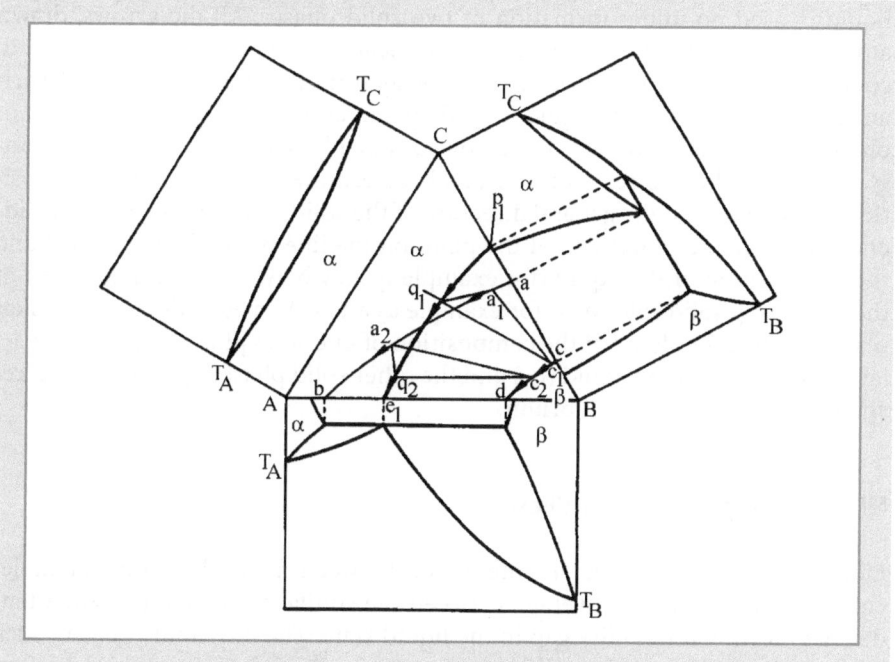

Fig. 4.43 Transition of a binary eutectic equilibrium to a binary peritectic equilibrium

Fig. 4.44
Spatial of the transition of
a binary eutectic equilibrium to a binary peritectic
equilibrium. Notations as in
Fig. 4.43, ---- line beneath
melting spheres

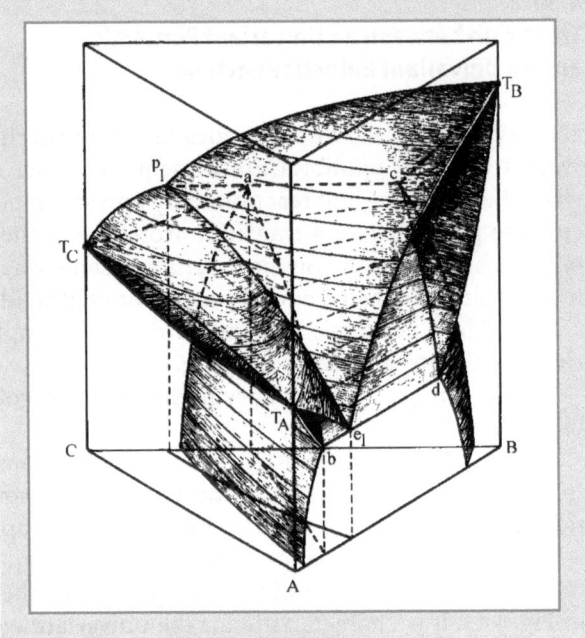

liquid is used up under formation of two solid phases. All the arrows, drawn at the corners of the tie triangle, which indicate the direction of the change in composition during the reaction, direct away from the triangle. At the univariant peritectic reaction the liquid is used up and at the same time the amount of one solid phase is decreased. The arrow for one phase directs into the tie triangle. Which of the two reactions occurs, is determined by the relative composition of the phases. If the general direction of the solidification process is considered (for example movement of the point on the line $p_1 - e_1$ from p_1 to e_1) and the composition of the liquid (for example q_2) lay between the composition of the coexisting solid solutions (for example a_2 and c_2) an eutectic reaction takes place. On the other hand, if the composition of one solid phase (for example a_1) lay between that of the liquid (q_1) and the other solid phase (c_1) than this corresponds to a peritectic equilibrium.

4.31
Miscibility Gap in the Liquid Phase

Liquid solutions, in which the interaction between dissimilar atoms or molecules is smaller than the average between the similar participants, have a tendency to form a miscibility gap in the liquid state. The miscibility gaps in such binary systems extend naturally into the ternary, by addition of a third component which is soluble in the binary liquid.

The simplest case occurs, when only one bounding binary system shows a miscibility gap in the liquid state. Figure 4.45 shows an isothermal section, which lays above the liquidus surfaces and only cuts the miscibility gap. In an analogous fashion to the bounding binary system A – B at T_1, where the liquids L_1 and L_1' are in equilibrium with each other, thus, $L_1 - L_1'$ is a tie line, in the ternary there are coexisting liquids, for example a and b. As the bounding binary systems A – C and B – C do not present a miscibility gap in the liquid, the miscibility gap originating in the base system A – B must close by necessity in the ternary system. With increasing additions of component C, the composition of the coexisting phases approach each other, until at the preset temperature (T_1) they coincide at the composition K_1. This point K_1, in analogy to the point K, where in the bounding binary system A – B the liquids in equilibrium with each other with increasing temperature finally become identical, also is called (partial) critical point.

In general the extent of the miscibility gap in the ternary decreases with increasing temperature, corresponding to the position of the critical point changes. The totality of the critical points, represents a line, which for a composition $x_C \to 0$ finally reaches the critical point K of the binary system A – B. It is called the critical curve.

If the two bounding binary systems, where the components are completely miscible in the liquid state, present eutectic equilibria, Fig. 4.46 represents the phase reactions in the simplest case. Let the line fgh be the trace, along which

Fig. 4.45
Isothermal cut through a ternary system with a miscibility gap terminating in the ternary in the liquid state

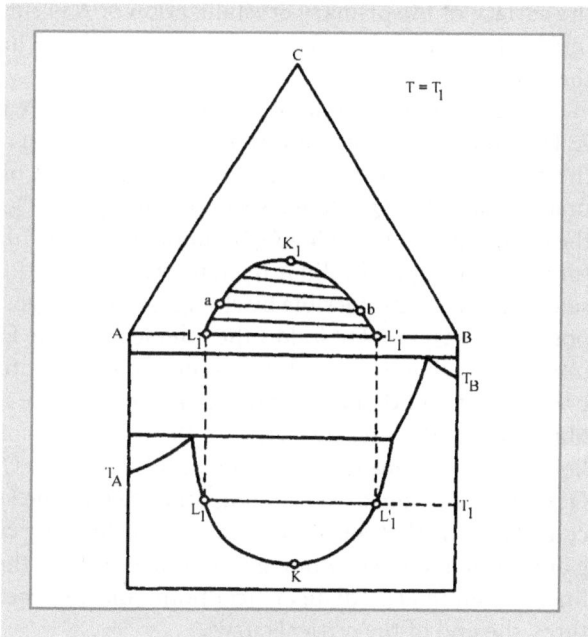

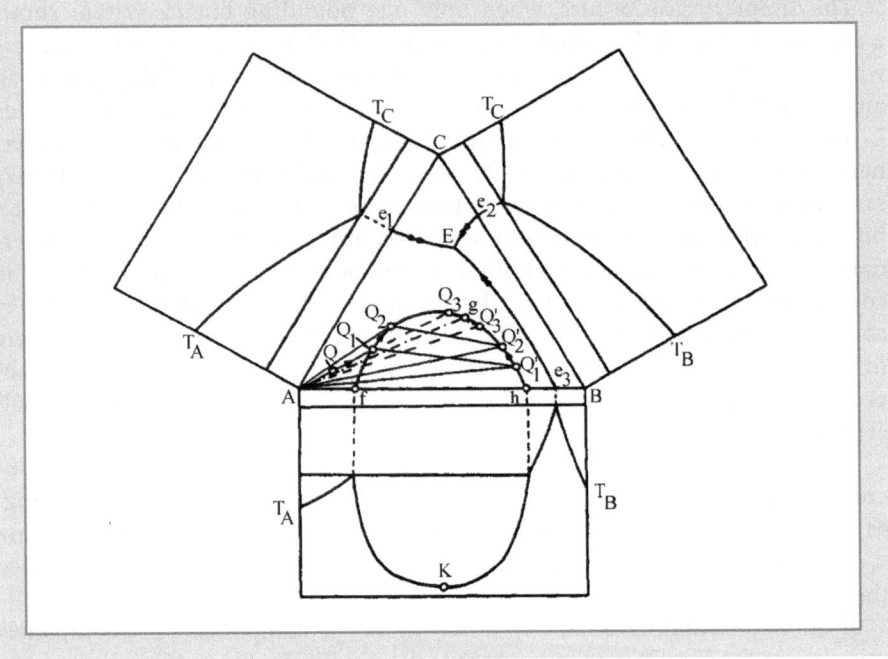

Fig. 4.46 Outlines of phase equilibria in a ternary system, in which one bounding system has a miscibility gap in the liquid state, -.-.-. critical knod

the surface of the primary crystallization of A is cut by the miscibility gap. This intersection line is not at constant temperature. The surface fghf is not an isothermal section.

During cooling a liquid of composition Q first reaches the liquidus surface of A. Due to the crystallization of A, the composition of the liquid proceeds along the extension of the tie line A – Q in the direction of the miscibility gap. This is finally reached at Q_1. Here a second liquid phase Q_1' appears. For this situation the compatibility triangle AQ_1Q_1' applies. During further cooling from the mixture of liquid Q_1 and solid A in an univariant reaction the liquid phase Q_1' separates, which is only possible by changing the composition of liquid Q_1. The compositions of Q_1 and Q_1' follow the intersection line fgh in the direction of g. This intersection line is, in analogy to the binary eutectic line, the place of all compositions of liquid phases, which participate in the univariant three phase equilibria. During this reaction, due to the change in composition of the coexisting liquids, for example from Q_1 to Q_2 and from Q_1' to Q_2', the starting composition Q reaches the interior of the compatibility triangle (AQ_2Q_2'). It follows from this consideration, that the temperature of the limiting curve fgh decreases from f to g and from h to g. The point g is assumed to be at the lowest temperature of this line. It is an intermediate critical point and represents towards lower temperatures, the end of the critical curve.

In the preceding case the end of the univariant reaction occurs, when as a result of the composition deplacement of the liquids, the point Q of the starting composition lays on the tie line A – Q_3 of the tie triangle AQ_3Q_3'. The quantity of the liquid Q_3' now become zero. Only phase A and liquid Q_3 are left. The primary crystallization of A, started in the region Q – Q' can now continue as a bivariant reaction during additional cooling. At another position of Q the first formed liquid can be used up first.

The primary crystallization of A still continues. When the starting composition of Q lays on the straight line A-g (critical tie line, see Fig. 4.46) both liquids are used up at the same time.

In three-component systems, based on the description in binary system, the reaction, in which two liquids and a solid phase participate is called monotectic three phase reaction.

4.32
Monotectic Four Phase Reaction

In a binary system two coexisting liquids cannot solidify eutectically, because in such a reaction four phases would be in equilibrium. According to the Gibbs phase rule the maximum number of coexisting phases in a two-component system at a constant pressure is P = 3 (invariant equilibrium). This possibility, however, exists in a three-component system, because based on by one larger number of components the degree of freedoms is also greater by one. Figure 4.47 represents the projection of such equilibria onto the base plane.

During cooling of liquid Q the liquidus surface of primary crystallization of B is reached. Due to the crystallization of B, the composition of Q is displaced along the extension of the tie line B – Q, until the line e_2-a of the binary eutectic crystallization of B and C is reached.

Fig. 4.47
Monotectic reaction of four phases

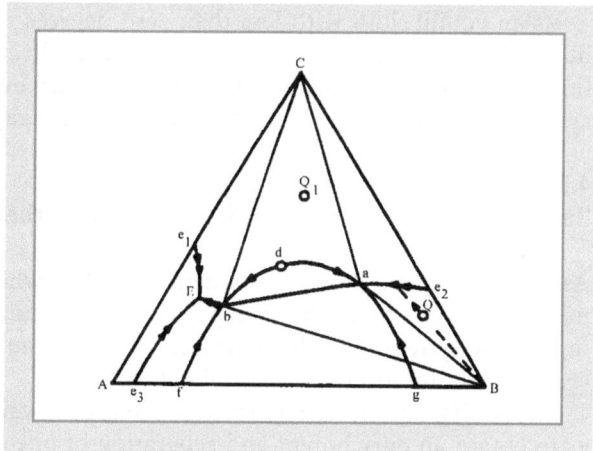

During continued cooling the composition of the liquid follows the eutectic line e_2 – a while simultanously B and C precipitate. If the composition a is reached, a second liquid appears, having the composition b. Now four phases are in equilibrium. During continued heat removal the invariant reaction, liquid L_a reacting to solid B, solid C and liquid L_b occurs:

$$L_a \rightarrow B + C + L_b \tag{4.21}$$

This is a monotectic four phase reaction. It is analogous one to the ternary eutectic reaction. The compositions of the three phases B, C and L_b, formed during heat removal, form a triangle, in which the composition of the used up liquid L_a lays.

After the liquid L_a is used up completely, three phases B, C and L_b remain. This corresponds to an univariant equilibrium. During temperature decrease, B and C precipitate simultaneously from liquid L_b. The composition of the liquid follows the line b-E, until the ternary eutectic is reached, where under simultaneous precipitation of A, B and C, the leftover liquid solidifies completely.

The lines f-b and g-a represent intersections of the liquidus surface of A resp. B with the miscibility gap. Along these line the temperature decreases in direction of b, resp. a. The line b-d-a is the intersection line of the miscibility gap with the liquidus surface of the primary crystallization of C. Point d represents the maximum temperature of the line b – d – a. The temperature decreases from d to a and from d to b. Points a and b lay at the lowest temperatures occuring along the line fbdag.

During cooling of a liquid of composition Q_1, primary C crystallizes. The liquid becomes poorer in C, until line b – d – a is reached. A second liquid phase is formed. With decreasing temperature and continued precipitation of C, the compositions of the two liquids change along the limiting line b – d – a, while the univariant three-phase equilibrium proceeds, until the composition a and b is reached. During continued heat removal the invariant monotectic four-phase reaction takes place, where in addition to B, C precipitates.

As an equilibrium with less than zero degrees of freedom (invariant equilibrium) cannot exist, a ternary eutectic crystallization from a mixture of two liquid phases is impossible, similarly as a binary eutectic in a two-component system cannot fall into the region of a miscibility gap in the liquid state.

4.33
Real Ternary Diagrams with Limited Solubility in the Liquid Phase

As an example of a three-component system, in which one of the bounding binary systems has a miscibility gap in the liquid state, while the other two bounding binary systems are characterized by simple eutectic fusion equilibria, the system lead-zinc-tin is shown in Fig. 4.48.

Miscibility gaps in ternary systems are used in various metallurgical processes to obtain an enrichment and separation of one component. As an example,

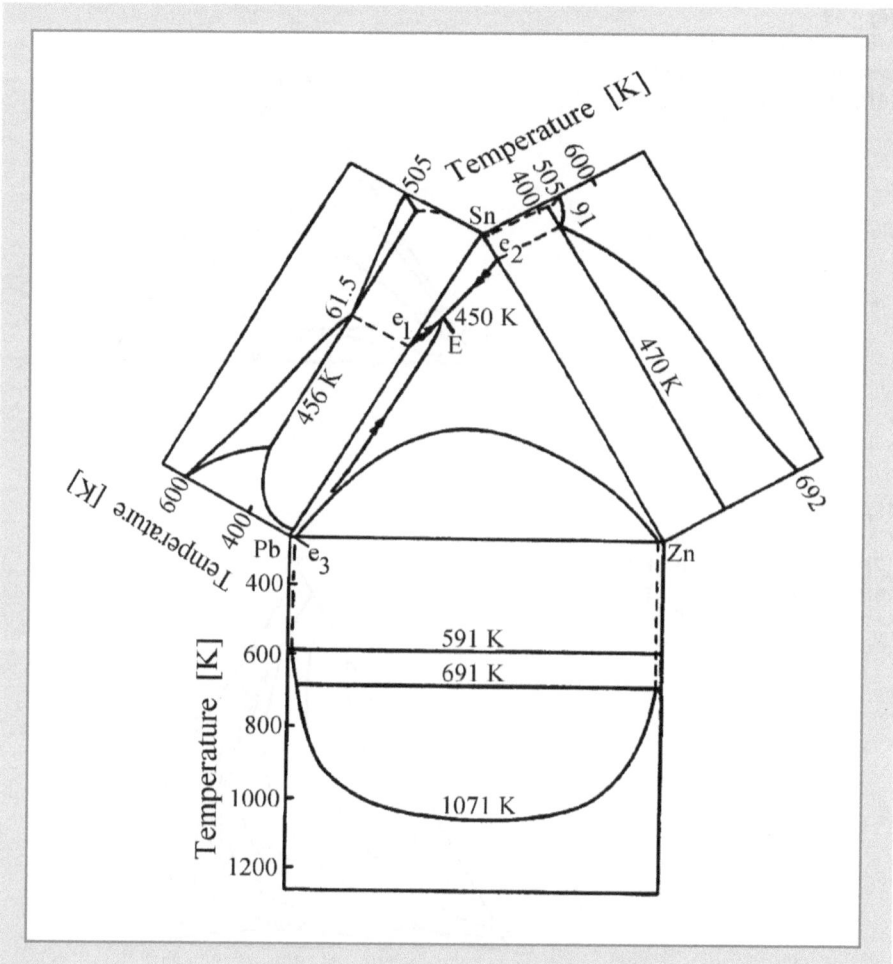

Fig. 4.48 Pb-Zn-Sn phase diagram (after [6]), concentration in wt.-%

Fig. 4.49 reproduces the miscibility gap in the system silver-lead-zinc. The lines representing other equilibria are omitted.

If zinc is added to a lead-silver mixture with the composition a, until the liquid alloy has the concentration b, the system splits into two liquid phases L_1 and L_1'. The silver content of L_1 is less, that of L_1' higher than in the original alloy. The zinc rich liquid L_1' has a lower density than the lead rich alloy L_1, and floats on top. The two liquid layers can easily be separated from each other.

If significantly more zinc is added, so that, for example, the concentration corresponds to the point c, the lead rich layer is indeed still silver poorer (L_2), but according to the lever rule the quantity of the silver rich layer (L_2') is increased, and its silver content decreased.

Fig. 4.49
Miscibility gap in the ternary
system Pb – Zn – Ag (after
[4]), concentration in wt.-%

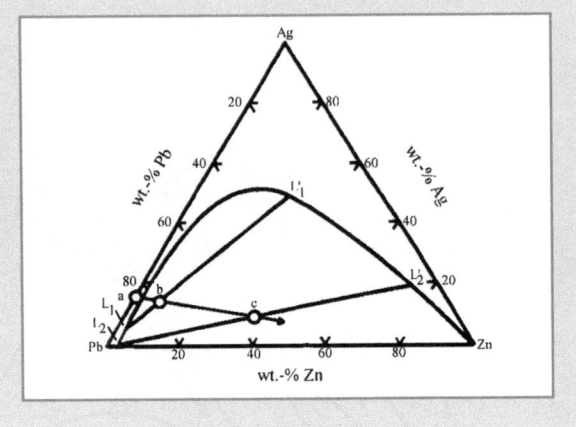

Fig. 4.50
Ga – Pb – Cd system (after
[7]), ——— Isotherms of the
miscibility gap, ---------- projection of the critical line

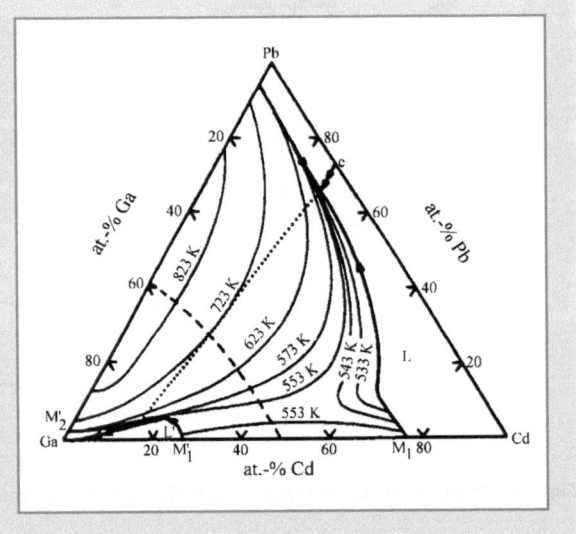

The most favorable zinc addition for an effective separation of silver from
lead is given by the shape of the miscibility gap and the position of the tie
lines.

As an example of a real diagram with a monotectic four-phase equilibrium
Fig. 4.50 reproduces the important equilibrium lines of the system gallium-cad-
mium-lead. In the bounding binary systems gallium – cadmium and gallium –
lead miscibility gaps appear in the liquid state, which extend over the whole ternary system. In the system cadmium-lead simple eutectic equilibria are present.
The ternary eutectic, as well as the binary eutectics of the gallium – lead and
gallium – cadmium systems are very close to 100 at.-% Ga. These invariant equi-

libria are therefore not represented in Fig. 4.50. On the basis of the isotherms of the miscibility gap it is obvious, that in the liquid-liquid two phase region extending in the ternary from the bounding binary system Ga-Pb to the bounding binary system Ga–Cd, a minimum in the critical temperature occurs at about 10 at.-% lead. Simultaneously, a contraction of the composition range of the miscibility gap takes place. This is an indication that, in this composition range the interaction parameter between dissimilar atoms is stronger, than in other parts of the miscibility gap.

4.34
Reaction Schemes

To obtain a rapid overview about neighboring phase fields and reaction steps in a ternary system, E. Scheil proposed the construction of a reaction scheme, where the phase equilibria, which follow each other during solidification are connected in a systematic way (see also [8]).

In the case of a ternary system with eutectics in the base systems and a ternary eutectic, the following univariant reactions of binary eutectic crystallization appear:

$$L \leftrightarrow \alpha + \beta \tag{4.22}$$
$$L \leftrightarrow \alpha + \gamma \tag{4.23}$$
$$L \leftrightarrow \beta + \gamma \tag{4.24}$$

L is the liquid, α, β and γ are the solid solutions of components A, B and C. This binary crystallization is followed by the ternary eutectic crystallization:

$$L \leftrightarrow \alpha + \beta + \gamma \tag{4.25}$$

As the final result of the solidification a mixture of the phases α, β, and γ remains.

The succession of the reactions is represented according to E. Scheil in the form shown in Table 4.3. From the representation it is immediately visible, which phase fields with univariant equilibria are contiguous to the plane of the four-phase equilibrium.

Similarly Table 4.4 shows the succession of the reactions in connection with an intermediate equilibrium.

The plane of the four-phase reaction

$$L + \alpha \leftrightarrow \beta + \gamma \tag{4.26}$$

is met on one hand by the univariant reactions

$$L + \alpha \leftrightarrow \beta \tag{4.27}$$

and

$$L + \alpha \leftrightarrow \gamma \tag{4.28}$$

Table 4.3 Scheme of reaction after E. Scheil for a ternary system with an eutectic four-phase equilibrium

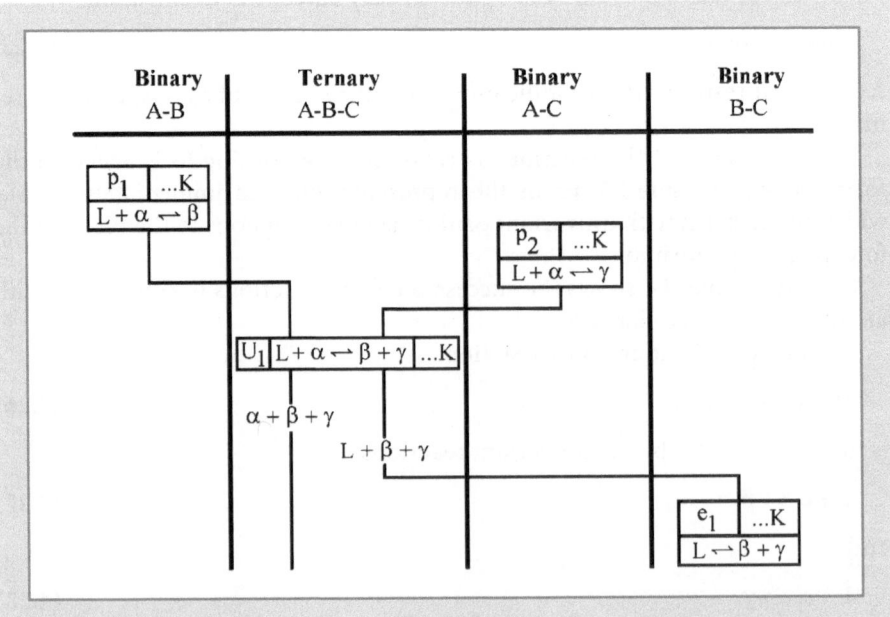

Table 4.4 Scheme of reaction after E. Scheil for a ternary system with a non-variant intermediate equilibrium

Table 4.5 Scheme of reaction after E. Scheil for a ternary system with a non-variant peritectic equilibrium

Binary A-B	Ternary A-B-C	Binary A-C	Binary B-C

$$\boxed{\begin{array}{|c|c|}\hline e_1 & ...K \\\hline L \leftrightharpoons \alpha + \beta \\\hline\end{array}}$$

$$\boxed{\begin{array}{|c|c|}\hline P_1 & L + \alpha + \beta \leftrightharpoons \gamma & ...K \\\hline\end{array}}$$

$\alpha + \beta +$

$L + \alpha + \gamma$

$$\boxed{\begin{array}{|c|c|}\hline P_2 & ...K \\\hline L + \beta \leftrightharpoons \gamma \\\hline\end{array}}$$

$$\boxed{\begin{array}{|c|c|}\hline P_1 & ...K \\\hline L + \alpha \leftrightharpoons \gamma \\\hline\end{array}}$$

as well as the three phase fields of the end products of the crystallization $\alpha+\beta+\gamma$, and that of the univariant reaction

$$L \leftrightarrow \beta + \gamma \tag{4.29}$$

Finally, Table 4.5 represents the reaction scheme for a system with a peritectic four-phase reaction. As in Tables 4.3 and 4.4 represented four-phase equilibria, also in this case four fields of three-phase equilibria, meet the four-phase plane.

4.35
Four-Component Systems

The same principles apply to the determination of the equilibrium conditions in a four-component system are used in two- and three-components systems. The variety of the possible equilibria and their combinations is, however, considerable. In addition, to represent the compositions in a four-component system, three dimensions are required. Only isothermal sections can be represented directly.

For a graphical representation of phase equilibria at a fixed temperature in quaternary systems, the equilateral tetrahedron (composition tetrahedron) is most suitable (see Fig. 4.51). The four corners A, B, C and D correspond to the pure components. The length of the edges is set as 100 at.-% (or 100 wt.-%) as is assigned to binary mixtures. The surfaces are the regions of the bounding ternary systems. At the interior of the composition tetrahedron lay the compositions of the quaternary mixtures. In Fig. 4.51 the position of one composition is shown. From the corner A the concentration are drawn: of B (distance b on the edge A – B), of C (distance c on the edge A – C), of D (distance d on the edge A – D). Construction of a parallelogram from the distances b, c and d yields the position of point Q in the composition tetrahedron. A few simple sections can be constructed in the composition tetrahedron. (see Fig. 4.52).

Fig. 4.51
Presentation of the con-
centration in a quaternary
system

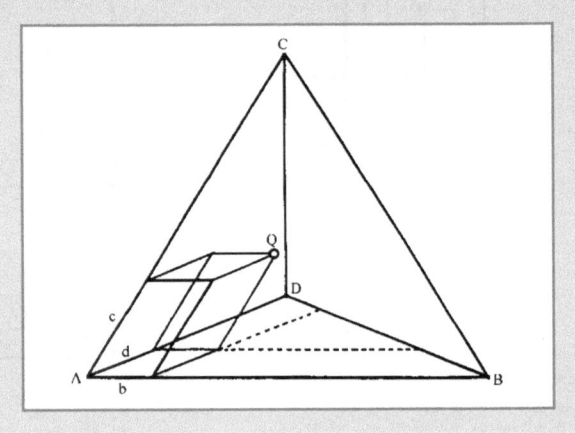

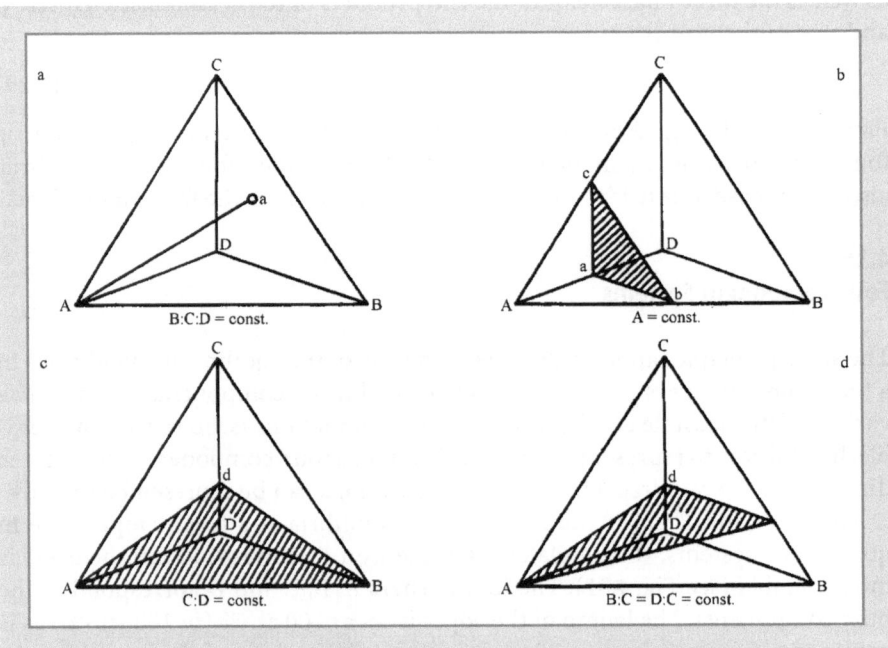

Fig. 4.52 Geometrically simple cuts through a tetrahedron of concentrations

a) Mixtures with a constant ratio of three components lay on a line, which passes through the corner of the fourth component. In the case represented in Figure 4.52a the ratio B:C:D = const. The concentration of A increases from a to A.

b) Figure 4.52b represents a section with constant concentration of A. The plane abc is parallel to the tetrahedron surface BCD, which is opposite the A corner.

c) Figure 4.52c shows a section, where the ration C:D = const. The surface is bound by the line of the binary system A-B, where the concentrations of A and B are not fixed. The lines A–d and B–d are intersections in the ternary bounding systems A–C–D and B–C–D for constant ratio C:D.

d) Figure 4.52d represents a section with B:C = const. and C:D = const. The line A–d corresponds to a constant ration C:D in the bounding ternary system A–D–C, and for the line A–e B:C = const. in the bounding ternary system A–B–C.

4.36
Simple Equilibria in Four-Component Systems

A two-phase equilibrium L ↔ α at a constant temperature is shown in Fig. 4.53 for a four-component system. Compared to an analogous equilibrium in a ternary system, the dimensions of the phase fields and the number of degrees of freedom is increased by one. The two-phase equilibrium at constant pressure is, in a quaternary system according to the Gibbs phase rule, a trivariant reaction. The phase field of the liquid is a space, also that of the solid solutions, and of the mixture L + α (see Fig. 4.53a). The tie lines between the surfaces def and ghi do not lie in a plane. Naturally they do not intersect.

With decreasing temperature the surfaces def and ghi displace themselves farther from A (see Fig. 4.53b). If all bounding binary systems are simple eutec-

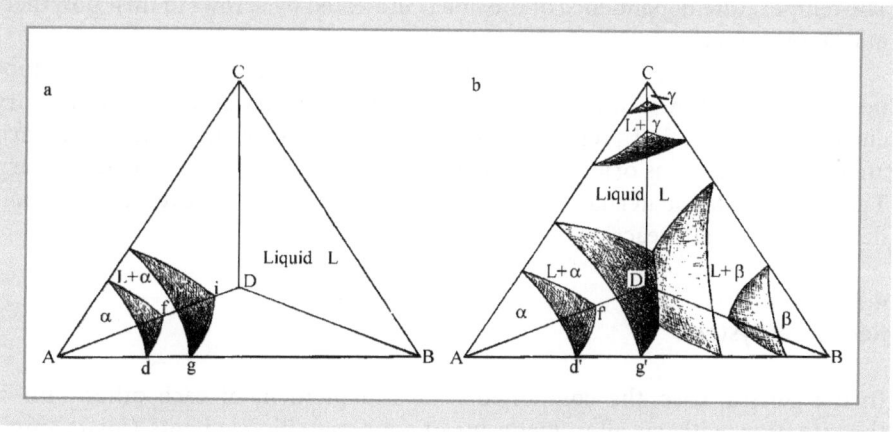

Fig. 4.53 Special presentation of liquidus-solidus spheres in a tetrahedron of concentrations

Fig. 4.54
Polythermic presentation of
eutectic crystallization in a
simple quarternary system
with a quaternary eutectic
E_Q

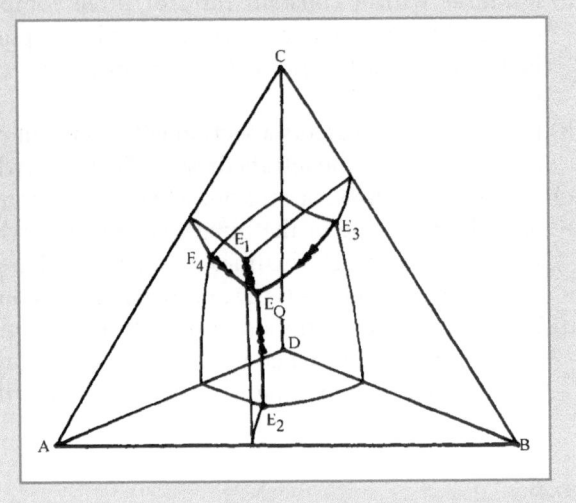

tic systems, and ternary eutectic points are present in all bounding ternary systems, then when two trivariant equilibria meet, bivariant reactions follow, where two components precipitate simultaneously. If two bivariant reactions meet, an univariant ternary solidification results. Finally the quaternary eutectic point can be reached, where the leftover liquid solidifies at constant temperature.

$$L \leftrightarrow \alpha + \beta + \gamma + \delta \qquad (4.30)$$

The graphical representation of such equilibria poses difficulties. Similarly as was done in ternary systems, to project onto the plane of the composition triangle the lines of univariant equilibria to obtain a rapid overview, the equilibrium lines in quaternary systems can be projected into the composition tetrahedron. The temperature dependence of the line is indicated by arrows in this "polythermal" representation.

A simple example for the case of an eutectic solidification in a four-component system is shown in Fig. 4.54. The bounding ternary systems have ternary eutectic points E_1, E_2, E_3 resp. E_4. With decreasing temperature, lines of univariant eutectic reactions originate at these points, and extend into the quaternary. They meet at the quaternary eutectic point E_Q, where the leftover liquid solidifies in an invariant reaction.

4.37
Reciprocal Systems

In the general case, the components are independent of each other. This is the situation with an alloy made up of four metallic elements (for example Cd – Bi – Pb – Sn). In systems with compounds, however, reactions can take place, which lead to special situations. A double reaction occurs during melting of two salts, whose anions and cations are different:

$$BaCl_2 + Mg(NO_3)_2 \leftrightarrow Ba(NO_3)_2 + MgCl_2 \tag{4.31}$$

The melt consists of four components. They are not independent from each other, but are connected through Eq. (4.31). The number of degrees of freedom is reduced by one compared to a system with four independent components. Four salts, which are connected by a double reaction are called reciprocal salt pairs.

In a reciprocal system AX – AY – BX – BY, where A,B are the cations and X, Y the anions of the salt, only three compositions have to be given, to know all the compositions; the fourth is given by the reaction

$$AX + BY \leftrightarrow AY + BX \tag{4.32}$$

The consideration of the phase equilibria can thus, proceed as with a three-component system. It is possible to represent the concentrations in a plane. Instead of a composition triangle a composition square is more appropriate.

In a system AX – AY – BX – BY, whose components are connected by a double reaction according to Eq. (4.32), the constituents form the corners of a square (see Fig. 4.55). It is convenient for neighboring corners to have a common part component. The reciprocal pairs are positioned diagonally. The sides of the square represent the totality of concentrations of the binary systems with one common part component (AX – AY, AY – BY, BY – BX, BX – AX), which cannot perform a double reaction. The inside of the square is the compositions region with double reaction. The point P is simply given by the two mole fractions

Fig. 4.55
Square of concentrations for a quaternary system AX – AY – BY – BX with double reaction

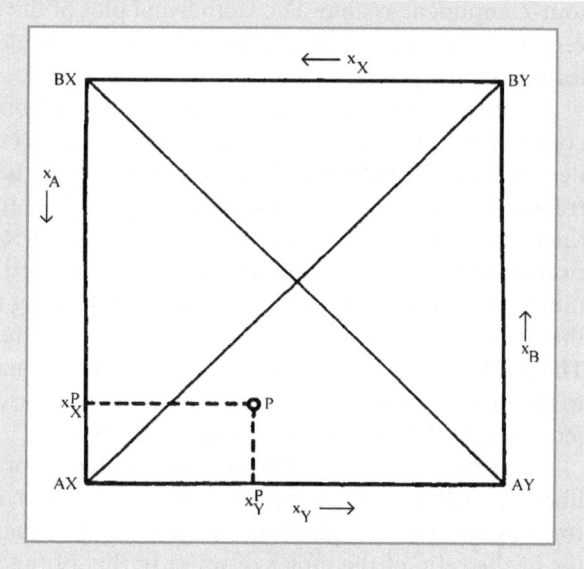

$x_x^P = 1 - x_A^P$ and x_y^P. To represent the phase equilibria at constant pressure, the temperature is plotted pendicularly to the composition square.

Naturally the sum of all mole fractions is 1. For the point P, shown in Fig. 4.55, which lays in the triangle AX – AY – BX, we have

$$x_{AY} = 0.4$$

and

$$x_{BX} = 0.2$$

Correspondingly the mole fraction of AX is $x_{AX} = 0.4$

The point P also lays in the triangle AX – AY – BY. Thus, the concentration of P can also be given as

$$x_{AX} = 0.6,$$
$$x_{BY} = 0.2,$$
$$x_{AX} = 0.2$$

The mole fractions of the cations and anions are naturally independent from the used triangle. They are, as can be easily seen

$$x_A = 0.8, x_B = 0.2, \text{ and } x_x = 0.6, x_y = 0.4.$$

4.38
Solubility of Reciprocal Salt Pairs in Water

For the representation of the concentrations of a reciprocal salt pair in water a plane is not sufficient. The special five-component system can be treated as a four-component system. The isothermal plot of the solubility equilibria already requires a spatial representation. As a rule, suitable projections are acceptable for certain purposes.

According to Löwenherz, to represent the solubility at a fixed temperature a composition pyramid with a square base surface is used (see Fig. 4.56). The corners of the base surface of the pyramid represent the pure salts (x =1). The mole fraction of these salts is plotted from the top O of the pyramid, which serves as the coordinate starting point. As shown in Fig. 4.56, to construct the composition point P, first the mole fraction of BX is plotted on the edge O – BX and the mole fraction of AX on the edge O – BX, and using the lines a and b, parallel to the above plots, the intersection Q is obtained. It represents a ternary solution. The line Q – P, starting at Q and parallel to the pyramid edge O – BY corresponds to the mole fraction x_{BY}. As the spatial representation is difficult, often the projection onto the square base surface is used.

According to Jänecke, for the representation of isothermal solubility equilibria of a solution of a reciprocal salt pair in water, the concentrations of water-free salt mixtures are plotted in a composition square. The mole fraction of water, or the ratio of the moles of water to that of the salt are plotted perpendicu-

Fig. 4.56
Isothermal pyramid of
concentrations of solutions
of a reciprocal pair of salts in
water (source: Löwenherz)

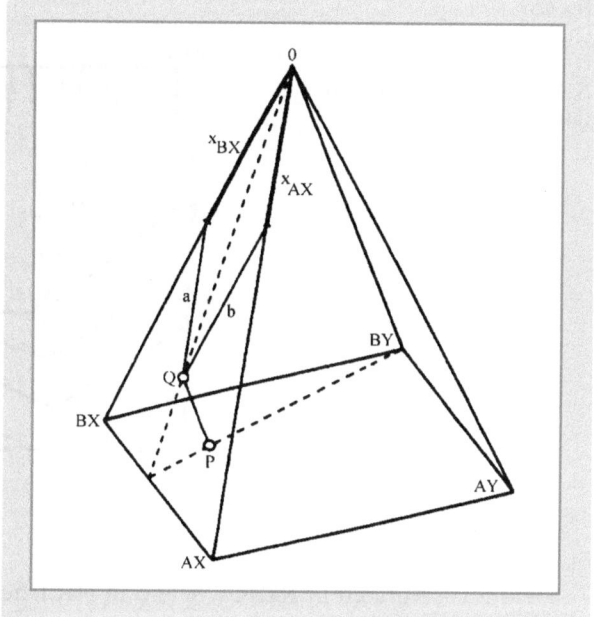

larly to this base plane. The concentrations of the quinary mixed phases lay in the interior of the square prism. The solubility of one salt is changed by the addition of another salt. The surfaces, which represent the totality of the solutions saturated in one salt, are curved. It is customary to show the course of the curvature in the projection on the base surface as isohypses. Similarly, one represents the position of the projection of the line saturated in two salts, which is the intersection of two curved surfaces, each saturated in one salt. The intersection of the lines representing two saturated solutions yield composition points, which are simultaneously saturated in three salts.

As an example, Fig. 4.57 represents the solubility relationships in the reciprocal salt system

$$NaNO_3 + KCl \leftrightarrow KNO_3 + NaCl \tag{4.33}$$

The water content of the saturated solutions is given as moles of water per mole of salt. The numerical values of these water concentrations are shown in the isohypses with equal concentration. Let us note, that the sides of the quadratic projection surface correspond to ternary systems ($NaCl - KCl - H_2O$, $KCl - KNO_3 - H_2O$, $KNO_3 - NaNO_3 - H_2O$ and $NaNO_3 - NaCl-H_2O$). Point A corresponds to a solution in the system $NaNO_3 - NaCl-H_2O$, which is simultaneously saturated in $NaNO_3$ and $NaCl$. The same applies to points B, C and D. From these points curves go into the base square, indicating solutions which are saturated in two salts. They divide the base surface into sub surfaces $\pi, \tau, \varrho, \sigma$, which represents fields saturated in one salt.

Fig. 4.57
Situation of solubility in a
reciprocal system of the salts
$NaNO_3 + KCl \leftrightarrow KNO_3 + NaCl$
(after [1])

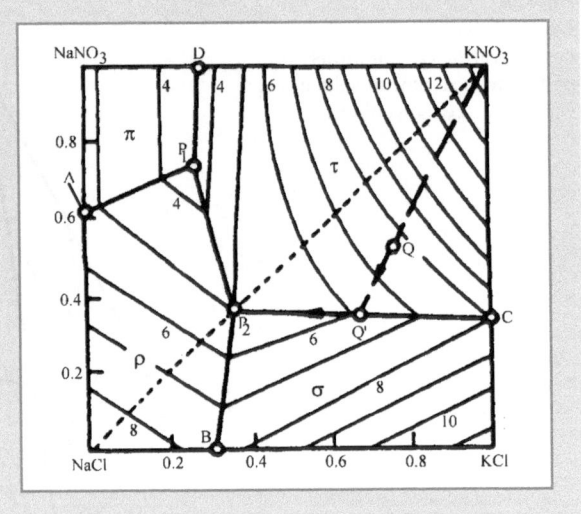

Field π is saturated in $NaNO_3$, field ϱ in $NaCl$, field σ in KCl and finally field τ in KNO_3. The isohypses of the water content are also indicated. Points P_1 and P_2 represents solutions, which are saturated in three salts, which belong to the neighboring fields on single saturation, π, σ, τ, resp. ϱ, σ, τ.

One can easily find the crystallization processes which occur during isothermal evaporation of water. The basics can be compared to the crystallization in a three-component system, when the temperature is decreased. As an example the isothermal evaporation of water from a starting solution Q is considered (see Fig. 4.57). KNO_3 crystallizes during loss of water, until the line $P_2 - C$ at Q' is reached. Now, under continued water evaporization, KNO_3 and KCl crystallize simultaneously.

The composition of the solution, moving along $Q' - P_2$, reaches P_2. Now KNO_3, KCl and $NaCl$ crystallize from the solution, until all the water is eliminated.

It must be remarked, that the reciprocal salt pair shown in Eq. (4.33) has gained significant importance in obtaining KNO_3 from $NaNO_3$ by reaction with KCl. The fields π and σ, as seen in Fig. 4.57, do not meet each other. Along the line $P_1 - P_2$ during water removal from a solution of $NaNO_3$ and KCl, a mixture of KNO_3 and $NaCl$ precipitates. In addition advantage is taken, during preparation of the "conversion nitrate" of the temperature dependence of the solubilities.

4.39
Comments to the Extent of Higher Order Systems

Heterogeneous equilibria have intensively been investigated since the beginning of this century. As example Table 4.6 gives a rough overview about the metallic systems already investigated, and those which have not.

Table 4.6
Number of possible poly-
nary systems and number of
investigated systems regarding 80 metallic elements as
possible components (after
[11])

Number of components n	Number of possible systems $\dbinom{80}{n}$	Approximate number of investigated systems
1	80	80
2	3 160	2 000
3	82 160	1 000
4	1 581 580	100
5	24 040 016	10
6	300 500 200	–
7	3 176 716 400	–

The number of possible combinations of the 80 metallic elements is considered up to seven-component systems.

The number of multi-component systems increases extremely rapid with the increase in the number of components. A comparison with the number of investigated systems shows that only about 70 % of the binary and 1 % of the ternary systems have been studied. With increasing number of components, the number of investigated systems decreases rapidly. It must be mentioned, that systems marked "investigated" are not always cleared up completely, but in many cases only small, very interesting parts were researched.

The number of possible n-component systems with a preset number of components does not increase continuously, but passes a maximum, as shown in Fig. 4.58. With 80 components the number n of systems increases until $n = 40$, and then decreases. Naturally, there is only one 80 component system, in which all combinations are contained in the bounding systems. The number of 40 component systems is in the order of magnitude 10^{23}.

The multitude of possible combinations becomes even larger, if one considers the other inorganic elements and compounds, and the endless number of organic compounds. Finally, the metastable phase equilibria have to be considered, which are often more important than the stable ones.

In summary, it must be stated, that in spite of an intensive research activity, only a small fraction, though an important fraction, of all possible stable and metastable heterogeneous equilibria has been cleared up.

Many, not very complex, metallic and non-metallic materials could be optimally developed for certain applications. It is known that small additions of other components can greatly influence the property of the material. In this respect, it is useful to carry out investigations of heterogeneous equilibria in multi-component systems. With increasing number of components the extent of research needed to understand a system increases greatly. The expenditure of

Fig. 4.58
Correlation between the
logarithm of number of
components of a system and
the number of possible systems, if 80 different elements
as components are available
(after [11])

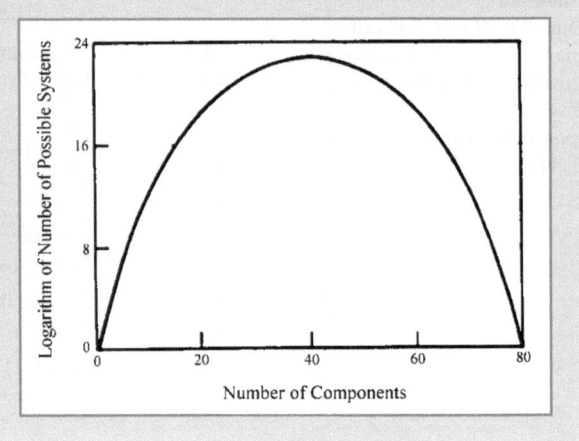

time increases rapidly, so that purely experimental investigations can no longer
be realized, or at least do not appear profitable.

The phase equilibria in a system are direct expressions of the thermodynamic properties of the mixture. It is thus, plausible, to investigate theoretically the
relationships between phase equilibria in higher order systems and the interaction parameters between atoms and molecules of the components. The necessary large calculations can rapidly be carried out with modern computer systems. The problem now is to set up suitable models, to adequately comprehend
the interaction parameters and the structure of the various phases and their effects on the phase equilibria (see Chap. 6).

References

Citations
[1] R. Vogel, "Die heterogenen Gleichgewichte", 2. Edition, Akademische Verlagsgesellschaft
Geest u. Portig, Leipzig (1959)
[2] W. Guertler, "Konstitution der ternären metallischen Systeme", M.W. Guertler, Berlin-
Dahlem (1956)
[3] L.S. Palatnik and A.I. Landau, "Phase Equilibria in Multicomponent Systems", Holt, Rein-
hart and Winston Inc., New York (1964)
[4] R. Vogel, "Einführung in die Metallurgie", Musterschmidt-Verlag, Göttingen (1955)
[5] B. Predel, H. Bankstahl and T. Gödecke, J. Less-Common Metals, 44, 39 (1976)
[6] W. Guertler and E. Anastasiadis, "Konstitution ternärer metallischer Systeme", Verlag
Wirtschaft und Kultur, E Jaster, Berlin (1956)
[7] W. Dreizier, F. Aldinger and G. Petzow, Z. Metallkde., 70 769 (1979)
[8] A. Prince, "Alloy Phase Equilibria", Elsevier Publishing Comp., Amsterdam (1966)
[9] G. Petzow, S. Steeb and G. Kiessler, Z. Metallkde., 54 (1963) 473
[10] G. Petzow and H.L. Lukas, Z. Metallkde., 61 (1970) 877

General References

A.M. Alper (Editor), "Phase Diagrams", Vol.I, Theory, Principles and Techniques of Phase Diagrams Vol.II The Use of Phase Diagrams in Metal, Refractory, Ceramic and Cement Technology, Vol.III, The Use of Phase Diagrams in Electronic Materials and Glass Technology. Academic Press, New York (1970)

W. Eitel, "Die heterogenen Schmelzgleichgewichte silikatischer Mehrstoffsysteme", J. Ambrosius Barth, Leipzig (1945)

R. Haase und H. Schonert, "Solid-Liquid Equilibrium", in: The International Encyclopedia of Physical Chemistry and Chemical Physics, Topic 13: "Mixtures, Solutions, Chemical and Phase Equilibria", Editor: M.L. McGlashan; Vol. l, Pergamon Press, Oxford (1969).

J. Hansen and F. Beiner, "Heterogene Gleichgewichte", W. de Grueter Verlag, Berlin (1974)

E.M. Levin and H.F. McMurdie (Editor: M.K. Reser), "Phase Diagrams for Ceramists, 1975 Supplement", The Amer. Ceramic Soc., Columbus, Ohio, USA (1975)

T. Mager, H.L. Lukas and G. Petzow, Z. Metallkde., 63 (1972) 877

G. Masing, "Ternäre Systeme", 2. Edition, Akademische Verlagsgesellschaft Geest u. Portig, Leipzig (1959)

J. Nyvlt, "Solid-Liquid Phase Equilibria", Elsevier Scientific Publ. Comp., Amsterdam (1977)

L.S. Palatnik and A.I. Landau, Zh. Fiz. Khim. 29 1748 and 2054 (1955); 30 (1956) 2399

G. Petzow and F. Aldinger, Z. Metallkde., 59 (1968) 145

CHAPTER 5

Phase Equilibria Including a Vapor Phase

5.1
Vapor-Liquid Equilibrium in a One-Component System

As already presented in Sect. 1.2, some of the atoms or molecules tied to a solid or liquid, can, at a fixed temperature, leave the condensed phase. In a closed system a certain vapor pressure p is established, which depends, according to Eq. (1.1b), on the temperature. Equation (1.1b) is only valid when the enthalpy of vaporization dH^V is independent of the temperature. This is not always true. In such cases, to represent the vapor pressure as a function of temperature, the following expression has been found applicable:

$$\log\left(\frac{p}{p_O}\right) = \frac{a}{T} + b\cdot\log\left(\frac{T}{T_O}\right) + c\cdot T + d \tag{5.1a}$$

The constant a has the dimension of [K], c the dimension [1/K]; b and d are dimensionless. Further we have $p_o = 1$ Pa, $T_o = 1$ K. In many tabular works, the – dimensionally not correct – simplified equation

$$\log p = \frac{a}{T} + b\cdot\log T + c\cdot T + d \tag{5.1b}$$

is used.

As an example, the values of the constants for solid NaCl, valid between room temperature and the melting point are (p in Pa):

$a = -1.658 \cdot 10^6; b = -120; c = -61 \cdot 10^{-3}; d = 1907$

At high pressure and low temperatures intermolecular interactions occur in gases. The ideal gas law

$$p\cdot V = R\cdot T \tag{5.2}$$

is no longer valid; V = molar volume. With increasing pressure the volume of real gases decreases faster than predicted by Eq. (5.2).

Figure 5.1 represents the volume of a liquid and of the coexisting vapor as a function of the temperature close to the critical point. With increasing temper-

Fig. 5.1
Specific volume of vapor
and liquid, which are in
equilibrium near the critical
point K

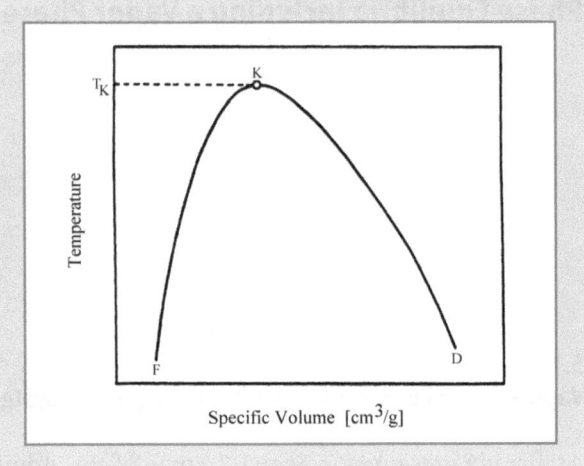

ature the volume of the vapor decreases along the line D – K, the volume of the liquid increases along the line F – K. If the critical temperature T_K is reached, liquid and vapor become identical (point K). At a temperature $T < T_K$ and fixed total volume the ratio of the specific volume of liquid and vapor is given by the lever rule. When the critical point is approached, the specific volumes of the liquid and vapor in equilibrium approach each other. Simultaneously the enthalpy of vaporization of the liquid decreases, becoming zero when the T_K is reached. As an example, the critical point of water is at $T_K = 638$ K and $p_K = 20.3 \cdot 10^6$ Pa.

5.2
Phase Equilibria Between Liquid and Vapor in Binary Systems, without a Miscibility Gap

In a two-component system A – B the total pressure P is formed additively from the partial pressure of the components p_A, p_B:

$$P = p_A + p_B \tag{5.3}$$

In an ideal binary system, where at a given temperature the vapor pressures of the components are equal, the composition of the vapor phase must equal that of the vaporizing liquid. In real systems this is not the case. The different vapor pressures of the components at a fixed temperature cause the composition of the vapor phase to differ from that of the liquid phase.

Figure 5.2 represents a simple p-x diagram of a binary melt. p_A^0 and p_B^0 are the vapor pressures of the pure components A and B at the fixed temperature T. In the region between the two drawn curves, liquid and vapor are in equilibrium.

If vapor with the composition $x_B^{D'}$ is compressed from D^0 to D', when p' is reached, condensation takes place. A liquid of composition $x_B^{L'}$ is formed. Be-

Fig. 5.2
P-x diagram for vapor-
liquid-equilibria in a simple
binary system without
miscibility gap

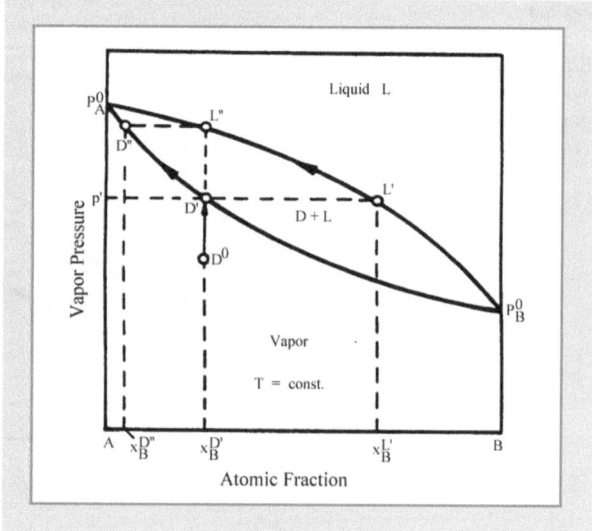

cause $x_B^{L'} > x_B^{D'}$, component A is enriched in the vapor. With increasing pressure the composition of the vapor phase follows the line $D' - D''$. The liquid phase is enriched in A during the continuing condensation, until the end point L'' is reached. Here, the last vapor (D'') condenses and only liquid is present.

The p-x diagram can be formally treated the same way as the T-x diagram of fusion equilibria. The lower curve is the line of the begin, the upper the end of the condensation. The curve of the end of the condensation represents the composition of the vapor phase at constant temperature. Similarly to melting point minima and to melting point maxima in T-x diagrams vapor pressure minima and vapor pressure maxima can occur in p-x diagrams. At these extreme positions, the curves of the beginning and end of the condensation touch each other. Vapor pressure minima are caused by strong interactions between the molecules of the components. Vapor pressure maxima on the other hand, are conditioned by demixing tendencies of the system.

A two-dimensional representation of the boiling process in a binary liquid can also be represented in a T-x diagram if $p = const$. Figure 5.3 represents schematically such a diagram, and for a binary system, to which the p-x diagram shown in Fig. 5.2 applies. $T_A^{V,0}$ and $T_B^{V,0}$ are the boiling temperature of the pure components. To high vapor pressures (see Fig. 5.2) low boiling temperatures are assigned, and the opposite (see Fig. 5.3). An increase in pressure signifies condensation, a temperature increase causes vaporization. In systems where a vapor pressure maximum occurs, a minimum in the boiling point is present. A vapor pressure minimum is equivalent to the appearance of a boiling point maximum. During heating, a liquid L^0 (see Fig. 5.3), when temperature T' is reached, vaporization takes place. The vapor is richer in A than the coexisting liquid. With con-

Fig. 5.3

T-x diagram for vapor-liquid-equilibria in a binary system without a miscibility gap, Fig. 5.2. gives the corresponding p-x diagram

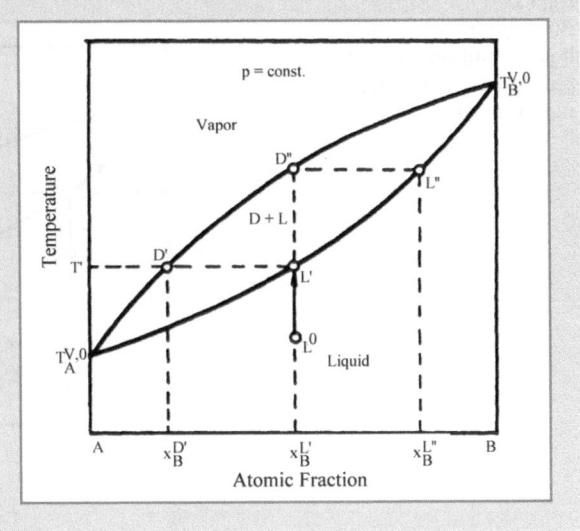

tinuing temperature increase with heat influx, both vapor and liquid become richer in B. The maximum concentration in B the liquid can reach, just before complete vaporization, is $x_B^{L''}$.

In the case of a large difference in the vapor pressure of the two components and correspondingly large difference in boiling temperatures, the two curves of the T-x diagram are greatly separated (see Fig. 5.4).

This is, for example, the case, if at room temperature one component is a gas, the other a liquid. The often interesting solubility of gases in liquids can also be described with a T-x diagram. The line indicating the dependence of the vapor composition on the temperature (condensation curve) starts at the boiling point of the component with the lower melting point, passes first very close to the T-axis on the side of the more volatile component. The vapor consists almost only of the more volatile component.

The line indicating the dependence of the composition of the liquid phase on the temperature (boiling curve) passes rapidly with increasing temperature to the T-axis on the side of the solvent (B-side). The solubility of a gas decreases at constant pressure with increasing temperature.

The fact that liquid and vapor do not have the same composition is used, to a great extent, to separate mixtures by fractionating distillation. During distillation a solution made up of liquids is heated at constant pressure, the vapor removed from the liquid surface, and condensed. The vapor is richer in the more volatile component. In the remaining liquid the concentration of the less volatile component increases. During distillation the temperature increases. To increase the yield, the distillation is interrupted after a certain fraction has been removed, and a separate distillation operation on the condensate and residue performed. This process can be repeated several times.

Fig. 5.4

T-x diagram of vaporization and condensation in a binary system with components deferring strongly in boiling point

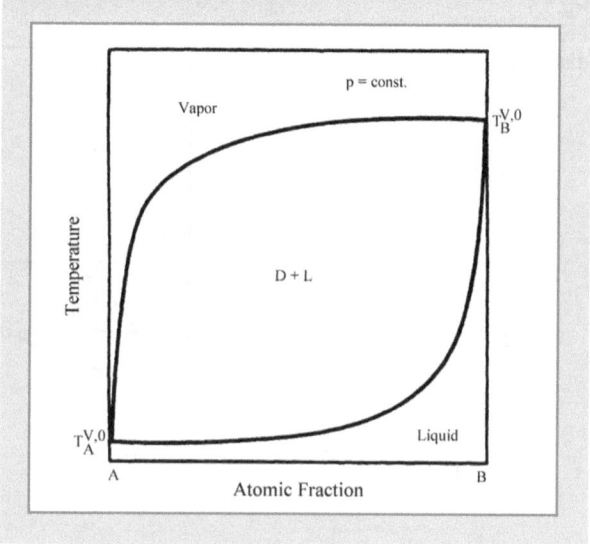

The principle of such a fractionating distillation can be seen in Fig. 5.5. The starting liquid has the composition x_B^0. If this solution is heated to the temperature $T^{(0)}$, and the vapor phase remains with the liquid in a closed container, a certain amount, determined by the lever rule, of vapor D is formed, which is richer in A. The distillation residue has a composition richer in B, that corresponds to x_B^0. If the vapor D is condensed, and the condensate heated to the temperature T_1, a certain amount of vapor is formed again, which has the composition D_1, which is richer in A than the original condensate. The condensation of D_1 and the equilibration of this condensate at T_2 yields a vapor still richer in A, and so on.

Similarly to this fractionating distillation, the liquid L can be subjected to a fractionating distillation, during which it becomes richer in B (see equilibria $D_1'-L_1'$ and $D_2'-L_2'$ in Fig. 5.5).

The effectiveness of each step at the fractionating distillation is greater, the more the boiling points of the components differ (see the T-x curve in Fig. 5.4).

With a large difference in boiling points of the components, a strong enrichment of the more volatile component in the vapor, and of the less volatile component in the residue occurs.

Solutions with a composition, which corresponds to a boiling point maximum or a boiling point minimum, are called azeotropic mixtures. Azeotropic solutions cannot be decomposed into the two components with fractionating distillation. In solutions at the composition of a boiling point maximum or boiling point minimum, liquid and vapor of the same composition are in equilibrium. Only one component can be obtained in the pure state. For example the solution of ethylalcohol-water exhibits at 95.6 wt.-% alcohol and 351.28 K a boiling point minimum. Therefore it is not possible to obtain water free al-

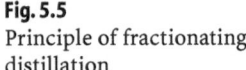

Fig. 5.5
Principle of fractionating
distillation

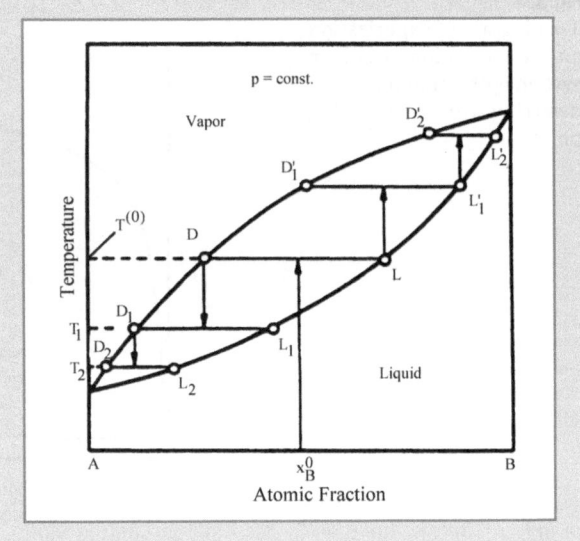

cohol through distillation. It is only possible to reach a water-alcohol solution
with 95.6 wt.-% alcohol. The boiling diagram of a system can be determined
by thermal analysis. The temperature of the liquid is registered as a function
of time, while heat is added. Care has to be taken, that the vapor formed does
not escape, but remains in contact with the liquid, until the latter disappears
completely. The begin and end of the vaporization can be determined from a
temperature-time-heating curve. Through determination of the composition of
vapor and liquid at constant temperature and constant pressure p-x and T-x
diagrams can be determined.

5.3
Gas-Solid Equilibria in a Binary System

Let us assume a vapor phase of two components A and B. Their total pressure is
composed additively from the partial pressures p_A and p_B according to Eq. (5.3).
The partial vapor pressures are proportional to the atom fractions x_A^D and x_B^D of
the components in the vapor. One has

$$\frac{p_A}{P} = \frac{x_A^D}{x_A^D + x_B^D} \tag{5.4}$$

Transforming and by the fact that $x_A + x_B = 1$ one obtains

$$P = p_A \cdot \frac{1}{x_A^D} \tag{5.5a}$$

Naturally by analogy:

$$P = p_B \cdot \frac{1}{x_B^D} \tag{5.5b}$$

From a vapor, which contains only component A, crystals of the solid phase A precipitate as frost at a fixed temperature, when the equilibrium pressure p_A^0 is reached. In this one-component system the total vapor pressure is equal to the partial vapor pressure $(x_B^D = 0, p_B = 0, p_A + p_B = p_A = p_A^0)$.

Also, if in addition to vapor A, vapor B is present, when thus, A-B vapor is given, crystals of A precipitate, when the partial pressure is $p_A = p_A^0$.

If the atomic fraction of component A in the vapor is $x_A^D \neq 1$, the total pressure P must be larger than p_A to have frost formation, and we must have

$$P^0 = p_A^0 \cdot \frac{1}{x_A^D} \tag{5.6}$$

If at a fixed temperature a binary vapor mixture with the composition x_A^D reaches the total pressure P^0, and in the case of equilibrium (no hindering due to nucleation difficulties), crystals of the pure component A precipitate and the volume decreases.

In an analogous fashion the same applies to material B (see Fig. 5.6). The hyperbolic frost curves of components A and B intersect at point M. This means, that at pressure p_M^0 and atomic fraction $x_B^{D,M}$ the vapor is simultaneously saturated in A and B. In this case a mixture of A and B crystallizes from the vapor.

The extension of the frost curves to higher pressures above p_M^0 (M–N and M–L) correspond to metastable conditions. For example, starting from the va-

Fig. 5.6
Situation of vapor-solid-equilibria in a binary system, in which the components in the solid state are not soluble in each other

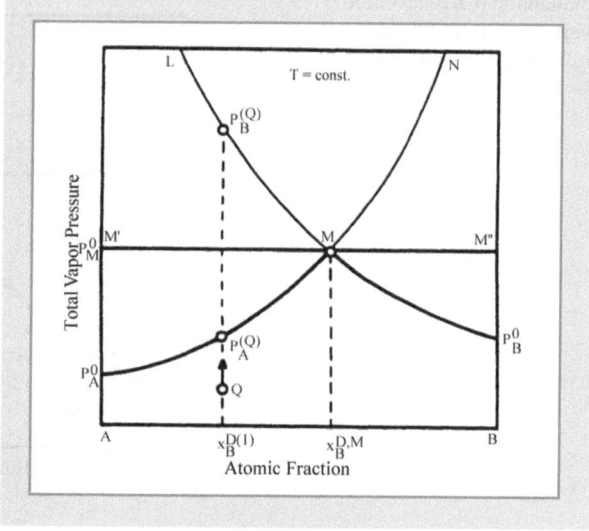

por mixture Q at atomic fraction $x_B^{Q(1)}$, the frost curve of component B at $p_B^{(Q)}$ can only be reached, if due to nucleation difficulties the precipitation of crystals of A at a much lower equilibrium pressure $p_A^{(Q)}$ of A has not taken place. In the case of equilibrium, frost formation only occurs along the line $p_A^0 - M - p_B^0$. The two frost curves end in point M. The precipitation of frost in the case of equilibrium is terminated when pressure p_M^0 is reached. At pressure $p > p_M^0$ only solid phases A and B are present. The line $M' - M - M''$ represents the boundary between the range of existence of the solid ($p > p_M^0$) and the range, where a gas phase can appear.

Figure 5.7 presents further details about the range of existence of the phases and phase mixtures. It shows schematically a p-x diagram, which represents the phase transitions during sublimation and frost formation in a binary system without solubility in the solid. For example, if a vapor mixture starting from point Q is compressed until the frost curve of A at p_A^{Q1} (point Q_1) is reached, A crystallizes; the vapor becomes poorer in A.

If the pressure is further increased, the composition of the vapor follows the line $Q_1 - M$. When M is reached, A and B crystallize simultaneously, while the pressure stays constant (p_M^0), until the total vapor phase is transformed into a mixture of the solid phases (invariant equilibrium at T = const.). Only when no more vapor is present, can the pressure be increased.

Same applies to the sublimation process. The pressure weighing upon the mixture of A and B (point Q_2) is decreased. If p_M^0 (point Q_3) is reached, then both components sublime together at constant pressure, until the total amount of B is transferred into the vapor. The left over component A can then vaporize further while the pressure is decreased, and the composition of the vapor fol-

Fig. 5.7
Schematic p-x diagram for vapor-solid-equilibria in a binary system without formation of solid solutions

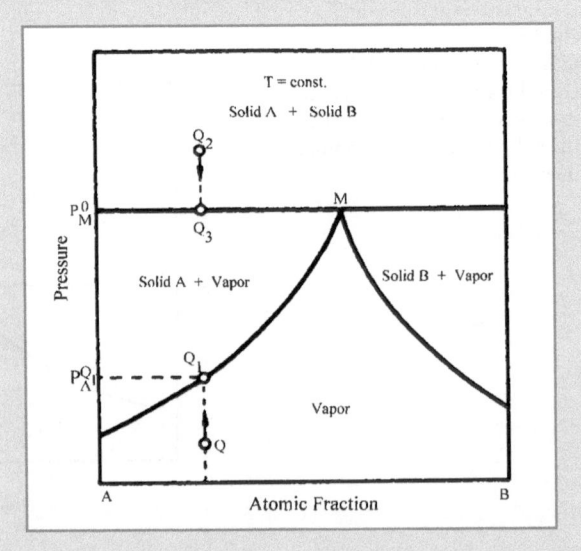

Fig. 5.8
T-x diagram to demonstrate
vapor-solid-equilibria in a
binary system without
formation of solid solutions

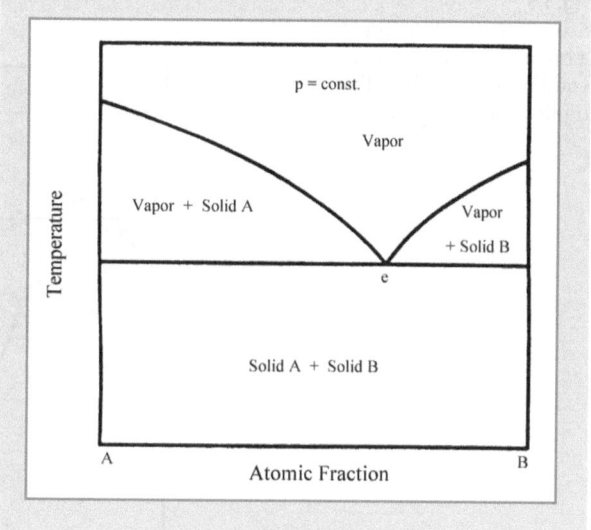

lows the line $M - Q_1$, until finally pressure p_A^{Q1} (point Q_1) is reached, and only the vapor phase is present.

The sublimation pressures increase with increasing temperature in a system. For a series of temperatures a band of analogous p-x diagrams result. From these, T-x diagrams for $p = $const., can be gained. Figure 5.8 shows a schematic T-x diagram, which can be assigned to a system, that can lead to the p-x diagram represented in Fig. 5.7. The phase equilibria in this T-x diagram are analogous to the fusion equilibria in an eutectic system. Vapor, that can be transferred in a mixture of solid A and B is called "eutectic vapor mixture".

5.4
Phase Equilibria in a Binary System in which Solid, Liquid and Gas can Appear

In a binary system at high temperatures the liquid phase can be expected to be present as a condensed phase, at lower temperatures only solid phases exist over the whole composition range. The condensed phases can be in equilibrium with vapor. Between the temperature regions, where these extreme conditions exist, in certain pressure-temperature-composition ranges liquid or solid phases, but also a mixture of both types can be in equilibrium with the vapor phase. A p-x diagram is present, which in its main features represents a combination of the p-x diagram for equilibria, in which vapor coexists with a binary, completely misci-ble liquid (see Fig. 5.2), and of a p-x diagram for binary solid-gas equilibria (see Fig. 5.7). An important extension of a three-phase equilibrium is added, where a solid phase, a liquid solution and vapor of the corresponding composition are in equilibrium. Figure 5.9 represents schematically such a p-x diagram.

Fig. 5.9

Schematic p-x diagram of a binary system with phase equilibria involving solid, liquid and vapor

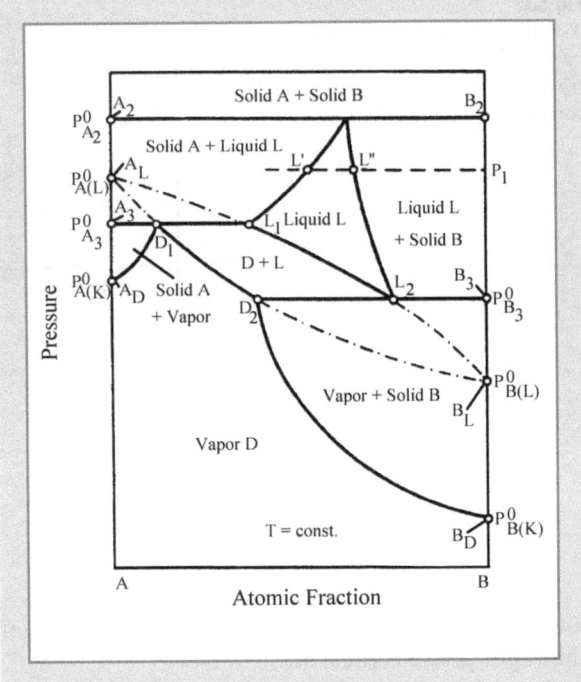

Assume that $p^0_{A(K)}$ and $p^0_{B(K)}$ are the pressures at which, at a given temperature T, the crystals A resp. B are in equilibrium with their vapors. First, only the equilibrium solid-vapor is considered, which starts at pure A (point A_D). The line, which represents the totality of the P-x points where this two-phase equilibrium exists, is $A_D - D_1$. Similarly, at B_D a phase equilibrium, pure solid B – pure gaseous B occurs. Addition of A displaces the equilibrium along the line $B_D - D_2$. At the given temperature, Ta, liquid solution exists in a P-x region, which is limited to lower pressures by the line $L_1 - L_2$. The liquid L is in equilibrium with vapor. The compositions of the liquids and the coexisting vapor depend naturally on the pressure.

Liquids, whose P-x values lay on the line $L_1 - L_2$, are in equilibrium with vapor, whose composition varies along the line $D_1 - D_2$. If the liquid L as a condensed phase, would be in equilibrium with vapor not only between the points L_1 and L_2, but over the whole composition range, than the boiling diagram would be given by the two lines $A_L - L_1 - L_2 - B_L$ and $A_L - D_1 - D_2 - B_L$. The dotted lines of the boiling diagrams in Fig. 5.9, $A_L - L_1$, $A_L - D_1$, and $L_2 - B_L$, $D_2 - B_L$ represent metastable equilibria. For example in the case of pure A, the equilibrium pressure $p^0_{A(L)}$ of liquid A (point A_L) is higher than the equilibrium pressure $p^0_{A(K)}$ of crystalline pure A. Molten A can transform into solid A, with a decrease in pressure, thus, a decrease in the Gibbs energy (see Eq. (1.4)). Only for a delay in nucleation of crystals of A could the sections $A_L - L_1$, $A_L - D_1$ of the boiling diagram be realized. The same applies to the section of the boiling diagram on the B rich side of the system.

In the case of a stable equilibrium the two phase equilibrium solid A – vapor at pressure p_{A3}^0 with the equilibrium vapor liquid. Here a three phase equilibrium $A_3 - D_1 - L_1$ exists. A vapor D_1 is in equilibrium with a saturated solution L_1, in which, at the bottom, solid A lays. The same applies to the binary equilibria on the B-side of the system and for the three-phase equilibrium $D_2 - L_2 - B_3$ at pressure p_{B3}^0.

The three-phase equilibria are invariant at a fixed temperature. If for example pressure is applied onto the coexisting phases A_3, D_1 and L, the reaction

$$A_3 + D_1 + L_1 \rightarrow A_3 + L_1 \tag{5.7}$$

proceeds. The volume decreases with decrease of the vapor fraction, until finally the vapor disappeares completely. Only then is an increase in pressure possible. The two-phase equilibrium $A_3 - L_1$ present now, is univariant, and therefore pressure dependent. The line $L_1 - \varepsilon$ represents the path of the composition of the liquid solution, which can be in equilibrium with solid A. The line $L_1 - \varepsilon$ and the analogous p-x line $L_2 - \varepsilon$ are drawn in Fig. 5.9 with slopes such that an increase in pressure decreases the solubility of A resp. B in the liquid. The sign of the slope of the p-x solubility line depends on the change of volume during precipitation of A resp. B from the solution. If the volume decreases during crystallization of the pure component in question, the solubility lines approach each other, based on the least resistance principle of Le Chatelier (see the Clausius-Clapeyron equation) as shown in Fig. 5.9, until, with increasing pressure, they finally intersect in point ε. A three-phase equilibrium occurs, where the pure solid components (points A_2 and B_2) coexist with liquid ε under pressure p_{A2}^0. Only, if at constant pressure p_{A2}^0 with decreasing volume the liquid phase ε completely disappears according to

$$A_2 + \varepsilon + B_2 \rightarrow A_2 + B_2 \tag{5.8}$$

can the pressure increase further. Above p_{A2}^0 neither the liquid, nor the vapor appear as an equilibrium phase.

If in a system the volume increases during crystallization of the components A and B from the liquid L, the solubility lines diverge with increasing pressure. The phase field of the liquid increases at the expense of the phase fields of the solids. However, it is possible that in a system one component crystallizes with decrease in volume and the other with increase in volume. Then the phase field of the liquid is displaced with increase in pressure correspondingly in a direction of higher A resp. higher B concentrations.

The combination of p-x sections at various temperatures yields a spatial representation, the p-T-x diagram. Through this three-dimensional diagram sections of constant pressure can be placed. Of interest is a T-x section for a pressure, where no vapor is present for example p_1 in Fig. 5.9. At the given temperature T, L' is the liquid in equilibrium with A, and L'' the liquid in equilibrium with B (see also Fig. 5.10). As a rule, the solubility increases with increasing temperature. Therefore at $T_1 > T$ the liquid L_1', coexisting with solid A, has a greater content of A than L'. Likewise is $x_B^{L1''} > x_B^{L''}$, where $x_B^{L''}$ is the atom fraction of B in the liquid L'' in equilibrium (at T) with solid B, and $x_B^{L1''}$ is the corresponding atom fraction of L_1'' at T_1. With decreasing temperature the field of the liquid ($L' - L''$) narrows, until – at the given pressure P_1 – the compositions of the two liquids become equal at a temperature $T_e < T$ ($L_e' = L_e''$). This is the eutectic point e, at which under the given pressure p_1, liquid e and solid components A and B are in an invariant three-phase equilibrium.

T-x phase diagram of this nature (at p = const. and absence of a gas phase) are of very great importance. They were already treated in detail in Chaps. 3 and 4.

The extent of the pressure dependence of solubility equilibria (for example liquidus lines in Fig. 5.11) are correlated, according to the Clausius-Clapeyron equation (see Chap. 2), with the change of the enthalpy ΔH^F, and the change in volume ΔV^F. In comparison to the change in enthalpy and volume of vaporization, the enthalpy and the volume of fusion is small. Correspondingly the influence of the pressure on the fusion equilibria is small. To obtain visible effects, pressures of the order of $> 10^9$ Pa are required. As ΔH^F is always positive – the enthalpy always increases while passing from the solid to the liquid state – the sign of ΔV_F depends on the variation with pressure of the solubility equilibria. If the extent of the pressure dependence of dissolution of A in the liquid differs from that of the dissolution of B in the liquid, then the eutectic composition is necessarily displaced with pressure. This can have significant consequences for the crystallization process, as in the rock formation from the magma. Figure 5.11 represents the case, where the fusion temperature of A increases more

Fig. 5.10

Schematic presentation of
T-x diagram generated from
an p-x diagram as given in
Fig. 5.9 at different temperatures and at a pressure p_1, at
which no vapor exists as an
equilibrium phase

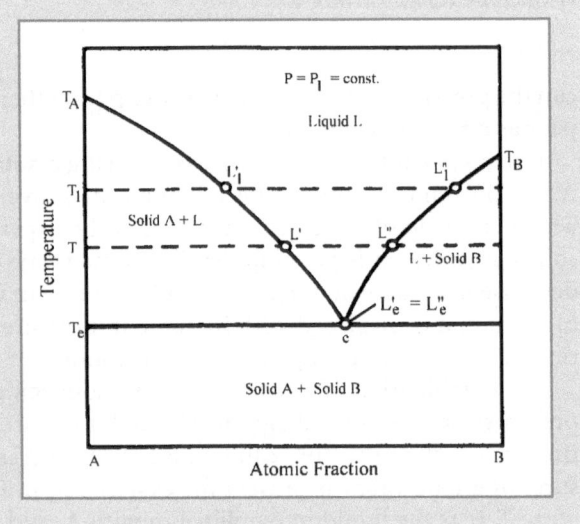

Fig. 5.11
T-x isobars at two different pressures to demonstrate the dependence of simple melting equilibria and the shifting eutectic concentration by application of pressure

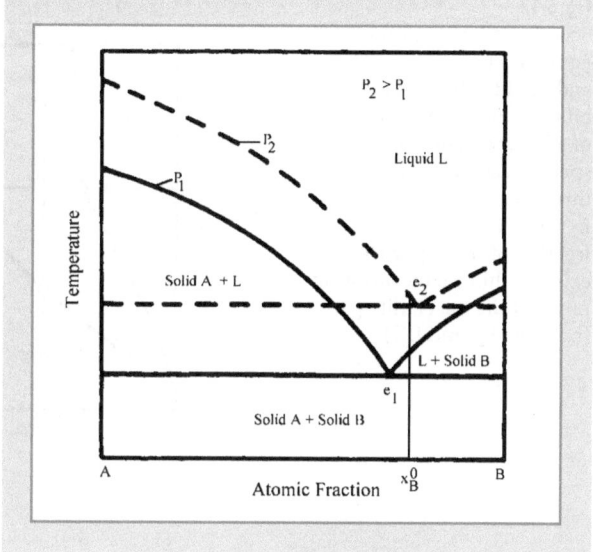

with increasing pressure than the melting point of B. As can be seen from the two isobars $(P_2 > P_1)$, the eutectic composition is displaced towards the B-side of the system with increasing pressure. From a liquid with the composition x_B^0, primary crystals of B precipitate at pressure p_1. At p_2 on the other hand, primary crystals of A appear.

5.5
Phase Equilibria with Participation of the Gas Phase, with Limited Solubility in the Liquid Phase

In a system with two liquid components, which are not completely miscible, an invariant three-phase equilibrium liquid L'-liquid L''-vapor D can occur.

The composition of the vapor, which coexists with the two liquid phases, can lay in the region of the miscibility gap, or correspond to a composition, where the condensed body is a single phase solution.

In the case presented below, the composition of the vapor lays in the region of the two phase mixture. The p-x diagram is represented schematically in Fig. 5.12.

On the A side of the p-x diagram, below p_2, the composition of the single phase liquid solution in equilibrium with vapor is given by the line $A_L' - L_2'$. The compositions of the vapor, which coexists with these solutions is given by the line $A_L' - D_2$. Similarly the part of p-x diagram on the B side, below the pressure of p_2 is given by the lines $B_L'' - L_2''$ and $B_L'' - D_2$. In point D lines of the vapor composition $A_L' - D_2$ and $B_L'' - D_2$ intersect. At the given temperature an invariant three-phase equilibrium $L_2' - D_2 - L_2''$ exists.

Fig. 5.12
Schematic p-x diagram for
equilibria between vapor
and liquid at limited solubility of the components
of a binary system. In the
special case shown in the
three-phase equilibrium
$L_2' - D_2 - L_2''$ the atomic
fraction of the vapor D_2
exists within a concentration
range of the miscibility gap
(between L_2' and L_2'')

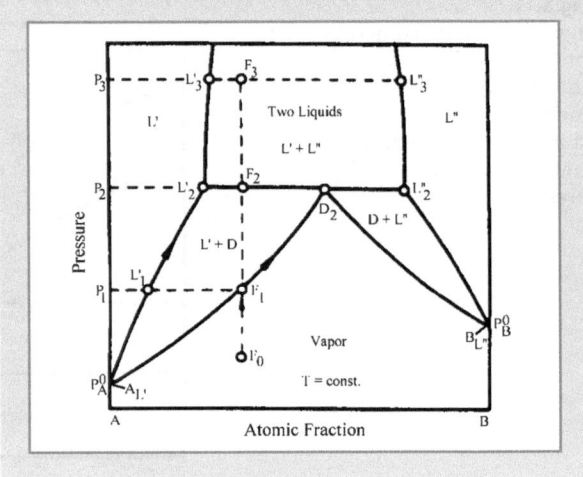

During compression of vapor F_0, when pressure p_1 belonging to F_1 is reached, condensation to liquid L_1' occurs, with a decrease in volume.

Further increase in pressure causes continuous condensation of vapor, while the compositions of liquid L' and vapor D change along the lines L_1'-L_2' and F_1-D_2 in the direction of the arrows. Is pressure p_2 finally reached, vapor D_2 condenses simultaneously to liquid L_2' as well as to liquid L_2''. This occurs at constant pressure, until the vapor phase, under continuous decrease in volume, disappears completely. Now only two liquids (L' + L'') are present at equilibrium, for example at p_3 the phases L_3' and L_3''. The vapor phase is no longer present. The lines $L_2' - L_3'$ and $L_2'' - L_3''$ and their extension over L_3' resp L_3'' to higher pressures represent the dependence of the mutual solubility of the liquids on pressure. The miscibility gap can close at higher pressure still in the liquid region (critical pressure), or reach the lines bounding the solid-liquid equilibria.

A system with a miscibility gap in the liquid phase, which corresponds to the p-x diagram represented in Fig. 5.12, can have a T-x diagram shown schematically in Fig. 5.13. During distillation of a two phase mixture of liquids with limited solubility at temperature T_a, vapor D_a is removed. As its composition lays within the miscibility gap, during condensation in the condenser a two phase mixture appears again. A separation of the components by capturing the vapor during distillation at T_a is not possible. A separation, however, is successful, if in a system with a miscibility gap in the liquid, the composition of the vapor corresponds to a composition which lays outside of the composition of the miscibility gap. In the first case a separation can also take place, but only then, when at T_a = const. vapor is removed, until one of the phases (L_a' or L_a'', depending on the overall composition of the starting two phase mixture) completely disappeared. The vapor formed with increasing temperature yields a single phase condensation product.

Fig. 5.13
Schematic T-x diagram
for vapor-liquid equilibria
in a binary system with a
miscibility gap and with a
p-x diagram, as shown in
Fig. 5.12

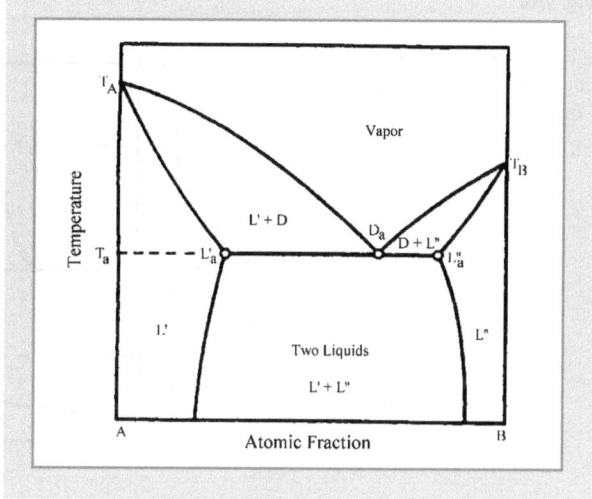

5.6
Vapor-Solid Equilibria with Solid Solution Formation

The vapor-solid equilibria in a system with complete solid solubility correspond completely to those between vapor and liquid with complete miscibility in the liquid phase. The p-x diagrams can also occur with a vapor pressure maximum or vapor pressure minimum, corresponding to a boiling point minimum resp. boiling point maximum.

If only a limited mutual solubility of the solid components is present, then analogous equilibrium conditions are found, as in the case of the vaporization of liquid with limited miscibility. With fixed temperature at a certain pressure a three-phase equilibrium occurs, where two saturated solid solutions and a vapor phase coexist.

5.7
Gas-Solid Equilibria in Systems with Compounds

The influence of a compound on the phase equilibria between solid and vapor will be described using metal-gas systems, where naturally, at a fixed pressure the boiling points of the components are very different. To simplify the situation, systems with one single compound will be considered.

Figure 5.14 reproduces schematically the p-x diagram of a system, which shows a compound β, melting incongruently. The component A for example, shall represent a metal, component B a gas (hydrogen, oxygen). At a given temperature T the vapor pressure of the pure metal is p_A^0. The vapor pressure of the pure gas B, p_B^0 is so high, that it falls outside of the pressure range, as indicated by

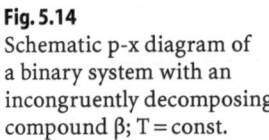

Fig. 5.14
Schematic p-x diagram of
a binary system with an
incongruently decomposing
compound β; T = const.

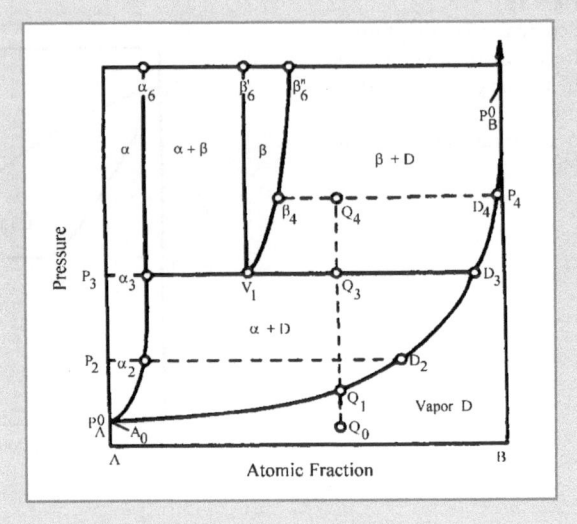

the arrow, given in Fig. 5.14. In the pressure range between p_A^0 and P_3 two-phase equilibria can appear. Here solid solution α and vapor D are in equilibrium. With increasing pressure the composition of the α-phase follows the line $A_0 - \alpha_3$, that of the vapor phase the line $A_0 - D_3$. Starting with vapor at point Q_0, condensation of the α-solid solution starts at Q_1. If the pressure is p_2, then α_2 and D_2 are in equilibrium. If, with increasing pressure, p_3 is reached, an invariant equilibrium with decrease in volume and constant pressure is formed with α_3, D_3 and the β-phase V_1. Only when α_3 is completely used up, can the pressure be further increased. Now only two phase equilibria appear. For example at P_4, β_4 and D_4 coexist. With further increase in pressure, the line representing the composition of the vapor coexisting with the β-phase, approaches the ordinate on the B-side of the diagram. At sufficiently high pressures, in addition of β, finally only the practically pure gas phase B ($p_B \gg p_A$) is present.

The line $V_1 - \beta_4 - \beta_6''$ indicates how high the concentration of the gaseous component in the β-phase at a given pressure is. The β-phase is only a stable phase at pressures above p_3; p_3 is thus, called the decomposition pressure of β.

Figure 5.15 shows a possible section at constant pressure across the complete p-T-x diagram, for a system with an incongruently vaporizing compound β, for which Fig. 5.14 represents a p-x diagram, a constant temperature section across the complete p-T-x diagram. The essential phase equilibria will be considered while heating an alloy with the atom fraction x_B^0.

At T_0 the β-phase β_0 is in equilibrium with vapor D. Heating to T_1, the β-phase V_1 decomposes into the solid solution α_1 and vapor D_1 in an invariant equilibrium. If the total amount of the originally present β-phase is decomposed, a heat input can cause a temperature increase above T_1. Only solid solution and vapor coexist. For example at T_2 the phases represented by points α_2

Fig. 5.15
Schematic T-x diagram for
vapor-liquid-solid equilibria
in a binary system with an
incongruently decomposing
compound

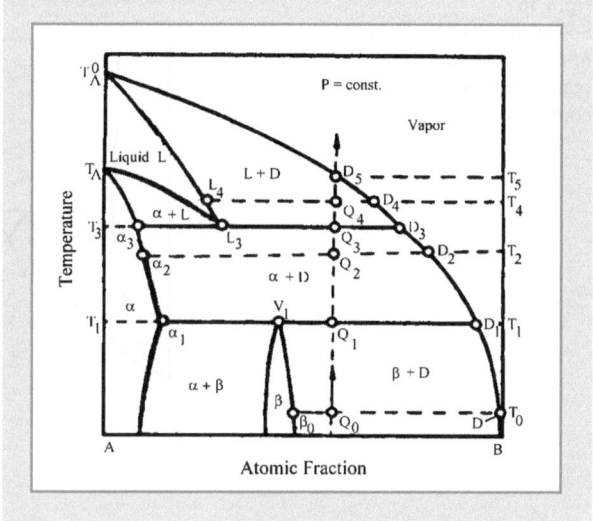

and D_2 are in equilibrium. A further heating up to T_3 is possible. Here the two
phase mixture $\alpha_3 + D_3$ reacts in an invariant reaction with the formation of a
liquid L_3. If, during this reaction, α_3 is completely consumed the temperature
can further increase. A two-phase equilibrium liquid + vapor are present. With
increasing temperature the compositions of liquid and vapor change along the
lines $L_3 - T_A^0$ resp. $D_3 - D_4 - D_5 - T_A^0$. If T_5 is passed, then the sample is completely
vaporized. T_A^0 is the boiling point and T_A the melting temperature of component
A (for example a high melting metal). The boiling point T_B^0 of component B (gas
component) lay way below the temperature T_0, outside of the range represented
in Fig. 5.15. T_1 is for the given pressure P = const. the decomposition temperature of β-phase. Here β decomposes into α-solid solution and a vapor, which
contains mostly the gas component B.

Figure 5.16 reproduces a p-x diagram for a system which possesses a congruently melting component, γ. In the case presented here, a reduction in pressure
will transform the phase γ, which exists at high pressures, into a vapor with the
same composition as the vaporizing solid.

It must be mentioned, that the vapor in the metal-gas system does not always consists of metal atoms A and gas molecules B_2. Molecular species can
be present, which contain both components. As an example MoO_3 and NbO_2
should be named. Such molecules can sometimes make up a large fraction of
the gas phase. This necessarily influences the frost line. The various atomic and
molecular species are connected through dissociation equilibria.

These equilibria in the homogeneous phase depend on the state variables
pressure, temperature and composition.

Fig. 5.16
Schematic p-x diagram
of a binary system with a
congruently vaporizing
compound γ

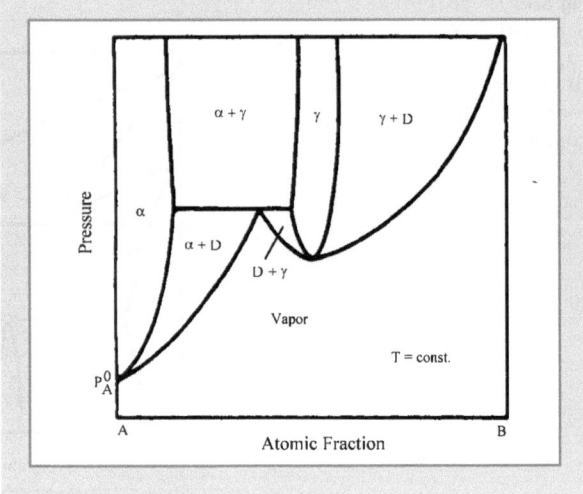

5.8
Heterogeneous Equilibria at Chemical Transport Reaction

In a large number of systems solid or liquid materials A react with a gas B with the formation of a gaseous chemical compound C:

$$aA + bB \rightarrow cC \tag{5.9}$$

a, b and c are the mole fractions of the materials participating in the reaction. The gaseous reaction product C can be removed from the reaction area, and can then in a reverse reaction, under different conditions (p, T), decompose into the starting material A and starting gas B.

A classical example is the Mond-Langer process for the winning of nickel. The raw nickel, obtained from the ores, is transformed, with carbon monoxide at $T_1 = 320$ K under normal pressure, into nickel carbonyl $Ni(CO)_4$. The latter can be removed and decomposed at $T_2 = 470$ K into pure nickel and CO. At the temperatures in question the vapor pressure of nickel is extremely low. The material transport does not take place through sublimation or distillation. It is a chemical transport reaction, where at different temperatures the different positions of a chemical heterogeneous equilibrium are taken advantage of. The situations can be described quantitatively with the mass action law.

In addition to the gain in Gibbs energy, which occurs during the formation of nickel carbonyl at T_1, and during the following decomposition at T_2, and which is the driving force of the transport reaction, the method of transport of the gaseous reaction partner from its place of origin to its place of decomposition is of importance for the effectiveness of the process. This transport can occur through simple gas flow, through diffusion or thermal convection.

Chemical transport reactions are used in many fashions in industrial processes. The purification of high melting heavy metals was made very simple by a method developed by van Arkel and de Boer. In the case of zirconium the process is based on the following reaction:

$$Zr + 2I_2 \leftrightarrow ZrI_4 \tag{5.10}$$

The formation of ZrJ_4 occurs at 670 K, its decomposition at 1670 K. A certain J_2 pressure is maintained in a quartz vessel, containing the raw zirconium and equipped with an incandescent wire. The raw zirconium is heated to 670 K. On the electrically heated incandescent wire pure solid zirconium is deposited.

This process can, among others, be also used for the purification of uranium. The uranium is deposited as a liquid on the hot wire, and drops into a collection vessel.

The transported substances in chemical transport reactions can often be collected as single crystals. This can occur also with substances, which cannot be obtained as single crystals using other methods. The preparation of whiskers, which are used in the production of composites, can be carried out by this process.

References

E. Fromm and E. Gebhardt (Editors), "Gase und Kohlenstoff in Metallen", Springer-Verlag, Berlin (1976)

W. Paul and D.M. Warschauer (Editors), "Solids Under Pressure", McGraw-Hill Book Comp., New York (1963)

R.G. Ross and D.A. Greenwood, "Liquid Metals and Vapours Under Pressure", in "Progress in Materials Science", Editors: B. Chalmers and W. Hume-Rothery, Vol. 14. No. 4, Pergamon Press, London (1969)

H. Schäfer, "Chemische Transportreaktionen", Verlag Chemie, Weinheim/Bergstrasse (1962)

CHAPTER 6

Thermodynamics

6.1
General

The combination of different substances into multi-component systems yields a large number of different phase diagrams. The complex variety of phase diagrams is, however, conditioned only by the interaction of a few factors. Experience shows, that certain types of phase diagrams can be found in very different groups of substances. Simple eutectic systems can, for example, be found in the combination of two metals, two salts or two organic compounds.

For the appearance of certain types of phase diagrams it is not the type and strength of the interatomic or intermolecular forces, but the difference in the bonding and structural conditions in the different phases of the system is responsible. This difference can be caused by differences in the valences, in the atomic radii of the components, by differences in the lattice stability or naturally by special electronic conditions. The variety of phase diagrams come about through a sensitive energy balance, which, as a rule, is only influenced by a few of the above named causes.

The atomic interactions express themselves in macroscopic energy quantities, which can be determined empirically or in simple cases, calculated from the properties of the participating atoms. The phase diagrams are connected to these quantities. The knowledge of these connections leads to a basic understanding of special phase equilibria in concrete systems. It makes it possible to answer questions relating to the stability of the individual phases under certain conditions, and permits the rapid calculation of phase diagrams from the given thermodynamic quantities of the system. Thus, an useful check of the experimentally obtained phase diagrams is possible. Similarly, parts of the equilibrium diagrams, which are difficult to approach experimentally, can be obtained easily. It is also possible to obtain a mathematical evaluation of very complex equilibria easily, rapidly and cheaply, in ternary and larger multi-component systems, if, for confirmation, certain critical points of the phase diagram are known. The number of needed experiments, which are very numerous in larger systems, can be reduced significantly. Also, predictions about equilibrium conditions under high pressures can be easily made. Finally, thermodynamic quantities can be calculated from known phase equilibria.

6.2
Basic Thermodynamic Concepts and Definitions

The supply or removal of heat Q or another form of energy changes the internal energy U of a system. The change dU of this state variable is independent of the way and means by which the energy is transferred. It is only given by the difference of the internal energy between the final state U_E and the starting state U_A:

$$dU = U_E - U_A \tag{6.1}$$

The increase of the internal energy is, according to the first law of thermodynamic, equal to the sum of the supplied energy quantities. If ΔQ is the supplied heat and dA the mechanical energy supplied to a system, then, as is well known:

$$dU = dQ + dA \tag{6.2}$$

If the heat is supplied at constant volume, than it serves only to increase the internal energy of the system.

If the heat is supplied under constant pressure p, a certain part of the heat is used up for the expansion of the volume V of the body. The energy given up by the system is counted negatively, the energy absorbed by the system is counted positively. The expansion work done by the system has also a negative sign.

As a further state variable the enthalpy or heat content is defined:

$$H = U + p \cdot V \tag{6.3}$$

For example, the enthalpy of vaporization ΔH^V of a substance is given by:

$$\Delta H^V = H_{Vapor} - H_{Liquid} = (U_{Vapor} + p \cdot V_{Vapor}) - (U_{Liquid} + p \cdot V_{Liquid}) \tag{6.4}$$

$$\Delta H^V = \Delta U + p \cdot \Delta V \tag{6.5}$$

ΔU is the change of the internal energy and ΔV the change in volume occuring during vaporization.

During vaporization a part of the supplied heat is transformed into volume work (expansion work) $p \cdot \Delta V$, which is done against the external pressure p. The volume change is significantly smaller during melting, changes of phases, mixing of liquid or solid substances and during reaction of two components to a compound than during vaporization. In these cases the change in enthalpy and internal energy is not large, because as a rule the transformation is carried out at a relatively low pressure ($p_0 = 1$ atm).

The enthalpy of a substance, as well as the internal energy are measured and reported against a standard state. In general the standard state is given by the temperature of 298.15 K (25 °C), pressure $p_0 = 1 = 10^5$ Pa, and the modification of the substance stable at this temperature and pressure. The following considerations are valid for the constant pressure p_0.

The experimental determination of the heat content at a temperature T can be carried out by dropping the substance from this temperature into a calori-

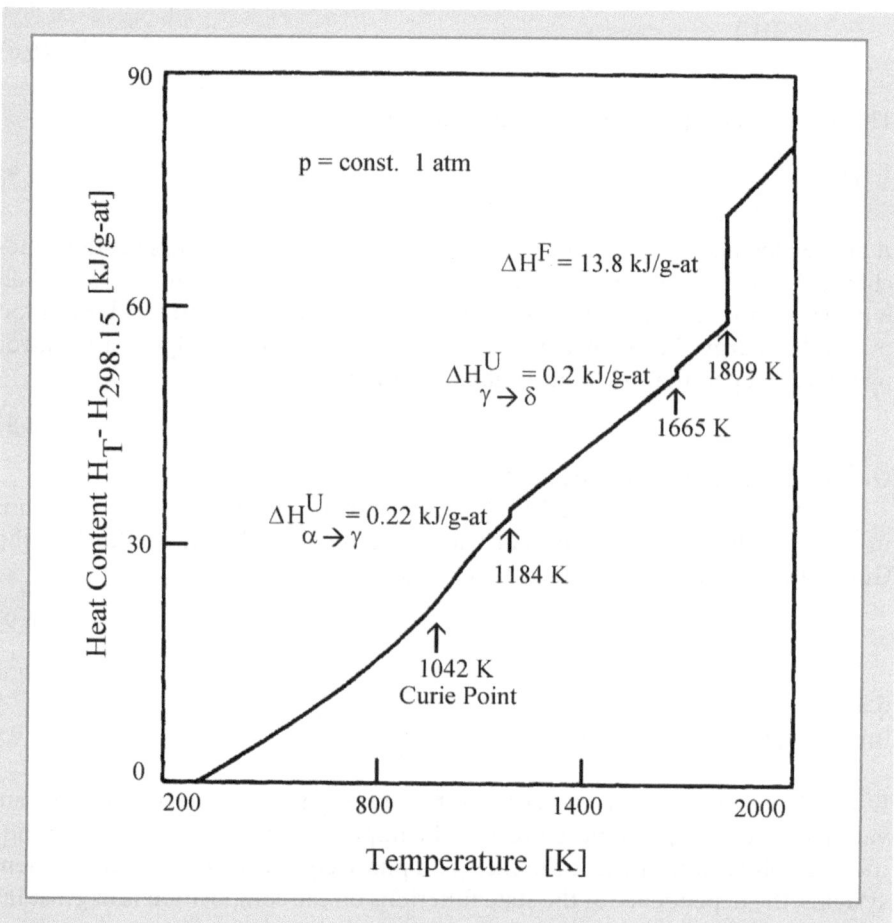

Fig. 6.1 Heat content of Fe [1]

meter at 298.15 K. The quantity of heat given up by one mole substance to the calorimeter is the enthalpy, referred to the standard state (H_T-$H_{298.15}$). Figure 6.1 represents the heat content curve of iron. The heat content naturally increases with increasing temperature.

The changes in structure (α-γ and γ-δ) present themselves as a jump in the heat content curve at the transformation temperature. The difference in heat content between the two modifications at the transformation temperature is the enthalpy of transformation. Same applies at the melting point and boiling point. The heat content curve shows a characteristic curvature at the transition from the ferromagnetic to the paramagnetic state, connected with the change in lattice energy.

The molar heat capacity at constant pressure is defined as the slope of the H – T curve:

$$C_p = \left(\frac{\partial H}{\partial T}\right)_p \tag{6.6}$$

The molar heat capacity at constant volume is

$$C_V = \left(\frac{\partial U}{\partial T}\right)_V \tag{6.7}$$

A ball, which has the possibility to move in a bowl (see Fig. 1.4), is in stable mechanical equilibrium, when it is at the lowest point of the container, that is, when it possesses the least amount of potential energy. In a somewhat similar fashion the Gibbs energy G is defined, which is connected to the enthalpy H and entropy S according to the Helmholtz-Gibbs equation:

$$G = H - T \cdot S \tag{6.8}$$

G and S are, as is H, state variables.

A system at constant p and T is in equilibrium, when it possesses a minimum in free enthalpy. This is equivalent with the statement, that at equilibrium the Gibbs energy of the system does not change:

$$dG = 0 \tag{6.9}$$

6.3
Integral Quantities of Mixing

Internal energy, enthalpy and entropy, the state functions, depend on the state variables pressure p, temperature T and composition x. When considering liquid and solid solutions, as is the case with phase equilibria in multi-component systems, the dependence of the state functions on the composition is of great interest.

Figure 6.2 represents the Gibbs energy as a function of composition (T = const.; p = const.) for a binary system A – B. G_A and G_B are the Gibbs energies of the components. If the components are present unmixed, as a heterogeneous mixture, then such mixtures, – depending on the content in A and B – have Gibbs energies which lay on a straight line connecting G_A and G_B (mixing rule). A solution of A and B can only be formed if the Gibbs energy of the system is decreased. In uncomplicated cases the amount of this decrease is smaller, when in a mole of substance the concentration of A or B is smaller, than in the middle of the composition range. The Gibbs energy as a function of composition is a hanging curve compared to the straight line of the mixing curve. It meets the ordinates at G_A and G_B with an infinite slope. The difference between the Gibbs energy of a mechanical mixture and the Gibbs energy of a solution of identical composition is called the Gibbs energy of mixing ΔG. Its dimension is J/g-atom. For phases which form spontaneously from the components the Gibbs energy of mixing necessarily always has a negative sign.

Fig. 6.2
Gibbs Energy G and Gibbs
Energy of Mixing ΔG_x as a
function of atomic fraction

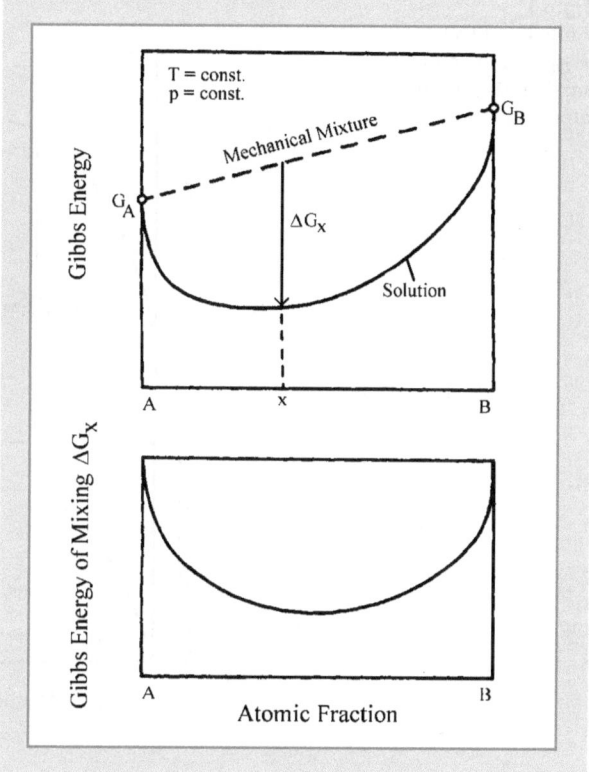

In an analogous fashion, the enthalpy of mixing ΔH is defined as the difference between the enthalpy of the solution and that of a mechanical mixture of
the same composition (see Fig. 6.3; H_A and H_B the enthalpies of the pure components A and B). This quantity ΔH can be determined directly with a calorimeter,
if nucleation is not hindered. For example, the two components can be mixed in
the calorimeter. The heat effect which appears, is ΔH. The enthalpies of mixing
for solid phases are, as a rule, determined using indirect methods.

The enthalpy of mixing, as well as the Gibbs energy of mixing, can be positive or negative. It is positive, if the interatomic interaction is reduced during the
mixing reaction, and it is negative, if an increase of the interaction occurs during the reaction. At small concentrations in the solution, ΔH is approximately
proportional to the concentration of the minority component. Thus, the $\Delta H - x$
curve meets the zero in the enthalpy of mixing-atom fraction curve with a finite
angle. The entropy of mixing is defined similarly:

$$\Delta S = S - (x_A \cdot S_A + x_B \cdot S_B) \tag{6.10}$$

where S_A and S_B are the entropies of the pure components and S the entropy of
the solution.

Fig. 6.3
Enthalpy H (relative to $H_{298.15}$) and enthalpy of mixing ΔH_x as a function of atomic fraction

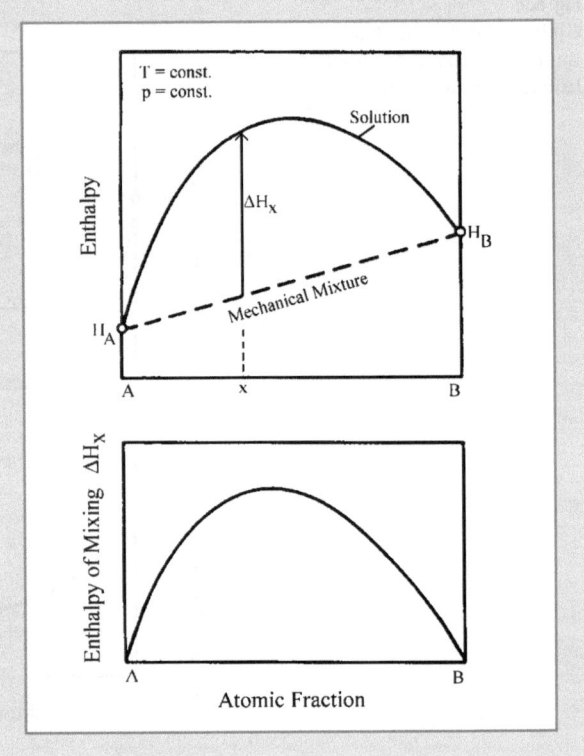

In the extreme case of negligible small change in the interatomic interactions during solution formation from the components, the change in entropy is given only by the fact, that the order existing before the reaction – the atoms of each components are separately strongly ordered – is changed into a random distribution of atoms in the solution. In this case, the entropy of mixing, based on statistical considerations, is given by the following expression, with R the gas constant:

$$\Delta S_i = - R \,[x_A \cdot \ln x_A + x_B \cdot \ln x_B] \tag{6.11}$$

In Fig. 6.4 the so-called ideal entropy of mixing is presented as a function of the atom fraction. The ΔS_i-x curve approaches with increased dilution the ordinate of the diagram with an infinite slope. This is the reason for the analogous behavior of the $\Delta G - x$ curve at $x_A \rightarrow 0$ and $x_B \rightarrow 0$.

The Gibbs-Helmholtz equation also applies naturally to the quantities of mixing:

$$\Delta G_x = \Delta H_x - T \cdot \Delta S_x \tag{6.12}$$

The subscript x signifies, that the quantities apply to the given concentration. ΔG, ΔH and ΔS are called integral quantities.

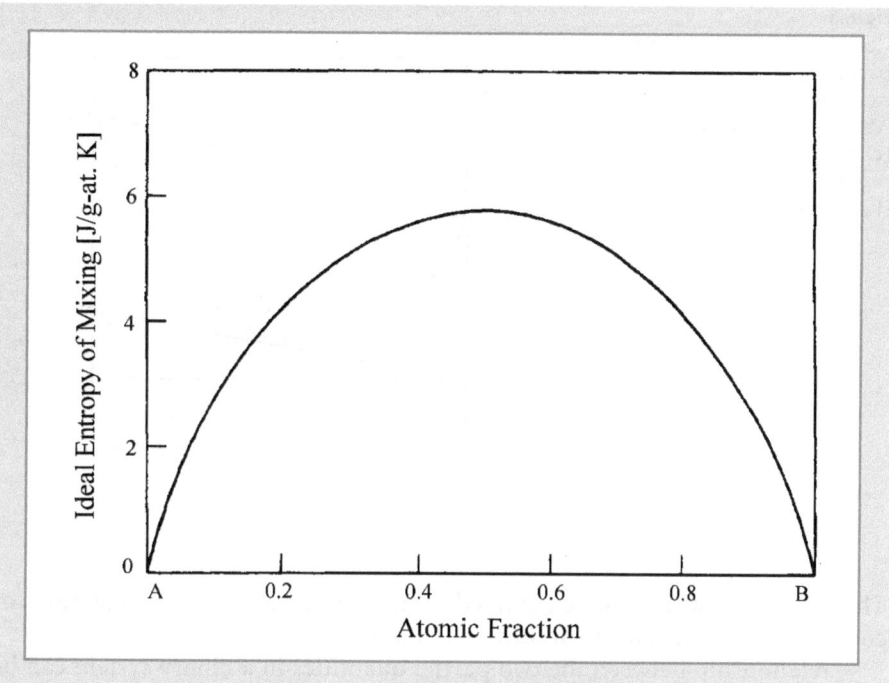

Fig. 6.4 Ideal entropy of mixing as a function of atomic fraction

6.4
Partial Quantities of Mixing

Partial quantities of the state functions are introduced to be able to follow the influence of the individual components on the thermodynamic properties of the system. For a clear understanding, they can be described as follows: they represent the change in the integral state function for the case that at constant pressure and temperature, formation of the solution occurs, without changing the composition of the solution. This can be realized in a thought experiment, when one mole of one component is added to an infinitely large quantity of solution with the composition x. The connection between partial and integral quantities is shown in Fig. 6.5 using the molecular volume as example.

V_A and V_B are the molecular volumes of the components in the two-component system A – B considered. V_x represents the molecular volume of the solution with the composition x. The partial volumes of the components are $\overline{V}_A$ and $\overline{V}_B$. Thus,

$$V_x = \overline{V}_A + y = \overline{V}_A + (\overline{V}_B - \overline{V}_A)\, x_B = \overline{V}_A\,(1 - x_B) + \overline{V}_B\, x_B \tag{6.13}$$

$$V_x = x_A\, \overline{V}_A + x_B\, \overline{V}_B \tag{6.14}$$

Fig. 6.5
Correlation of integral and
partial values shown as an
example for molar volume.
ΔV = volume of mixing
(change of the molar volume
occuring during mixing of
the components)

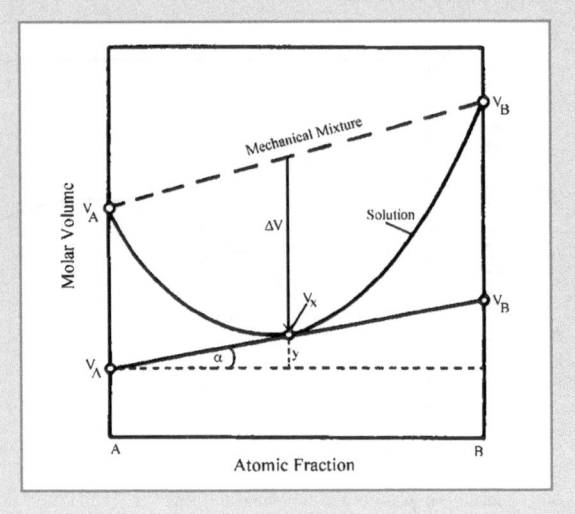

The partial quantities can be obtained from the integral ones by calculations or
construction of the tangent according to Fig. 6.5.

A relationship between the two partial quantities in a binary system can be
obtained as follows, according to Fig. 6.5

$$V_x = \overline{V}_A + y = \overline{V}_A + x_B \cdot tg\alpha = \overline{V}_A + x_B \cdot \frac{\partial V_x}{\partial x_B} \tag{6.15}$$

From

$$x_A + x_B = 1 \tag{6.16}$$

follows

$$d(x_A + x_B) = 0$$
$$dx_A = -dx_B \tag{6.17}$$

From Eq. (6.15) results:

$$\overline{V}_B = V_x + (1 - x_B) \cdot \frac{\partial V_x}{\partial x_B} = V_x - x_A \cdot \frac{\partial V_x}{\partial x_A} \tag{6.18}$$

For the second component, in an analogous fashion

$$\overline{V}_B = V_x + (1 - x_B) \cdot \frac{\partial V_x}{\partial x_B} = V_x - x_A \cdot \frac{\partial V_x}{\partial x_A} \tag{6.19}$$

Differentiation of V_A and V_B with respect to x_A yields

$$\frac{\partial \overline{V}_A}{\partial x_A} = (1 - x_A) \cdot \frac{\partial^2 V_x}{\partial x_A^2} \tag{6.20}$$

$$\frac{\partial \overline{V}_B}{\partial x_A} = -x_A \cdot \frac{\partial^2 V_x}{\partial x_A^2} \tag{6.21}$$

Division of Eq. (6.20) by Eq. (6.21) and rearranging

$$x_A \cdot \frac{\partial \overline{V}_A}{\partial x_A} + x_B \cdot \frac{\partial \overline{V}_B}{\partial x_A} = 0 \tag{6.22a}$$

or

$$x_A \cdot d\overline{V}_A + x_B \cdot d\overline{V}_B = 0 \tag{6.22b}$$

This is the Gibbs-Duhem equation. With it one can calculate from one partial quantity the corresponding other quantity. If for instance $\overline{V}_A$ is known, and $\overline{V}_B$ has to be calculated, one must integrate within the limits $x = 0$ and x_B. One obtains

$$\overline{V}_A = - \int_{\overline{V}_B \text{ at } x_B=0}^{\overline{V}_B \text{ at } x_B=x_B} \frac{x_B}{x_A} d\overline{V}_B \tag{6.23}$$

The same applies to the partial Gibbs energies, for the enthalpies and entropies, as well for the partial quantities of mixing $\overline{\Delta G}$, $\overline{\Delta H}$, $\overline{\Delta S}$ and $\overline{\Delta V}$.

6.5
The Ideal Solution

If during formation of a solution from two components no change in interatomic interaction occurs, than $\Delta H = 0$, and the change in molecular volume during mixing of the components (volume of mixing) is $\Delta V = 0$. In this case the change in Gibbs energy is fixed only by the change in entropy:

$$\Delta G_i = -T \cdot \Delta S_i \tag{6.24}$$

If absolutely no change in bonding conditions takes place, a completely random distribution of the atoms occur. The change in entropy is given by the ideal entropy of mixing (Eq. (6.11)). Solutions of this nature are called ideal solutions. In only a few cases are they realized approximately.

6.6
The Model of the Regular Solution

During formation of most real liquid and solid solutions, a positive or negative change in the enthalpy and volume takes place. ΔH and ΔV are often in a different way, dependent on the composition. J.H. Hildebrand has developed a model, which, based on a simple relationship between atomic interaction parameters and the quantities of mixing, makes it possible to describe the thermodynamic properties of a system, which as a rule corresponds much better to the reality, than the model of the ideal solution. It is assumed for such a so-called regular solution that even when $\Delta H \neq 0$, the atoms in the solution are distributed completely randomly. The entropy of mixing must only be determined by the ideal entropy of mixing.

Furthermore, it is assumed that only pairwise interactions (quasi chemical approximation) are present. In a binary system A – B interactions between next neighbors should only be important. The interaction energies of an A – A pair, a B – B pair and an A – B pair are w_{AA}, w_{BB} and w_{AB}. As the character of the interaction parameters considered here is of a bonding nature, they are always negative quantities. Furthermore it is assumed that the atomic interactions are composition independent.

The accounting of the different sorts of pair bondings with a random distribution of atoms, and the balance of total change in the bonding relationships – during mixing A – B bonds with w_{AB} at the expense of A – A and A – B pairs in the unmixed starting substances – yields the following expression for the integral molar enthalpy of mixing:

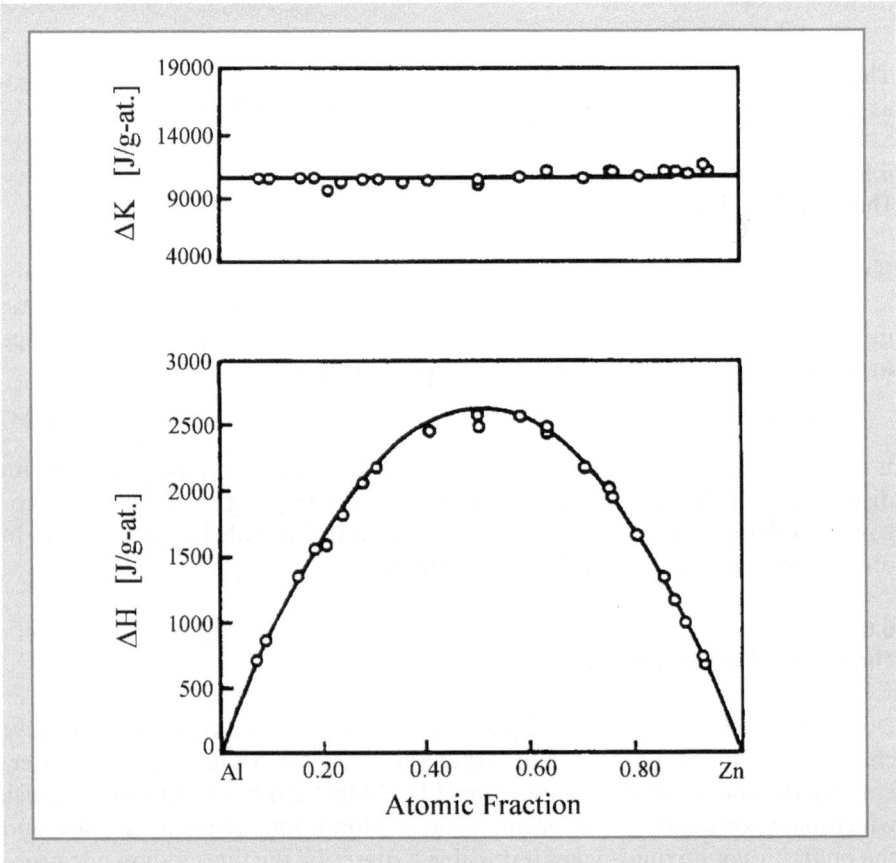

Fig. 6.6 Enthalpy of mixing and value K of liquid Al-Zn alloys as a function of the atomic fraction at T = 1226 K (after [2])

$$\Delta H = K \cdot x_A \cdot x_B \tag{6.25}$$

The quantity K, which for a given system is a constant, contains the Loschmidt number, the coordination number and the interaction parameters w_{AA}, w_{BB} and w_{BB}.

According to the model of the regular solution the ΔH-x curve is a parabola of second order, symmetric around $x = 0.5$. This represent a good approximation for many binary systems. As an example Fig. 6.6 represents the $\Delta H - x$ curve for liquid aluminum-zinc alloys.

The regularity of a solution can be proven in an especially sensitive fashion using the composition dependence of the enthalpy of mixing, when the quantity

$$K = \frac{\Delta H}{x_A \cdot x_B} \tag{6.26}$$

is plotted as a function of the mol fraction. As can be seen in Fig. 6.6, the quantity K has a constant value within experimental accuracy for liquid aluminum-zinc alloys.

A consideration of the relationship between integral and partial enthalpy of mixing, as was carried out based on Fig. 6.5, yields in a simple fashion the following expressions for the partial enthalpies of a regular solution:

$$\overline{\Delta H_A} = K \cdot (1 - x_A)^2 \tag{6.27}$$

$$\overline{\Delta H_B} = K \cdot (1 - x_B)^2 \tag{6.28}$$

For the relationship between ΔH, ΔH_A and ΔH_B one obtains in analogous fashion to Eq. (6.14)

$$\Delta H = x_A \cdot \overline{\Delta H_A} + x_B \cdot \overline{\Delta H_B} \tag{6.29}$$

Similar relationships apply naturally to the other quantities of mixing.

The model of the regular solution is, in general, a good approximation when the enthalpies of mixing are small. This is an indication of a small change in interatomic interaction during alloy formation, which guaranties that the assumption of a random distribution of the atoms is mostly correct.

6.7
Real Solutions and Excess Functions

In real solutions the composition dependence of the enthalpy of mixing can deviate greatly from that postulated by the regular solution model. Likewise the entropy of mixing often deviates strongly from the ideal entropy of mixing. This is not only due to the deviation from a random distribution of the atoms, but also to a change of the atomic vibration frequencies during solution formation. The vibration frequencies of an atom in a substance are connected directly with its heat capacity and entropy. If during formation of a solid solution, a change in the heat capacity ΔC_p occurs, then the corresponding change in entropy, still neglecting the possible presence of magnetic and electronic effects, is given by:

$$\Delta S_V = \int_0^T \frac{\Delta C_p}{T} dT \tag{6.30}$$

ΔS_v is the vibrational contribution to the total entropy of mixing.

It is customary to divide the entropy of mixing ΔS of a real solution in a part, which is determined by the ideal entropy of mixing, and in a remainder, the so-called excess entropy ΔS^{ex}, which is caused by specific properties of the system.

$$\Delta S = \Delta S_i + \Delta S^{ex} \tag{6.31}$$

As an example, Fig. 6.7 reproduces the enthalpies of mixing and entropies of mixing of liquid silver-antimony alloys as a function of composition.

One can observe significant deviations from the behavior of a regular solution. The ΔH-x does not resemble a parabola. In addition, the enthalpies of mixing of silver rich solutions have a negative sign, those of antimony rich solutions a positive one. The regular solution model requires $\Delta S^{ex} = 0$.

In an analogous fashion to the enthalpy of mixing, the Gibbs energy of mixing is divided into an ideal part ΔG_i, given by Eq. (6.24) and in ΔG^{ex}, the excess Gibbs energy of mixing

Fig. 6.7
Enthalpy of mixing ΔH and excess entropy ΔS^{ex} of liquid Ag – Sb alloys (after [1])

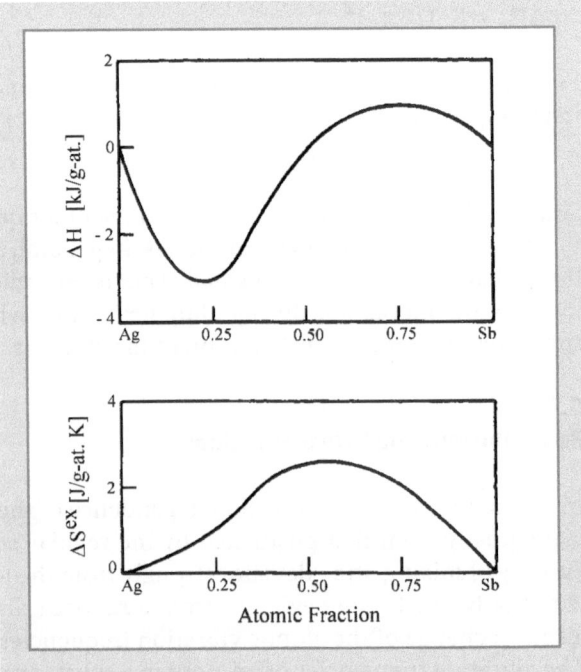

$$\Delta G = \Delta G_i + \Delta G^{ex} \tag{6.32a}$$

In general the excess functions are defined as the difference between the real mixing property, and the "ideal" mixing property, as can be expected for an ideal solution with identical composition, identical temperature and identical pressure. Accordingly, the enthalpy of mixing ΔH and the change in volume ΔV are excess properties, because for ideal solutions $\Delta H = 0$, and $\Delta V = 0$. The Helmholtz-Gibbs equation as applied to excess properties is

$$\Delta G^{ex} = \Delta H - T \cdot \Delta S^{ex} \tag{6.32b}$$

Partial mixing properties of real solutions are also divided into "ideal" parts and excess properties. For the partial entropy of mixing of component A of a binary solution A – B one obtains

$$\overline{\Delta S}_A = \Delta S_{i,1} + \Delta S_A^{ex} \tag{6.33}$$

As for all integral properties the integral entropy of mixing ΔS is composed additively, based on the composition, from the partial properties $\overline{\Delta S}_A$ and $\overline{\Delta S}_B$ (see Eq. (6.29)):

$$\Delta S = x_A \cdot \overline{\Delta S}_A + x_B \cdot \overline{\Delta S}_B \tag{6.34}$$

The expression for the ideal integral entropy of mixing is shown in Eq. (6.11).

Comparison of Eq. (6.34) with Eq. (6.11) shows that the ideal partial molar entropy of mixing for the components is given by:

$$\overline{\Delta S}_{A(i)} = -R \ln x_A \tag{6.35a}$$

and

$$\overline{\Delta S}_{B(i)} = -R \ln x_B \tag{6.35b}$$

The excess partial molar entropy of mixing is according to Eqs. (6.33) and (6.35) for example for component A given by:

$$\Delta S_A^{ex} = \overline{\Delta S}_A - \overline{\Delta S}_{A(i)} = \overline{\Delta S}_A + R \ln x_A \tag{6.36}$$

One has

$$\Delta S^{ex} = x_A \Delta S_A^{ex} + x_B \Delta S_B^{ex} \tag{6.37}$$

likewise

$$\Delta G^{ex} = x_A \Delta G_A^{ex} + x_B \Delta G_B^{ex} \tag{6.38}$$

6.8
Analysis of Experimental Thermodynamic Data

As could be shown, even in a five-component systems only binary interaction (interaction between neighboring atoms) is necessary to describe the energetic properties of the alloy. Thus, we may look at the factors which determine binary phase diagrams and accordingly binary interactions.

In binary systems we have a large variety of intermetallic phases and correspondingly also phase diagrams. This indicates that in metallic, and to a lesser extent in ceramic and polymeric systems, the stability relationships are governed by the combinations of different factors. Therefore it is not surprising, that certain relationships can be found, when experimental thermodynamic data are analysed. It is possible that one can deduce the cause of these relationships.

Alloy formation from the components can create a significant change in the binding character. The different size of the atoms also plays an important role, as was found by Hume-Rothery in 1934. Alpaut and Heumann [3] pointed out that a change in crystal structure has a noticeable influence on alloy formation. They showed that the effect of the three factors can be separated in their influence on the enthalpy of mixing of solid phases. They are additive, and can be determined separately. This facilitates the analysis of the thermodynamic relationships in a system. Let ΔH_B be the effect caused by changes in binding, ΔH_V the effect caused by the difference in atomic radii and ΔH_S the effect of the structure difference on ΔH, the total value of the enthalpy of formation, then we have

$$\Delta H = \Delta H_B + \Delta H_V + \Delta H_S \tag{6.39}$$

and similarly for the excess entropy of formation

$$\Delta S^{ex} = \Delta S_B + \Delta S_V + \Delta S_S \tag{6.40}$$

These relationships also apply to liquid alloys. It must be noted, that in the liquid case the difference in structure cannot be obtained easily, as it is not easily separated. The distortion, caused by the atomic size difference, can be large and calculable.

6.9.
Influence of the Atomic Size Difference

To demonstrate the influence of the size difference on ΔH even in the liquid alloys, in Fig. 6.8 the maximum ΔH of liquid Pb-alloys is plotted as a function of the relative difference d (see Eq. (6.41)) of the atomic volumes of some systems with demixing tendency:

$$d = \frac{|V_A - V_B|}{0.5(V_A + V_B)} \cdot 100 \tag{6.41}$$

Fig. 6.8
Maximum integral enthalpies of mixing in liquid lead alloys with tendency towards immiscibility, as a function of the relative atomic volume difference d between the components

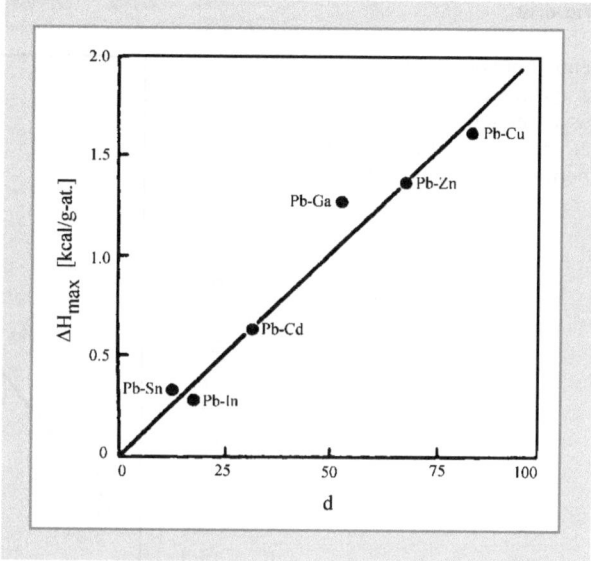

Fig. 6.9
Maximum integral excess entropies of liquid lead alloys with tendency towards immiscibility, as a function of the relative atomic volume difference d between the components

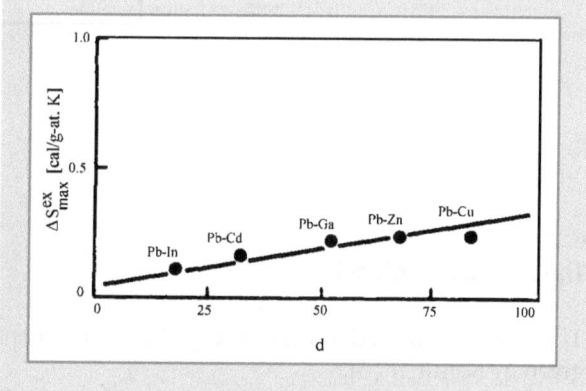

A similar dependence can be found by plotting the maximum of the excess entropy of mixing ΔS^{ex}_{max} as a function of d (see Fig. 6.9).

Extrapolation to $d \to 0$ gives the influence of the mean chemical interaction between the components, which is, in the case of simple demixing systems, almost zero. It is $\Delta H_{max} \neq 0$ for $d = 0$ in systems with components having compound forming tendency, like shown for liquid Hg-systems in Fig. 6.10. By extrapolation to $d = 0$ there results $\Delta H_V \approx -0.7$ kJ/g-at. For details see [3–5].

Fig. 6.10
Maximum integral
enthalpies of mixing of
some liquid mercury alloys
as a function of the relative
atomic volume difference d
between the components

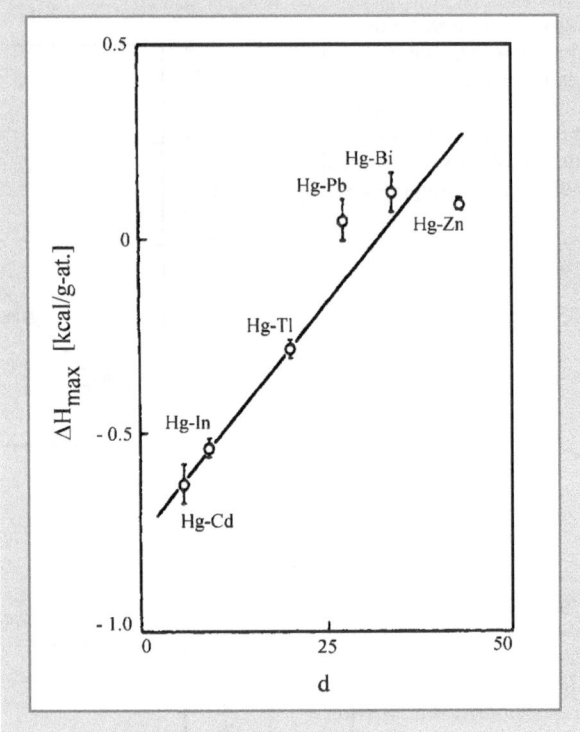

6.10
The Association Model

The association model was developed over the years, and the best description is given by Sommer [6].

The experimentally determined properties of compound-forming liquid alloys show specific concentration and temperature dependences, which are caused by the existence of chemical short range order in these alloys. The concentration and temperature dependences of thermodynamic mixing functions are explained on the basis of homogeneous equilibria reactions within an association model for a simple description of chemical short range order. The ability of this model to predict thermodynamic properties of a binary liquid alloy from experimental results and to extrapolate it to ternary liquid alloys is demonstrated.

6.10.1
Basic Formulae of the Association Model

The use of homogeneous equilibrium reactions for a simple description of CSRO (chemical short range order) in liquid alloys with a compound forming tendency enables a description of the different kinds of concentration and temperature dependences of the integral and partial thermodynamic mixing functions. The short range ordered volume parts are summarily described as associates with a well-defined composition, while the rest of the atoms are regarded to be randomly distributed. The associates are in a steady, dynamic equilibrium with the non-associated atoms, which is governed by the mass action law. The lifetime and the spatial arrangement of the short range ordered regions do not enter into the model. Particulars about the basis of this model, the development and application by other authors may be taken from [6–9]. The expressions for ΔH and ΔS for binary alloys with the formation of a single association type while neglecting volume effects are [7, 8].

$$\Delta H = \frac{n_{A_1} n_{B_1} C^{reg}_{A_1,B_1}}{n} + \frac{n_{A_1} n_{A_iB_j} C^{reg}_{A_1,A_iB_j}}{n} + \frac{n_{B_1} n_{A_iB_j} C^{reg}_{B_1,A_iB_j}}{n}$$
$$+ n_{A_iB_j} \Delta H^0_{A_iB_j} \tag{6.42}$$

and

$$\Delta S = -R(n_{A_1} \ln x_{A_1} + n_{B_1} \ln x_{B_1} + n_{A_iB_j} \ln x_{A_iB_j}) + n_{A_iB_j} \Delta S^0_{A_iB_j} \tag{6.43}$$

where n_{A1} and n_{B1} moles free A and B atoms are in equilibrium with n_{AiBj} moles of associates having the composition A_iB_j (i, j = 1, 2 ..., n). x_{A1}, x_{B1} and x_{AiBj} are the molar fractions of the assumed species for 1 mol of a binary alloy. The equilibrium value for $n_{A_iB_j}$ is determined by a mass action law with an association constant defined by

$$K_{A_iB_j} = \exp[-(\Delta H^0_{A_iB_j} - T\Delta S^0_{A_iB_j})/RT] \tag{6.44}$$

where $\Delta H^0_{A_iB_j}$ and $\Delta S^0_{A_iB_j}$ represent the enthalpy and entropy of formation of the associates. For the activity coefficient as one obtains from Eqs. (6.42)–(6.44)

$$\gamma_A = \exp[(C^{reg}_{A_1,B_1}(1-x_{A_1})x_{B_1} + C^{reg}_{A_1,A_iB_j}(1-x_{A_1})x_{A_iB_j}$$
$$- C^{reg}_{B_1,A_iB_j} x_{B_1} x_{A_iB_j})/RT] \tag{6.45}$$

$$\gamma_B = \exp[(C^{reg}_{A_1,B_1}(1-x_{B_1})x_{A_1} + C^{reg}_{B_1,A_iB_j}(1-x_{B_1})x_{A_iB_j}$$
$$- C^{reg}_{A_1,A_iB_j} x_{A_1} x_{A_iB_j})/RT] \tag{6.46}$$

The expression of excess molar heat capacity is given by [6, 9]

$$\Delta C_p(x,T) = nT[R(-i\ln x_{A_1} - j\ln x_{B_1} + \ln x_{A_iB_j}) - \Delta S^0_{A_iB_j}]^2$$

$$[2(C^{reg}_{A_1B_1}[ij + (i+j-1)\{-jx_{A_1} - ix_{B_1} + (i+j-1)x_{A_iB_1}\}]$$

$$+ C^{reg}_{A_1,A_iB_j}[-i + (i+j-1)\{x_{A_1} - ix_{A_iB_j} + (i+j-1)x_{A_1}x_{A_iB_j}\}]$$

$$+ C^{reg}_{B_1,A_iB_j}[-j + (i+j-1)\{x_{B_1} - jx_{A_iB_j} + (i+j-1)x_{B_1}x_{A_iB_j}\}])$$

$$+ RT[\frac{i^2}{x_{A_1}} + \frac{j^2}{x_{B_1}} + \frac{1}{x_{A_iB_j}} - (i+j-1)^2]]^{-1} \qquad (6.47)$$

with $n = n_{A1} + n_{B1} + n_{AiBj}$. The temperature dependences of the partial enthalpies are influenced by the stoichiometry of the associates. For $i, j > 1$, ΔH^0 exhibits no temperature dependence [7]

$$\Delta \overline{H}^0_A = C^{reg}_{A_1,B_1} \qquad (6.48)$$

$$\Delta \overline{H}^0_B = C^{reg}_{A_1,B_1} \qquad (6.49)$$

For $i = j = 1$, one obtains for ΔH^0 using Eqs. (6.42)–(6.44)

$$\Delta \overline{H}^0_A = C^{reg}_{A_1,B_1} + \frac{K_1}{1+K_1} RT^2 \partial \ln K_1 / \partial T \qquad (6.50)$$

$$\Delta \overline{H}^0_B = C^{reg}_{A_1,B_1} + \frac{K_2}{1+K_2} RT^2 \partial \ln K_2 / \partial T \qquad (6.51)$$

where

$$K_1 = K_{A_iB_j} \exp[(C^{reg}_{A_1,B_1} - C^{reg}_{B_1,A_iB_j})/RT] \qquad (6.51a)$$

and

$$K_2 = K_{A_iB_j} \exp[(C^{reg}_{A_1,B_1} - C^{reg}_{A_1,A_iB_j})/RT]. \qquad (6.51b)$$

For $i = 1, j > 1$. Eq. (6.50) is valid for ΔH^o_A and Eq. (6.49) is valid for ΔH^o_B. For $i > 1$, $j = 1$, ΔH^o_A is temperature independent (Eq. (6.48)) and ΔH^o_B is temperature dependent (Eq. 6.51)). It can be shown that the association model predicts an S-shaped temperature dependence for ΔH^0 with two limiting values at high and low temperatures [10, 11].

For alloys with high associate fractions and high values for the enthalpy and entropy of formation per atom of the associates, ΔH^0 and ΔS^0 generally cannot be regarded as independent of temperature. The following relationship can be used for a temperature T_2 if ΔH^0 and ΔS^0 are known at T_1 ($T_2 > T_1$) [9,12]

$$\Delta H^0_{A_iB_j}(T_2) = \Delta H^0_{A_iB_j}(T_1) + A(T_2 - T_1) \tag{6.52}$$

$$\Delta S^0_{A_iB_j}(T_2) = \Delta S^0_{A_iB_j}(T_1) + A\ln\frac{T_2}{T_1} \tag{6.53}$$

The association model enables the calculation of thermodynamic functions of ternary systems, if the model parameter of the respective basic binary systems are known and no additional association reaction occurs. Then enthalpy and entropy of the ternary alloy are given by the following expression [13]

$$\Delta H = \frac{n_{A_1}n_{B_1}C^{reg}_{A_1,B_1}}{n} + \frac{n_{A_1}n_{A_iB_j}C^{reg}_{A_1,A_iB_j}}{n} + \frac{n_{B_1}n_{A_iB_j}C^{reg}_{B_1,A_iB_j}}{n}$$

$$+\frac{n_{B_1}n_{C_1}C^{reg}_{B_1,C_1}}{n} + \frac{n_{B_1}n_{B_uC_v}C^{reg}_{B_1,B_uC_v}}{n} + \frac{n_{C_1}n_{B_uC_v}C^{reg}_{C_1,B_uC_v}}{n}$$

$$+\frac{n_{A_1}n_{C_1}C^{reg}_{A_1,C_1}}{n} + \frac{n_{A_1}n_{A_kC_l}C^{reg}_{A_1,A_kC_l}}{n} + \frac{n_{C_1}n_{A_kC_l}C^{reg}_{C_1,A_kC_l}}{n}$$

$$+n_{A_iB_j}\Delta H^0_{A_iB_j} +n_{B_uC_v}\Delta H^0_{B_uC_v} +n_{A_kC_l}\Delta H^0_{A_kC_l} \tag{6.54}$$

$$\Delta S = -R(n_{A_1}\ln x_{A_1} +n_{B_1}\ln x_{B_1} +n_{C_1}\ln x_{C_1}$$
$$+n_{A_iB_j}\ln x_{A_iB_j} +n_{B_uC_v}\ln x_{B_uC_v} +n_{A_kC_l}\ln x_{A_kC_l})$$
$$+n_{A_iB_j}\Delta S^0_{A_iB_j} +n_{B_uC_v}\Delta S^0_{B_uC_v} +n_{A_kC_l}\Delta S^0_{A_kC_l} \tag{6.55}$$

where n_{A_1}, n_{B_1}, n_{C_1} are the number of moles of free A, B and C atoms in equilibrium with n_{AiBj}, n_{BuCv} and n_{AkCl} moles of associates of the composition A_iB_j (i, j = 1, 2, ... n), B_uC_v (u, v = 1, 2, 3, ...), A_kC_l (k, l = 1, 2, 3, ... n) and the x_{A_1}, x_{B_1}, x_{C_1}, x_{AiBj}, x_{BuCv} and x_{AkCl} are the mole fractions of the respective species in 1 mole of ternary alloy.

6.10.2
Application to Liquid Binary and Ternary Alloys

It is necessary to define the stoichiometry of the associates to calculate the thermodynamic functions using the expressions given above. The results of all measured structure sensitive properties for a given liquid alloy should give indications for maximum CSRO at the same concentration. This information can be used to fix the composition of the associates. The model parameters are then obtained by fitting measured integral or partial values of ΔH and ΔG by the method of least squares [9, 11]. The model parameters of some exemplary systems, which are discussed in the following paragraph, are given in Table 6.1.

Table 6.1 Model parameters of some binary systems

	Sn-Te	Li-Pb	Cu-Ce	Ce-Mg	Cu-Mg
Associate A_iB_j	SnTe	Li_4Pb	Cu_3Ce	Mg_3Ce	Cu_2Mg
$\Delta H^\circ_{A_iB_j}$	-53.8	-143.7	-81.5	-56.4	-40.5
$\Delta S^\circ_{A_iB_j}$	-0.0103	-0.0238	0.0466	-0.026	-0.0115
$C^{reg}_{A,B}$	-28.2	-60.7	-46.0	-35.5	-26.1
C^{reg}_{A,A_iB_j}	11.6	0	-60.5	0	0
C^{reg}_{B,A_iB_j}	-6.3	-64.3	-28.7	-25.1	-12.2

Figure 6.11a [8] shows the calculated thermodynamic properties of the system Sn – Te [6] and Fig. 6.11b shows the phase diagram [14]. The experimental data for ΔH are from [15].

Figure 6.12a shows the calculated thermodynamic properties of the system Li – Pb [6] and Fig. 6.12b shows the phase diagram [14]. The experimental data for ΔH are from [16].

Figure 6.13a [14, 15] shows the calculated thermodynamic properties of the system Ag – Te [7, 8] and Fig. 6.13b shows the phase diagram [14]. The experimental data for ΔH are from [15] and for ΔG are from [58]. This system shows a miscibility gap; Sommer [10] calculated the miscibility gap: The top occurs at 25 at.-% Te and T = 1353 K.

Figures 6.14 und 6.15 show the calculated thermodynamic properties of the ternary system Cu – Mg – Ce along two composition lines [6]. The experimental data for ΔH are from [13].

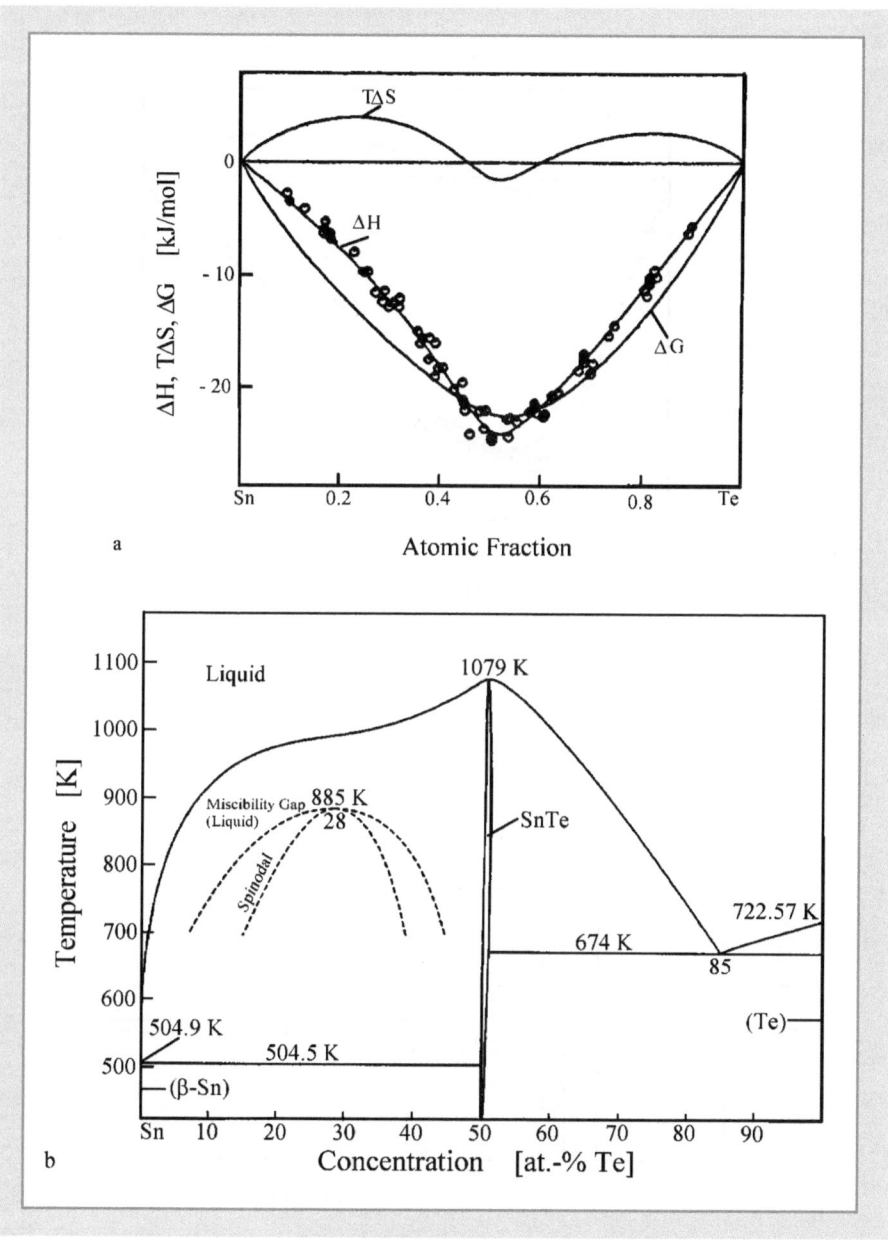

Fig. 6.11a Enthalpy, entropy and Gibbs energy of mixing of liquid Sn – Te alloys at 1140 K cal-culated using the association model (ooo experimental [8]), **b** Sn – Te phase diagram [14]

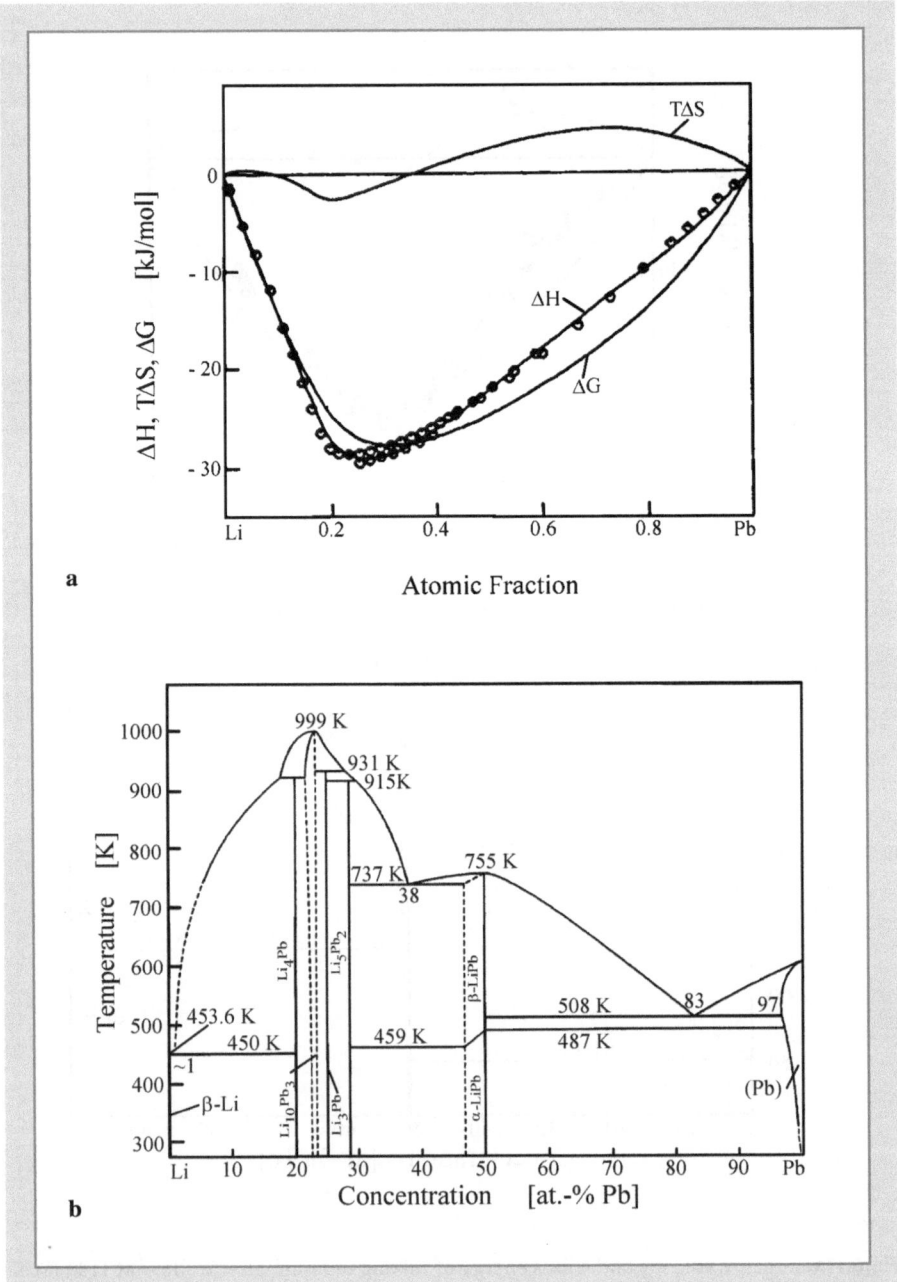

Fig. 6.12a Enthalpy, entropy and Gibbs energy of mixing of liquid Li-Pb alloys at 1000 K calculated using the association model (ooo experimental [16]), **b** Li-Pb phase diagram [14]

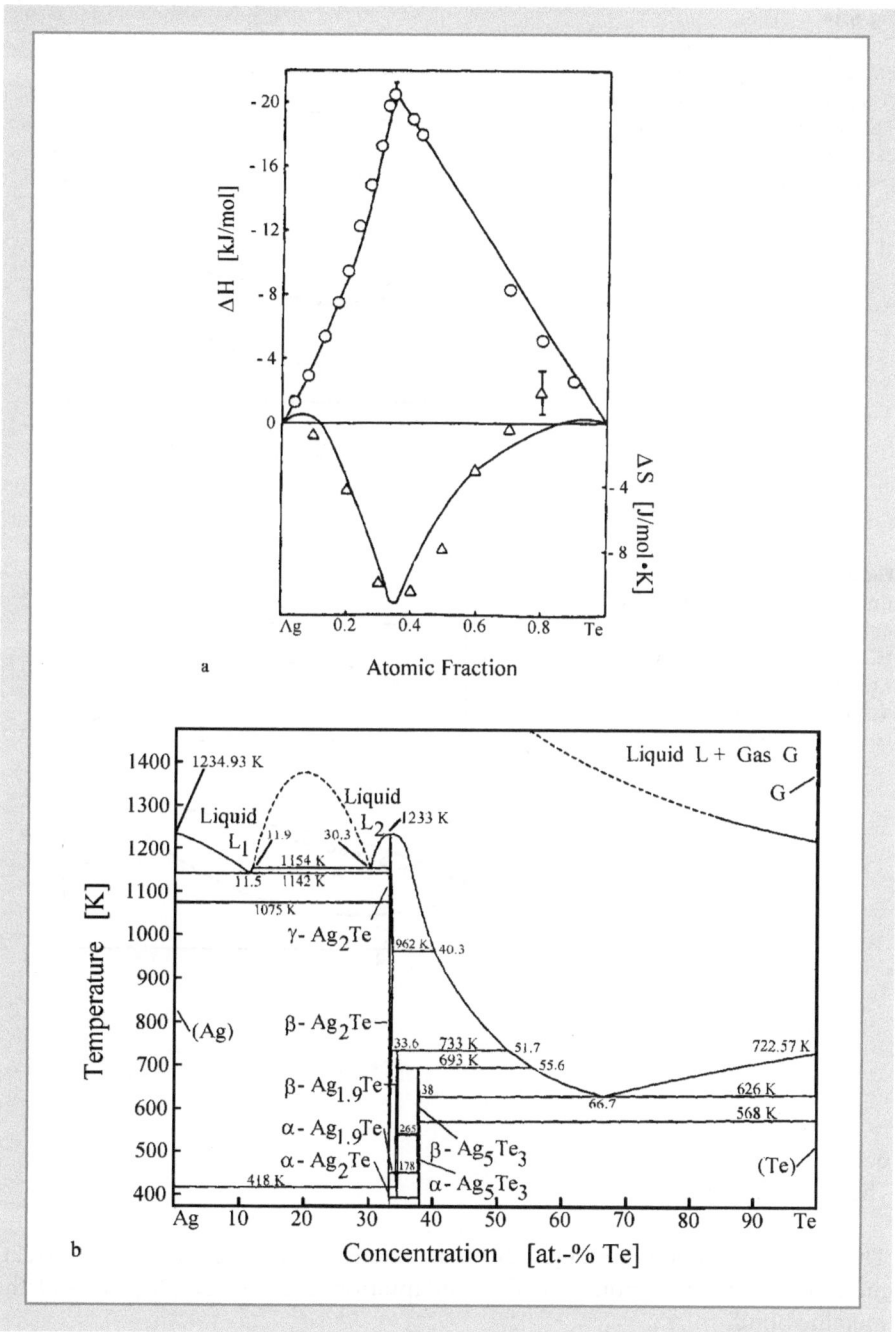

Fig. 6.13a Thermodynamic data for liquid Ag-Te alloys [7]. ------- calculated, ooo experimental, **b** Ag-Te phase diagram [7, 14]. Enthalpy and entropy of mixing of liquid Ag-Te alloys at 1281 K calculated using the association model [7, 8] (ooo [15], △△△ [58] experimental)

Fig. 6.14
Enthalpy, entropy and Gibbs energy of mixing of liquid $(Cu_{80}Mg_{10})_x Ce_{(1-x)}$ alloys at 1125 K calculated using the association model
(··· experimental [13])

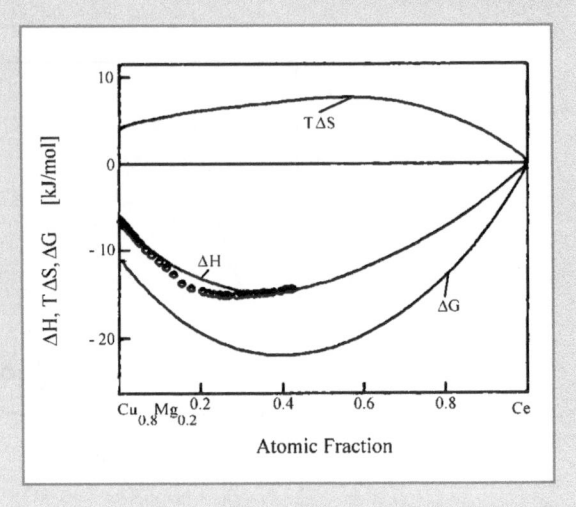

Fig. 6.15
Enthalpy, entropy and Gibbs energy of mixing of liquid $(Ce_{33}Cu_{67})_x Mg_{(1-x)}$ alloys at 1131 K calculated using the association model
(··· experimental [13])

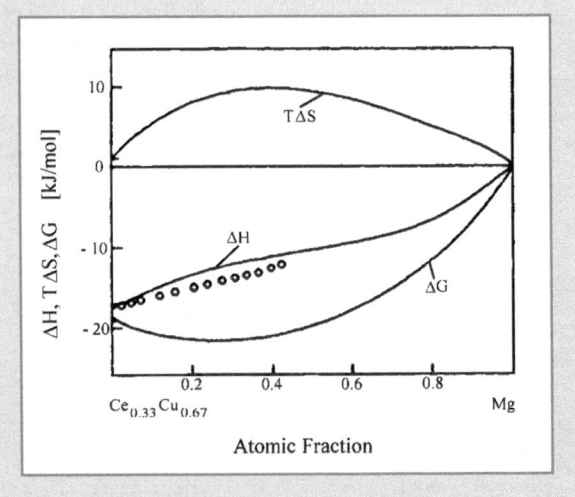

6.11
The Hoch-Arpshofen Model

The Hoch-Arpshofen Model [17, 18] is an extension of Guggenheim's [19] treatment of solutions, combined with an adaptation of Pauling's [20] ideas of the metallic bond.

Guggenheim [19], when treating regular solutions and superlattices speaks of "treatment of quadruplets of sites, forming regular tetrahedron" and "triplets of sites, forming equilateral triangles". Guggenheim, however, always treats the strength of the A – B bond the same, irrespective of what other atoms are present in the complex. In this model the strength of the A – B bond depends on the number of B atoms to which the A atom bonds or vice versa. In Pauling's [20] description of a metallic bond the bond number is defined as the number of bonding electrons divided by the number of neighbors that the specific atom bonds. In metallic copper, with one bonding electron and 12 neighbors, the bond number is 1/12. This is a one electron bond, which moves from one neighbor to the other. In our model this idea is applied to ionic materials (ceramics), and van der Waals type forces, both in an attractive and repulsive mode. This is not such an extravagant idea, as all bonds are due to the behavior of electrons.

Let us look at [17] a quaternary system with four components, A, B, C, and D and their mole fractions x, y, z, and u where three-atom "complexes" are present in the A – B binary and [18] a system with three components A, B, and C, and their mole fractions x, y, and z, where four-atom "complexes" are present in the A – B binary.

Let atom A have the lower binding capacity. In the binary A – B system the maximum interaction (in absolute term) occurs at $x < 0.5$. We have only to look at the A-B interaction in the multi-component system: the C and D atoms only dilute the system (the A – C interaction comes from the A – C binary system), and we do not consider ternary interactions.

The discussion below shows the effect of the A-B binary in the multi-component system. A similar calculation must be carried out for all the other binary systems A – C, A – D, B – C, B – D, and C – D, etc.

Looking at the case with four components and three-atom complexes (Fig. 6.16a) the possible combinations are given in Table 6.2 and ΔH has been calculated, the effect of the binary A – B system in the quaternary, where W is the interaction parameter.

$$\Delta H = W3xy[1 + (1 - y)] \tag{6.56}$$

Looking at the case with three components and four-atom complexes (Fig. 6.16b) we have given the possible combinations in Table 6.3 and have calculated ΔH, the effect of the binary A – B system in the ternary

$$\Delta H = W4xy[1 + (1-y) + (1-y)^2] \tag{6.57}$$

Thus, in general

$$\Delta H = Wnxy[1 + (1-y) + (1-y)^2 + (1-y)^3 + \ldots + (1-y)^{\{n-2\}}] \tag{6.58a}$$

or

$$\Delta H = Wnx[1 - (1-y)^{\{n-1\}}] \tag{6.58b}$$

The other binary systems A – C, … C – D,.. similarly contribute (obviously with different n, and their mol fractions) to the multi-component system.

The excess entropy of mixing, ΔS^{ex}, and the excess Gibbs energy of mixing, ΔG^{ex}, can behave similarily, thus, we use

$$\Delta G = Wnx[1 - (1-y)^{\{n-1\}}] \tag{6.59}$$

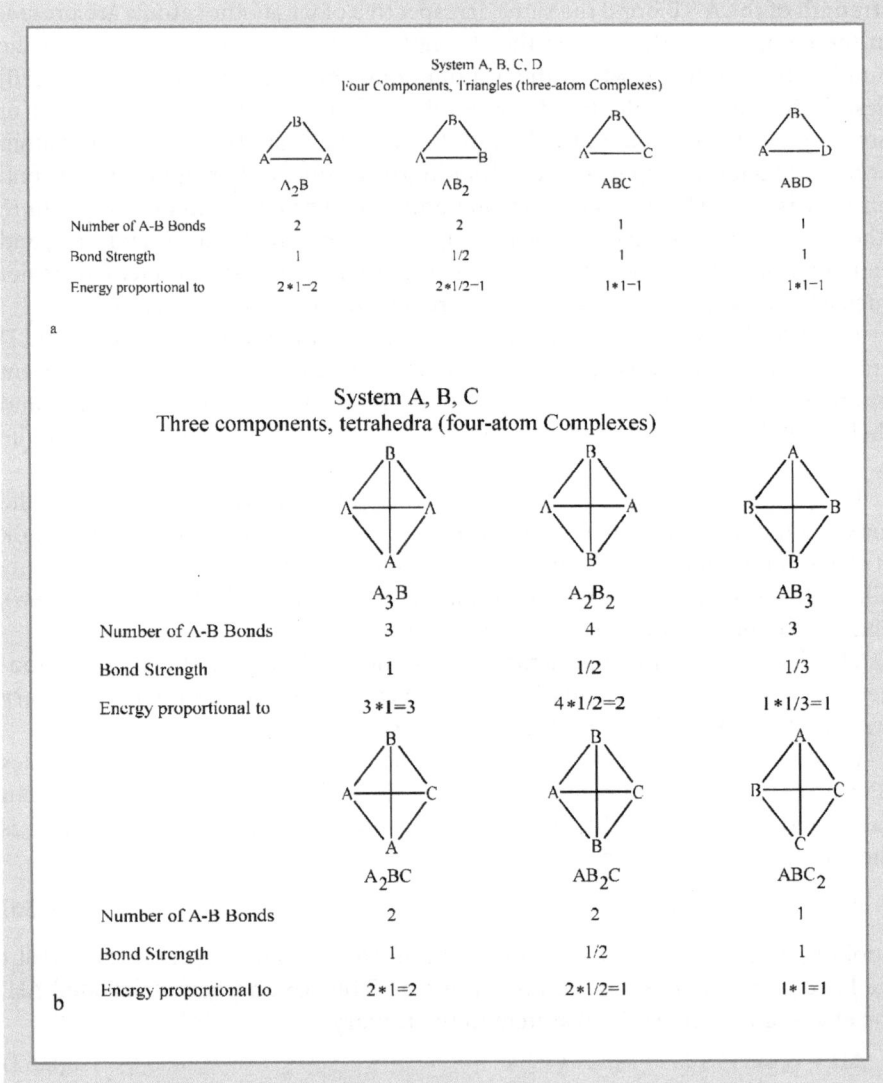

Fig. 6.16 Assemblage of atoms in ternary and larger systems in two-, three- and four-atoms complexes – triangles **(a)** and tetrahedra **(b)**

The partial quantities

$$\overline{\Delta G_x} = Wn[1 - (1-y)^{\{n-1\}} - xy(n-1)(1-y)^{\{n-2\}}] \tag{6.60}$$

$$\overline{\Delta G_y} = Wnx(n-1)(1-y)^{\{n-2\}} \tag{6.61}$$

$$\overline{\Delta G_z} = \Delta G_u = -Wnxy(n-1)(1-y)^{\{n-2\}} \tag{6.62}$$

Table 6.2 Complexes in a Four-Component System, with Three-Atom Complexes and Calculation of the Total Interaction. ▯ = Number

Complex	▯ of A–B Bonds	▯ of B Atoms One A Atome Bonds	Total Interaction	▯ of Complexes Proprotional to
AAB	2	1	$2 \cdot 1 = 2$	$3x^{201}y$
ABB	2	2	$2 \cdot 1/2 = 1$	$3xy^2$
ABC	1	1	$1 \cdot 1 = 1$	$6xyz$
ABD	1	1	$1 \cdot 1 = 1$	$6xyu$

and $x + y + z + u = 1$
we obtain $H_m = W3xy[1 + (1 - y)]$

Table 6.3 Complexes in a Three-Component System, with Four-Atom Complexes and Calculation of the Total Interaction. ▯ = Number

Complex	▯ of A–B Bonds	▯ of B Atoms One A Atome Bonds	Total Interaction	▯ of Complexes Proprotional to
AAAB	3	1	$3 \cdot 1 = 3$	$4x^3y$
AABB	4	2	$4 \cdot 1/2 = 2$	$6x^2y^2$
ABBB	3	3	$3 \cdot 1/3 = 1$	$4xy^3$
AABC	2	1	$2 \cdot 1 = 2$	$12x^2yz$
ABBC	2	2	$2 \cdot 1/2 = 1$	$12xy^2z$
ABCC	1	1	$1 \cdot 1 = 1$	$12xyz$

and $x + y + z = 1$
we obtain $H_m = W4xy[1 + (1 - y) + (1 - y)^2]$

Equations (6.59)–(6.61) also apply to the binary system A – B, if one replaces y by (1-x) then $x + y = 1$ and $z = u = \ldots = 0$.

$$\Delta G = Wn[x - x^n] \tag{6.63}$$

$$\overline{\Delta G}_x = Wn[1 - nx^{\{n-1\}} + (n-1)x^n] \tag{6.64}$$

$$\overline{\Delta G}_y = Wn(n-1)x^n \tag{6.65}$$

The partial quantities represented by Eqs. (6.64) and (6.65) do not change sign when the composition changes from $x = 0$ to $x = 1$; the sign of $\overline{\Delta H}_x$ and $\overline{\Delta H}_y$ is determined by the sign of W.

Table 6.4 Comparison of the Effect of a Binary System in the Center of a Ternary System, Using Different Extension Methods

n	Present Model Eqs. (15)	No Extension Eqs. (2)	Kohler [12] Eqs. (20)	Eqs. (2) Eqs. (15)	Kohler Eqs. (15)
2	0.1111	0.1111	0.1111	1.0000	1.0000
3	0.1852	0.1111	0.1667	0.6000	0.9000
4	0.2346	0.0864	0.1944	0.3684	0.8289
5	0.2675	0.0617	0.2083	0.2308	0.7788
6	0.2894	0.0425	0.2153	0.1469	0.7438
7	0.3041	0.0288	0.2188	0.0947	0.7194
8	0.3138	0.0194	0.2205	0.0617	0.7026
9	0.3203	0.0130	0.2214	0.0404	0.6910
10	0.3247	0.0087	0.2210	0.0267	0.6831

ΔS^{ex} (the excess entropy of mixing) also has the same form as Eq. (6.60), but it can have another value for n, and the weaker "bonding" component may be B, thus, it is possible that

$$\Delta S^{ex} = Ur(y-y^r) \tag{6.66}$$

where U is the entropy constant, again independent of composition.

Though we talk about "complexes", the ideal Gibbs energy of mixing is, as in Guggenheim [19]

$$\Delta G_{id} = RT(x_1 \ln_{x_1} + x_2 \ln_{x_2} + x_3 \ln_{x_3} + ...) \tag{6.67}$$

In a binary system, the maximum number of interactions for the mixing function ΔG, is two, one attractive (negative), the other repulsive (positive). In general

$$\Delta G = W_1 n(x-x^n) + W_2 m(y-y^m) \tag{6.68}$$

and

$$W_1 > 0 \text{ and } W_2 < 0 \text{ or vice versa} \tag{6.69}$$

we have as a maximum, two parameters for ΔH and two parameters for ΔS^{ex}, thus, for a binary system ,we need a maximum of four parameters. No ternary or larger system has yet been found, in which ternary or larger interaction parameters are needed.

ΔG^{ex} is a combination of ΔH and ΔS^{ex}, the above equations only apply to it, if ΔS^{ex} has the same form (n and m) as ΔH or is zero.

The model always mixes metal atoms or cations, i.e. B_2O_3 is always treated as $BO_{1.5}$ and Na_2O as $NaO_{0.5}$. This obviously requires a change in composition, before the model can be applied. Also we treat in one program, only components with the same crystal structure, i.e. solid FCC, BCC, or liquid systems.

If in a binary system a solid compound extends to much higher temperatures into the liquid than the melting points of the two components, we apply an extended Schottky-Wagner [21] model. In the Schottky-Wagner (5) model, a solid compound, such as AB_2 (s), dissociates according to

$$AB_2(s) \rightarrow A(s) + 2B(s) \tag{6.70}$$

The amount of dissociation is given by the chemical equilibrium of the above reaction. Schottky and Wagner [21] assumed ideal behavior: the activity coefficients of A(s) and B(s) were assumed to be one, and that of $AB_2(s)$ is necessarily one because the concentration of $AB_2(s)$ is close to unity.

We extended the model by postulating: (a) The compound exists in the liquid phase. (b) The phase diagram can be separated into two partial diagrams $A - AB_2$ and $AB_2 - B$. (c) In the partial systems $A - AB_2$ and $AB_2 - B$ there is interaction in the liquid between A and AB_2 as well as AB_2 and B according to the model described above. Thus, different activity coefficients are obtained for A(L) and B(L) when $AB_2(L)$ decomposes according to

$$AB_2(L) \rightarrow A(L) + 2B(L) \tag{6.61}$$

(d) in the partial diagram $A - AB_2$ one obtains activity coefficients for A(L) and $AB_2(L)$ from the phase diagram. The activity of B(L) in the $A - AB_2$ partial system can be obtained from Eq. (6.72) using the Gibbs energy change for that reaction and the activity coefficients of A(L) and $AB_2(L)$. Similar procedures applies for the activity of A in the partial binary system $AB_2 - B$. (e) It is possible that the compound $AB_2(L)$ is present, but $AB_2(s)$ is only stable at low temperatures and thus, not in equilibrium with the liquid.

In the shorthand notation 3, (A) means, that n = 3, and x, the atom or compound with the lower binding capacity is A. In a binary system we may find 3, (A) and 6, (B) indicating that in the A – B system, in the composition range around A, A is the weaker bonding component, with the complex A_2B, and close to B, B is the weaker bonding component, with the complex AB_5. In these cases one of the interaction parameters will be negative, the other one positive.

In all our calculations all quantities are divided by R, the gas constant, thus, ΔH and ΔG^{ex} are expressed in kK (kiloKelvin), ΔS^{ex} and C_p are dimensionless. This simplifies the tabulation of the data, as only maximal two significant figures are needed before the decimal point, and we use maximal four figures after the decimal point.

6.12
Difference in Heat Capacity between Liquid and Solid $C_{p(L-S)}$

The Gibbs energy of fusion $G_{(L-S)}$ at temperatures below and above the melting point, and thus, the calculated phase diagram, or thermodynamic data obtained from the phase diagram are greatly influenced by the difference in heat capacity between liquid and solid, $C_{p(L-S)}$.

All liquids can theoretically be transformed into non-crystalline solids. When we lower the temperature of a liquid below the melting point, a temperature will be reached where the entropy of the liquid will become less than that of the solid $S_{(L-S)}/R < 0$. This was already noticed by Kauzmann [22], and is known as the "Kauzmann paradox". At this temperature, known as the theoretical glass transition temperature or T_g, the liquid upon cooling transforms into an amorphous solid, having properties very close to the corresponding crystalline solid. On the other hand [23, 25], above the melting point, there is a temperature above which again $S_{(L-S)}/R < 0$, or a disordered liquid has a lower entropy than an ordered solid of the same composition, which is not thermodynamically acceptable. This is the theoretical "jelly" transformation temperature, T_j [23, 24]. If one carries out alloy stability calculations at high temperatures, using a low melting alloying element such as aluminum, solid aluminum becomes the stable form, falsifying the evaluations.

Hoch [24] treated the behavior of a one-component system from very low to very high temperatures. The heat capacity data of solids $C_{p(S)}$ and liquids $C_{p(L)}$ of elements and compounds, where accurate data are available, can be represented by [24, 25]

$$C_{p(S)}/R = 3 F (O_D(s)/T) + bT + dT^3 \tag{6.72}$$

$$C_{p(L)}/R = 3 F (O_D(L)/T) + bT + hT^{-2} \tag{6.73}$$

where $F (O_D/T)$ is the Debye function, b is the electronic contribution, and d and h are the anharmonic contributions. In our calculations we assume that at the elevated temperatures $3F (O_D(s)/T)$ and $3F (O_D(L)/T)$ are equal.

Figure 6.17 shows the heat capacity curves for a hypothetical material: $T_{mp} = 3000\,K$, $H_{mp}/R = 3\,kK$ (Kilo Kelvin), $d = 1.111 \times 10^{-10}\,K^{-4}$ and $h = 5.4 \times 10^7\,K$.

Figure 6.18 shows the specific heat, entropy, enthalpy and Gibbs energy difference liquid-solid. $C_{p(L-S)}/R$ is almost linear, $G_{(L-S)}/R$ crosses the zero line three times, $S_{(L-S)}/R$ and $H_{(L-S)}/R$ cross the zero line twice. The transition curves in Fig. 6.17 can be described and were constructed by an exponential function which was used in other rapid changes [26, 27]:

$$\ln (x/(1-x)) = E + FT \tag{6.74}$$

where x is the fraction of the phase forming and (1-x) the fraction of the phase disappearing. The constants are set by the boundary conditions, when the transition begins, $x = 0.01$ and when it is complete $x = 0.99$.

$C_{p(L-S)}$ was calculated (24) for various materials where heat capacity or heat content data for solid and liquid were available, using Eqs. (6.73) and (6.74). 25 elements and compounds were investigated, from low melting Pb to high melting UO_2. For each material T_g, the theoretical glass transition temperature was calculated, where $S_{(L-S)}$ becomes zero, below the melting point. Below this point,

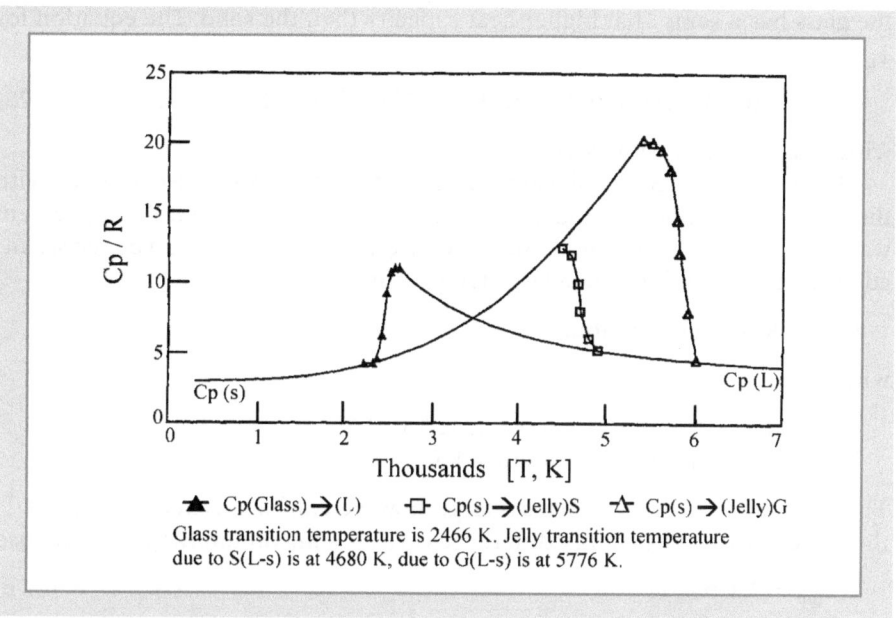

Fig. 6.17 C$_p$ of solid, glass, liquid, glass-liquid and solid-"jelly" transition. Melting point kK, at Mpt C$_{p(S)}$/R = 6, C$_{p(L)}$/R = 9. Hfus 3 kK

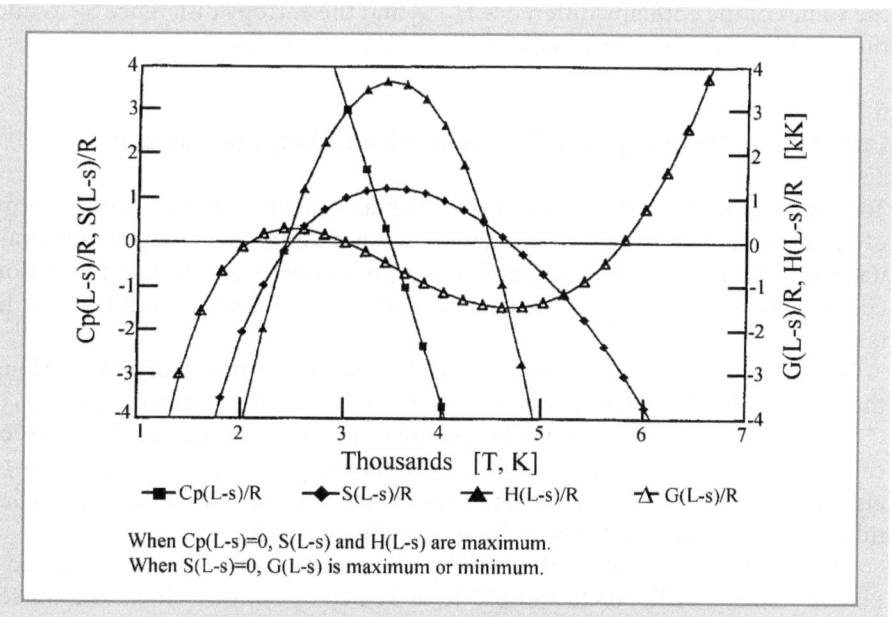

Fig. 6.18 Specific heat, entropy, enthalpy and Gibbs energy difference liquid-solid

the glass has a somewhat higher heat capacity than the solid. The equation for T_g is

$$T_g/T_{mp} = (0.4018 \pm 0.1601) + (8.53 \pm 2.68) \cdot 10^{-5} \cdot T_{mp}; R^* = 0.5228 \qquad (6.75)$$

with R^* as the regression coefficient.

The calculated theoretical glass transition temperatures were compared with the experimental ones, for materials where they were measured. The agreement was very good. We also found [25], that $C_{p(L-S)}$ in all these cases can be represented by a linear equation over a large temperature range,

$$C_{p(L-S)}/R = e + fT; \text{ per atom,}$$

with

$e/T_{mp} = (7 \pm 3)$ and $(e/T_{mp})/f = -(1.081 \pm 0.186)$,
T_{mp}, the standard melting point, in kK.

The uncertainty in the two terms is large: however, we have a check because at T_g the value of the Gibbs energy difference $G_{(L-S)}$ is a maximum [24]. Thus, we use

$$e/T_{mp} = (7 \pm p \cdot 3) \qquad (6.76)$$

$$(e/T_{mp})/f = -(1.081 \pm p \cdot 0.186) \qquad (6.77)$$

and we adjust p so that the above condition is met. The above uniform equation for $C_{p(L-S)}$ is good for phase diagram calculations where $C_{p(L-S)}$ enters into the values of the enthalpy difference $H_{(L-S)}$ and the entropy difference $S_{(L-S)}$ and their combination diminishes the effect of $C_{p(L-S)}$ on $G_{(L-S)}$.

6.13
Calculation of Thermodynamic Functions in Multi-Component Systems

To calculate thermodynamic functions, such as enthalpy, entropy, Gibbs energy of mixing or activities of a component in a multi-component system, one evaluates the binary data of Hultgren et al. [1b] to obtain n, m and the interaction coefficients, using regression analysis, and changing n and m until the best fit is obtained.

In the ternary Ag-Au-Si system Fig. 6.19 shows the enthalpy of mixing along the line (Au/Si = 1) as a function of x_{Ag} from the work of Hassam et al. [28].

In the quinary system Cd – Ga – In – Sn – Zn Figs. 6.20 and 6.21 results of calulations using the Hoch-Arpshofen model are presented. There are shown enthalpies of mixing along the line connecting Zn, resp. Cd, to the equimolar mixture of other components of the systems in questions.

Ptak et al. [29] measured the activity of Zn in the quinary system Bi – Cd – Pb – Sn – Zn at 714 K, 805 K and 877 K, at 124 compositions. Hoch and Moser [30] calculated thermodynamic values. There is a good agreement with data calculated on the basis of Hultgren [1].

Fig. 6.19
Enthalpy of mixing of the
Ag-Au-Si liquid alloy with
x_{Au}/x_{Si}: **** experimental
results at 1,423 K, ----- calculated with the Kohler model,
-.-.-. calculated with the
Hoch-Arpshofen model

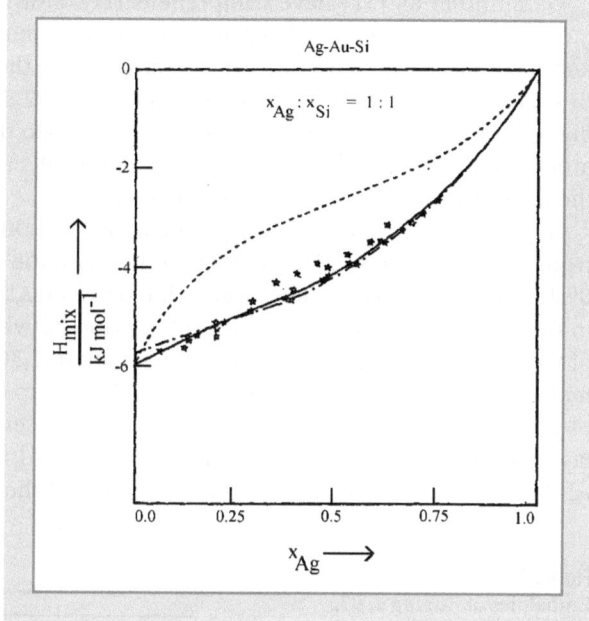

Fig. 6.20
Enthalpy of mixing of the
Cd-Ga-In-Sn-Zn liquid alloys
obtained by addition of Zn
to the equimolar Cd-Ga-In-
Sn solution at 730 K;
□□□ experimental,
+++ calculated (HA model)

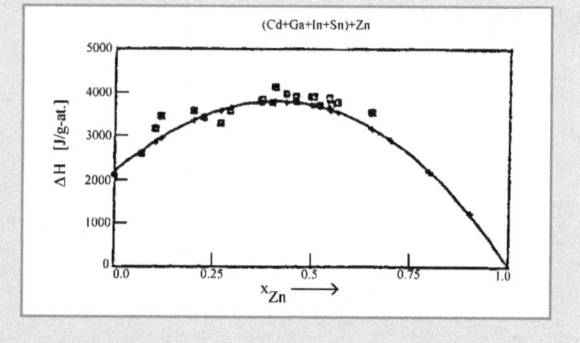

Fig. 6.21
Enthalpy of mixing of the
Cd-Ga-In-Sn-Zn liquid alloys
obtained by addition of Cd
to the equimolar Ga-In-Sn-
Zn solution at 730 K,
□□□ experimental;
+++ calculated (HA model)

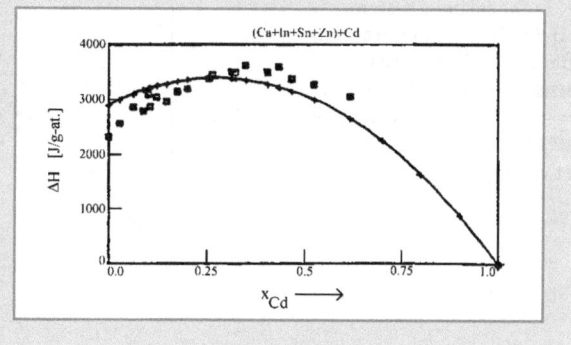

Gambino et al. [31] have comprehensively shown that the Hoch-Arpshofen model is a useful tool to also calculate thermodynamic data in higher component systems. As an example results obtained in the seven-component system Bi – Cd – Ga – In – Pb – Sn – Zn are shown in Fig. 6.22. There, the enthalpies of mixing for liquid alloys along the line connecting Pb to the equimolar mixture of all other components is shown. The calculated results are in good agreement with the data obtained by calorimetric experiments.

Figure 6.23 shows the liquidus line at about 1,300 K in the system Al-Au-Si from the calculations of Hoch [32] together with the experimental values of Lozovskiy and Politova [33]. The interaction between Al and Au plays the major role in this system: the strong negative interaction between Al and Au pushes Si out of the liquid, and thus, the liquidus line at about 1,300 K moves to lower Si content. The most stable compound in the Al-Au binary is Al_2Au, the strongest complex in the liquid is 3, (Al), and the lowest Si concentration of the liquidus line lays on the straight line connecting Si with x, Al = 0.67. By evaluation of the Al-Au phase diagram Hoch [32] also calculated the enthalpies of formation of

Fig. 6.22
Enthalpies of mixing at 973 K of the system Bi-Cd-Ga-In-Pb-Sn-Zn. ●●● experimental; — calculated

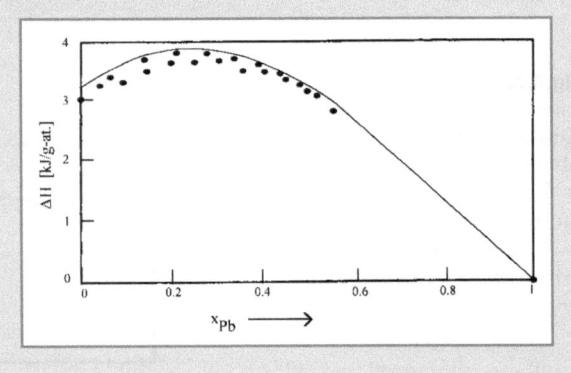

Fig. 6.23
The silicon-liquidus at 1373 K in the ternary system Al-Au-Si. Al-Au interaction parameter W is 3, (Al). ■■■ calculated: W (Au-Al) = -3.144 kK, ♦ ♦ ♦ calculated: W (Au-Al) = 0, × Lozovskiy et al. [33], experimental

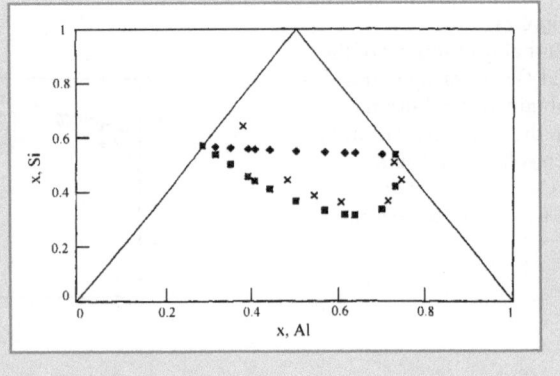

the binary Al-Au intermediate compounds which are shown in Table 6.5, together with the data of Kubaschewski and Alcock [34].

Using the Hoch-Arpshofen model, Hoch [19] treated the phase diagram $Rb-I_2$ (Fig. 6.24). It has been determined by [35] and can be separated in two parts $Rb - RbI$ and $RbI-I_2$. Thermodynamic data are available from Kubaschewski et al. [34]. The partial diagram Rb-RbI shows a miscibility gap. Thermodynamic data arrived at from the miscibility gap on the one hand and from the RbI-liquid equilibria on the other hand are in good agreement.

In the $RbI-I_2$ part there exists a non-congruently melting compound RbI_3 and an eutectic between RbI_3 and I_2. From RbI-liquid and I_2-liquid equilibria Gibbs energy of formation of solid RbI_3 from solid RbI and I_2, has been calculated. The

Table 6.5
Enthalpy of formation of binary Al-Au compounds

Compound	$\Delta H_{form} \cdot R$ [34] [kK]	$\Delta H_{form} \cdot R$ phase diagram [kK]
Al_2Au	-5.066 ± 0.168	-4.726 ± 0.138
AlAu	-4.655 ± 0.252	-3.956 ± 0.125
$AlAu_2$	-4.194 ± 0.252	-3.029 ± 0.125
Al_2Au_5		-2.586 ± 0.086
$AlAu_4$		-1.781 ± 0.064

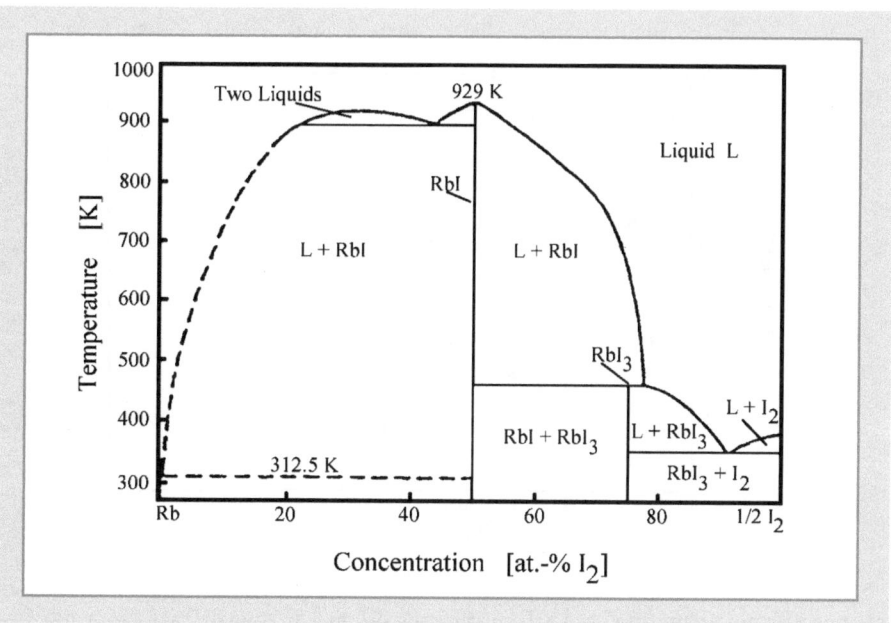

Fig. 6.24 Phase diagram Rb-I (sub-systems: $Rb - RbI$ and $RbI - I_2$) [35, 36]

thermodynamic activities of the components of the Rb-I_2 system are plotted in Fig. 6.25. Activities of Rb and I_2 are very low, as can be seen in Fig. 6.26.

On the basis of Hoch-Arpshofen model the thermodynamic data of the quaternary system $CaO - Al_2O_3 - MgO - SiO_2$ has been discussed by Hoch [37]. Some results are summarized in Table 6.6 and compared there with the results compiled by Kubaschewski et al. [34]. The agreement between the data of the two different sources [34] and [37] are rather good.

The method has also been applied to aqueous solutions of inorganic salts [39, 40]. In contrast to the accepted method [41] of representing aqueous solutions as solutions of ions, we treat the solubility and activity of molecules. The references [42, 43] give the activity coefficient of the ions as a function of the molarity. Figure 6.27 shows the activity coefficients of $CaCl_2$ as a function of mole fraction.

In this binary solution the "straight line" is a regular solution. At very low concentrations, where ionisation plays an important role, an additional term m, (H_2O) is needed where m is a coefficient with a value between 200 and 1,000 depending on the available experimental data. Here we have a complex made up of 199 moles of water, and one mole of $CaCl_2$, the latter being ionised, and wanting to spread out as much as it can. When treating ternary and larger solutions (two or more salts in water) we need the interaction between the salts. This is obtained from the solid binary phase diagrams given in [38].

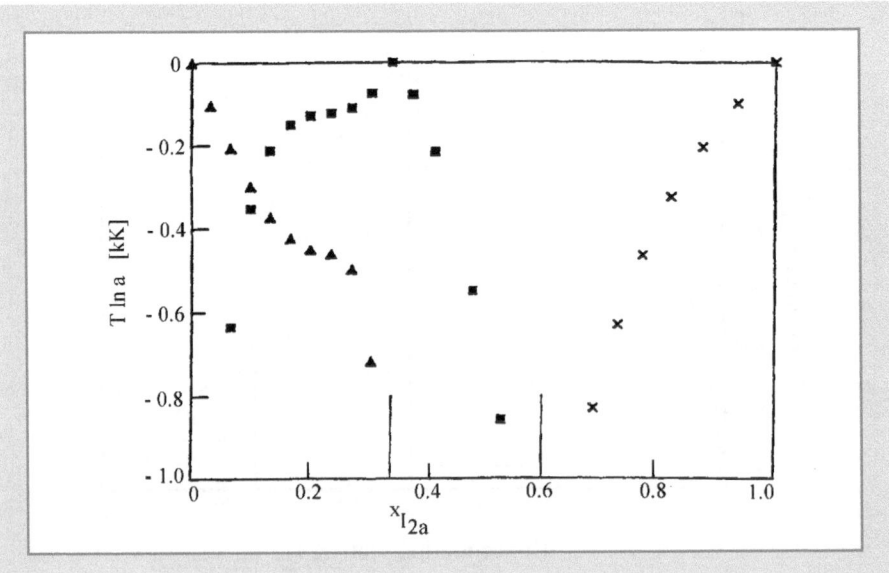

Fig. 6.25 Activity of Rb, RbI, and I_2 at 1000 K in the $Rb - I_2$ system. Compound RbI is at $x_{I_2} = 0.334$; RbI$_3$ is at $x_{I_2} = 0.6$. ■■■ RbI, ▲▲▲ Rb, ×××I_2

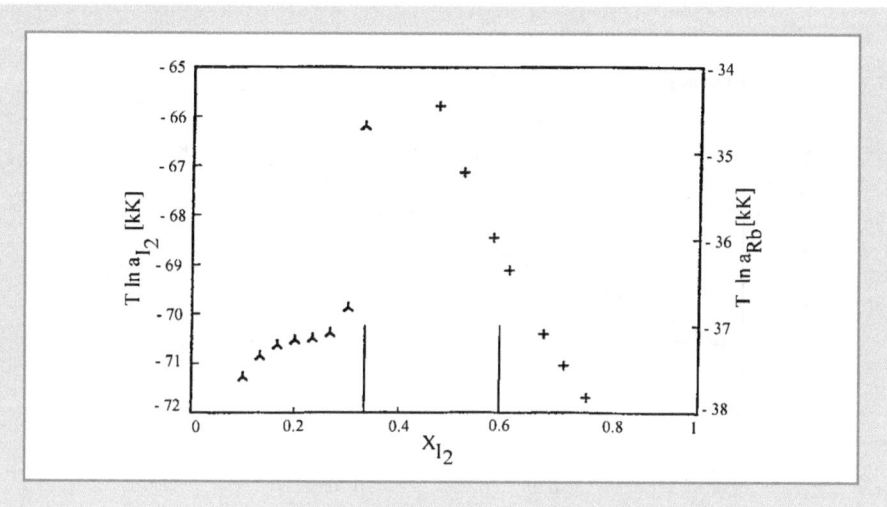

Fig. 6.26 Activity of Rb, and I_2 at 1000 K in the $Rb-I_2$ system. From the liquid equilibrium $Rb + \frac{1}{2} I_2 \rightarrow RbI \, \Delta G \, (RbI) = -355.5 = 20$ kK. Solid compounds are RbI at $x_{I_2} = 0.333$ and RbI_3 at $x_{I_2} = 0.6$. +++ Rb, ⋏ ⋏ ⋏ I_2

Table 6.6 Enthalpy of formation of ternary and quaternary system $Al_2O_3 - CaO - MgO - SiO_2$

Phase	Hoch-Arpshofen model [37] $\Delta H_{form} \cdot R$ in kK	Kubaschewski et al. [34] $\Delta H_{form} \cdot R$ in kK
Diopside	-2.282 ± 0.125	-4.595 ± 0.137
Monticellite	-2.147 ± 0.076	-4.723 ± 0.142
Akermanite	-3.221 ± 0.180	
Merwinite	-3.489 ± 0.086	-4.607 ± 0.091
Anorthite	-3.176 ± 0.161	-4.941 ± 0.116
Gehlenite	-4.012 ± 0.177	-2.980 ± 0.134
Cordierite	-1.451 ± 0.038	-4.117 ± 0.109
Saphirine	-1.876 ± 0.033	
Grossularite	-4.426 ± 0.194	-5.926 ± 0.123
Pyroxene		-3.031 ± 0.143

Figure 6.28 shows the saturation curves for the $Na_2SO_4 - NaCl - H_2O$ system at 298 K. The solubility of $Na_2SO_4 \cdot 10H_2O$ first decreases with the addition of NaCl. Then it increases, because the activity of H_2O decreases significantly. Finally the solubility of $Na_2SO_4 \cdot 10H_2O$ reaches that of Na_2SO_4, and Na_2SO_4 becomes the stable precipitate.

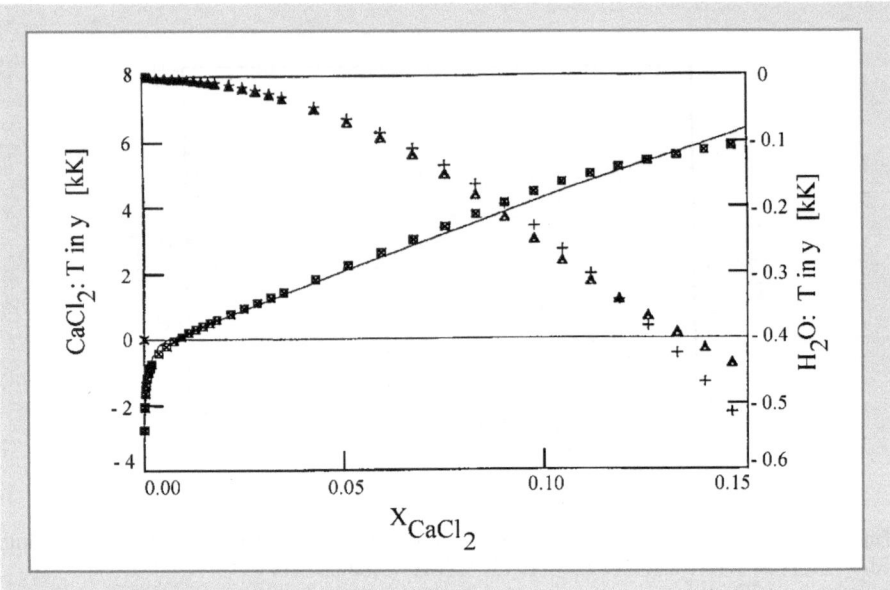

Fig. 6.27 Activity coefficient T ln y of CaCl$_2$ and H$_2$O at 298 K. CaCl$_2$: standard state infinite dilution. ■■■ CaCl$_2$: litt, — CaCl$_2$: calculated, ▲▲▲ H$_2$O: litt. +++ H$_2$O: calculated, H$_2$O: calculated with coefficients obtained with CaCl$_2$

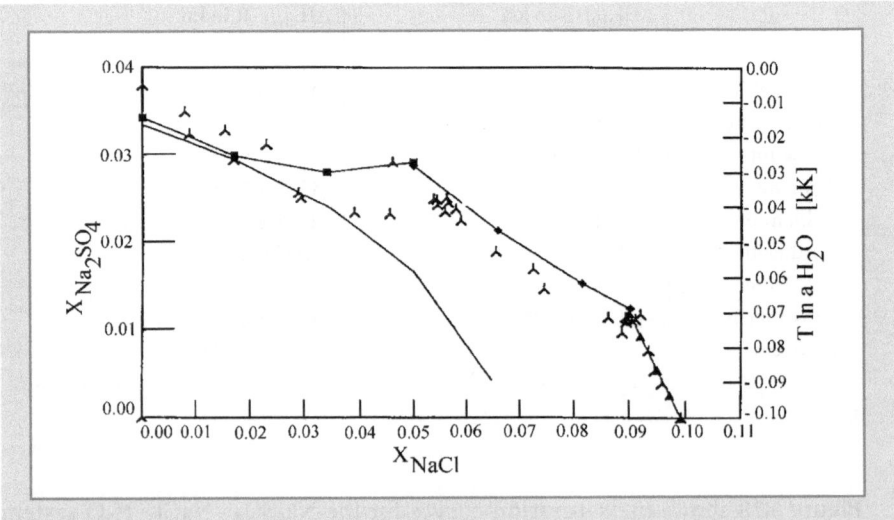

Fig. 6.28 Saturation curves for Na$_2$SO$_4$–NaCl-H$_2$O at 298 K. T ln a H$_2$O (with Na$_2$SO$_4$ · 10H$_2$O) is also plotted. ■■■Na$_2$SO$_4$ · 10H$_2$O, ♦♦♦ N$_2$SO$_4$, ▲▲▲ NaCL, — T ln a H$_2$O, ⅄⅄⅄ Linke [44], 1/10 Na$_2$SO$_4$ + H$_2$O → 1/10 Na$_2$SO$_4$ · 10H$_2$O, Handbook: T ln a H$_2$O = –0.063 ± 0.101 kK; Linke: T ln a H$_2$O = –0.058 kK

6.14
The Thermodynamic Activity

The measure of the deviation of a real solution from the behavior of an ideal solution is often expressed by the thermodynamic activity. In general:

$$\overline{\Delta G}_A = \overline{\Delta H}_A - T \cdot \overline{\Delta S}_A \tag{6.78}$$

In the case of an ideal solution $\overline{\Delta H}_A = 0$ and $\overline{\Delta S}_A = \overline{\Delta S}_{A(i)}$, from which with Eq. (6.35) follows

$$\overline{\Delta G}_{A(i)} = -T \cdot \overline{\Delta S}_{A(i)} = R \cdot T \cdot \ln x_A \tag{6.79}$$

In a formal fashion the deviation of a real solution from an ideal solution of identical composition can be expressed by replacing the atom fraction x_A with an "effective concentration", a_A, of component A, the so-called thermodynamic activity of A:

$$\Delta G_A = R \cdot T \cdot \ln a_A \tag{6.80}$$

Consideration of the vaporization equilibrium of the pure component A and of the component A from the A-B solution, shows that the thermodynamic activity depends, in the following fashion, on the partial pressure p_A of component A above the solution A-B and the vapor pressure p_A^0 of the pure component A at the same temperature:

$$a_A = \frac{p_A}{p_A^0} \tag{6.81}$$

A plot of a_A against x_A at constant temperature yields activity isotherms. For the ideal solution $a_A = x_A$. The corresponding activity isotherm is a straight line (Raoult's line), Raoult's law (see Fig. 6.11). In real systems the activity isotherms deviate from Raoult's law. The deviation can be positive ($a > x$) or negative ($a < x$). In the case of positive deviations the partial pressure p_A of component A is greater than the partial pressure $p_{A(i)}$ of an ideal solution at identical T and x. The atoms vaporize easier from the real solution, than from an ideal solution with the same temperature and composition. This means that the interatomic interactions are weaker in this real solution than in an ideal solution with the same state variables. For an ideal solution, where, according to its definition, during formation no change in the interaction parameters occur, and thus, a=x, we have:

$$w_{AB} = \frac{w_{AA} + w_{BB}}{2} \tag{6.82}$$

In the presence of a solution with $a_A > x_A$ accordingly

$$|w_{AB}| < \frac{|w_{AA}| + |w_{BB}|}{2} \tag{6.83}$$

The probability that an A – B pair is found at a certain place in the solution is less than can be expected in an ideal solution with completely random distribution of atoms. On the other hand, the probability is greater to find A – A and B – B pairs than in an ideal solution of identical composition. Such a solution indicates a tendency towards demixing.

On the other hand, if, during solution formation, an increase in interatomic interactions occurs, which means that

$$|w_{AB}| > \frac{|w_{AA}| + |w_{BB}|}{2} \tag{6.84}$$

and thus, the probability to find an A – B pair in such a solution is greater, than in an ideal solution with random distribution of atoms. A solution with tendency to compound formation is present. One has a < x.

As an example, Fig. 6.29a shows an activity isotherm of the components of liquid alloys of the system copper-lead. The deviation from Raoult's law is positive. In Fig. 6.29b the activity isotherms with negative deviation from Raoult's law are presented from the liquid alloys in the Cu – Al system.

The tendency towards demixing resp. compound formation in these systems can immediately be demonstrated based on the phase diagrams. In the copper-lead system a miscibility gap occurs in the liquid phase < 1253 K. On the other hand, in the system copper-aluminum compounds are present in the solid state. It must be emphasized, that the tendency towards demixing or compound formation present in the liquid based on thermodynamic activities, does not always lead to a miscibility gap or compound formation in the solid. For this to occur, in addition to the tendencies appearing in the liquid, other factors are of importance.

The composition dependence of the thermodynamic activities is governed by limiting laws for $x_A \to 0$, and $x_A \to 1$. For $x_A \to 1$, a_A follows Raoult's law, as can be easily shown and immediately observed in Fig. 6.29. For solutions with infinite dilution in B (dissolved substance), Raoult's law is valid for A (solvent):

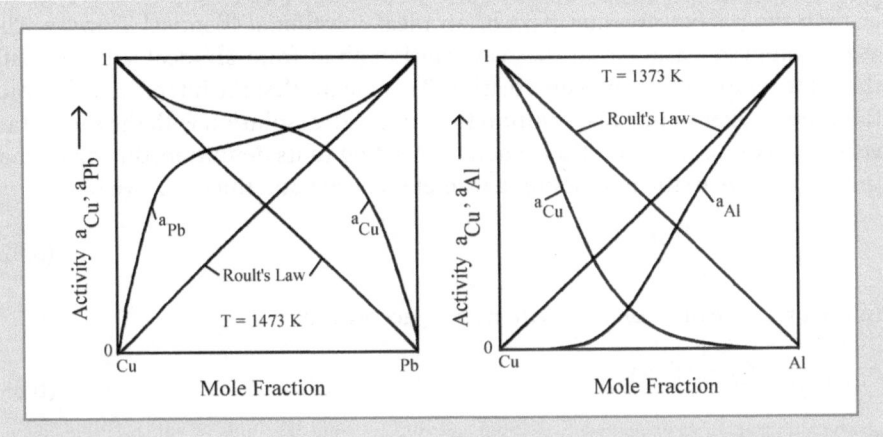

Fig. 6.29 Activity isotherms for liquid alloys of the system Cu-Pb **(a)** and Cu-Al **(b)** (after [1b])

$$a_A = x_A \tag{6.85}$$

For $x_A \rightarrow 0$, Henry's law is valid for component A:

$$a_A = C \cdot x_A \tag{6.86}$$

C is a constant, whose value depends on the change of the interatomic interactions during solution formation. At small atom fractions of the solvent A, a_A varies linearly with the atomic fraction. For an ideal solution $C = 1$. In this case Henry's law and Raoult's law are identical.

The activity coefficient f_A of component A in a solution A – B is defined as:

$$f_A = \frac{a_A}{x_A} \tag{6.87}$$

Naturally, for an ideal solution $f_A = 1$. The activity coefficient is a measure for the deviation of a real solution from ideal behavior. It is connected with the partial Gibbs energy of mixing (see Eqs. (6.79) and (6.80)):

$$\begin{aligned}
\overline{\Delta G}_A &= R \cdot T \cdot \ln a_A = R \cdot T \cdot \ln (x_A \cdot f_A) \\
&= R \cdot T \cdot \ln x_A + R \cdot T \cdot \ln f_A \\
&= \overline{\Delta G}_{A(i)} + \overline{\Delta G}_A^{ex}
\end{aligned} \tag{6.88}$$

$$\overline{\Delta G}_A^{ex} = R \cdot T \cdot \ln f_A \tag{6.89}$$

The activity coefficients of the two components of a binary solution are connected by the Duhem-Margules equation. In this case, it is

$$\ln f_A = - \int\limits_{\ln f_B \text{ at } x_B = 0}^{\ln f_B \text{ at } x_B = x_B} \frac{x_B}{x_A} \cdot d\ln f_B \tag{6.90}$$

The Gibbs energy G of a substance is temperature dependent, and if S is the entropy of the substance, one has

$$\frac{\partial G}{\partial T} = -S \tag{6.91}$$

The same is valid for the Gibbs energy of mixing, for example

$$\frac{\partial \overline{\Delta G}_A}{\partial T} = -\overline{\Delta S}_A \tag{6.92}$$

By consideration of Eq. (6.78) the following relation can be derived

$$\frac{\overline{\Delta H}_A}{R} = \frac{\partial \ln a_A}{\partial \left(\dfrac{1}{T} \right)} \tag{6.93}$$

A plot of $\ln a_A$ versus $1/T$ at constant composition yields, in the simplest case, a straight line the slope of which can be used to calculate the partial enthalpy of mixing of the component in question. Figure 6.30 represents a $\ln a_A$-1/T plot of the thermodynamic activity of gallium in liquid gallium-cadmium solutions. Straight lines result over the whole composition range. If, as it can occure in some real solutions, the nature of binding between atoms changes with variation of temperature. Thus deviation from Eq. (6.93) is possible. The consequence is curvature of the $\ln a_A$-1/T lines. ΔH is then temperature dependent.

As one can conclude from Fig. 6.30, and is immediately visible in Fig. 6.31, the activity isotherms approach Raoult's law with increasing temperature. With in-

Fig. 6.30
Thermodynamic activities of Ga in liquid Ga-Cd alloys in dependence on temperature (after [16])

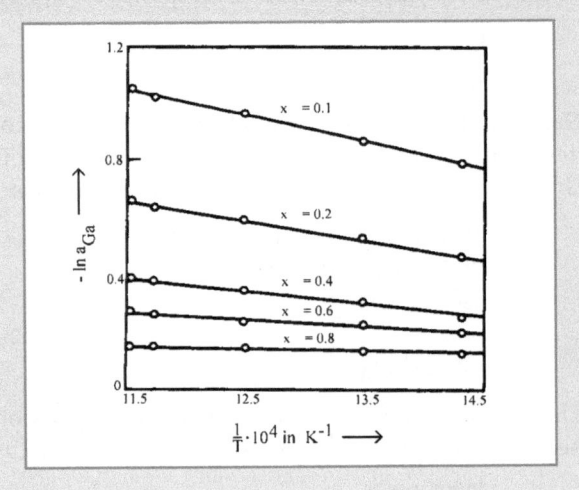

Fig. 6.31
Thermodynamic activities of Ga in liquid Ga-Cd alloys as a function of atomic fraction at different temperatures (after [16])

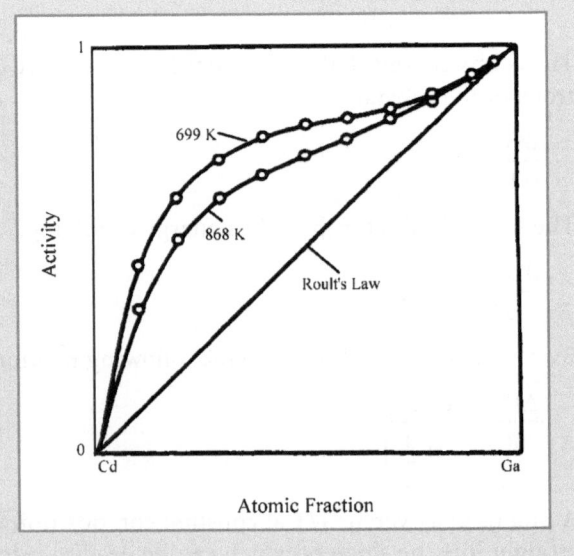

creasing temperature atomic distances in the solution increase, and the interatomic interactions become weaker. The individual binding effects between dissimilar atoms are reduced. The solution approaches ideal behavior.

6.15
General Comments about Experimental Methods to Determine Thermodynamic Mixing Properties

Thermodynamic mixing properties are directly connected to the interatomic
interactions and to the arrangement of the atoms in a solution. They can therefore, yield useful information about the structure of the solid solutions. In addition, they fix phase equilibria. A general method to obtain thermodynamic mixing properties does not exist. To obtain sufficiently accurate basic data, special
investigative methods are needed, which depend on the individual peculiarities
of the system studied.

To obtain the change in the state properties during formation of a solution, the
starting and end states of this reaction must be set unequivocally, and must be attainable. The attainment of equilibrium conditions is tied to kinetic possibilities.
In systems with high melting components (silicates, salts, metals, etc) equilibrium
takes place only at temperatures way above room temperature with a sufficiently high reaction rate for the experiment in question. For the determination of the
thermodynamic properties of mixing of such systems methods have been developed, which can be applied at 1000 K and higher. In this case only a limited number
of construction materials are available to set up the experimental procedure.

6.16
The High-Temperature Calorimeter

The determination of enthalpies of mixing can be carried out directly. Figure 6.32 represents the essential parts of an isothermal high-temperature calorimeter, which can be used for the direct determination of enthalpies of mixing of liquid solutions.

The two metallic blocks, 1 and 2, are placed in a suitable furnace. They contain two crucibles 3 and 4, situated one above the other, and which contain the
liquid components to be mixed. Lifting the stopper 5 clears the opening in the
upper crucible 4. The liquid from crucible 4 flows into crucible 3. A stirrer, 6,
guaranties thorough mixing.

To determine the change of heat during the mixing process, the temperature
difference ΔT between the liquid in crucible 3 and the metal block is plotted as
a function of time, using the two thermocouples 7 and 8.

A ΔT-time curve results, as shown schematically in Fig. 6.33. The mixing
process begins at $t = 0$, and is finished at $t = t_1$.

The temperature difference deviates from the basis line during the course of
the mixing process, but returns after the completion of the reaction ($t > t_1$) ac-

Fig. 6.32
Principle of the construction
of an isothermic high-temperature calorimeter for
direct determination of
enthalpies of mixing.
1, 2 massive metallic block,
3, 4 crucible, *5* thermocouple,
protection tube of which is
used as a stopper, *6* stirrer,
7, 8 thermocouples,
9 heating coil for calibration,
10 heat isolation

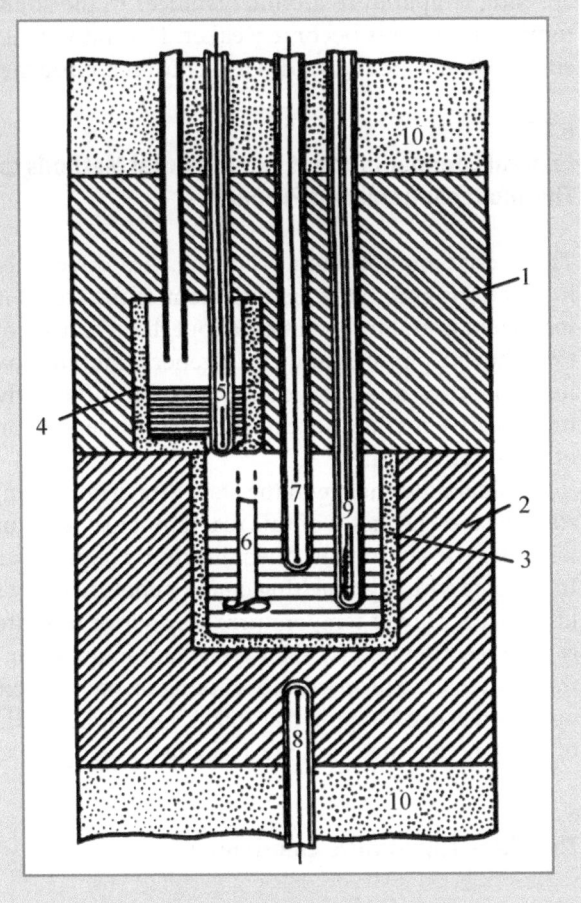

Fig. 6.33
Temperature-time diagram
of a mixing process in an
isotherm calorimeter

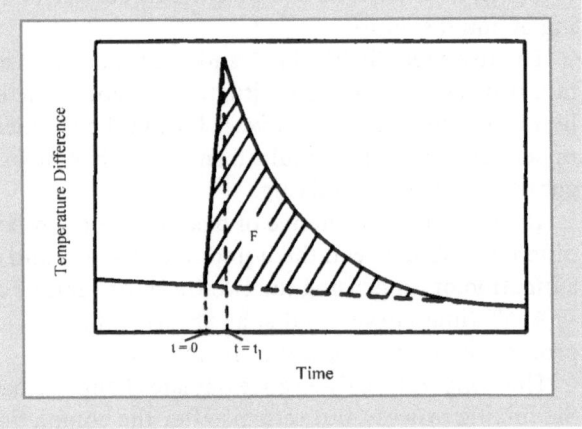

cording to Newton's temperature compensation law. The surface F between the real ΔT-time curve and the (interpolated) basis line is a measure for the change in enthalpy during formation of the solution. The calibration of the calorimeter is carried out by imitating such a thermal effect with accurately fixed Joule heating, using the heating spiral 9 in the sample. A thermocouple placed in the stopper 5 and thermocouple 7 serve to control the temperature of the liquids before mixing. The evaluation of the measurement results is naturally very simple, if the components are at the same temperature before the mixing.

The enthalpy of mixing of solid solutions can also be determined by measuring the heat effect during direct mixing of the components. For that purpose the solid components are finely pulverized, mixed and pressed into pellets. The pellets are heated in the calorimeter. When the temperature, where rapid diffusion occurs, is reached the solid solution can be formed within a short time, within which the change in enthalpy of the system occurs. It is registered by the calorimeter.

Indirect methods for the determination of enthalpies of mixing and formation have been proven to be useful. To these belong the so-called solution calorimetry. The solid solution is introduced into a calorimeter, which contains a solvent, in which the sample to be studied rapidly dissolves. In a second experiment the solution process is repeated under identical conditions, with the pure components. The difference between the two enthalpies of solution represents the enthalpy of formation of the solid solution phase.

6.17
Partial Vapor Pressure Measurements

Partial Gibbs energies of mixing can be obtained according to Eqs. (6.80) and (6.81) from partial pressure measurements. Numerous methods have been developed to determine vapor pressures. Two examples for the determination of very low partial vapor pressures at high temperatures are given below.

Figure 6.34 represents schematically the experimental arrangement for the so-called dew-point method. A solution, A-B, is enclosed at one end of an evacuated, transparent tube, 1. Let component A have a significantly higher vapor pressure than B. At the temperature T_1 practically only A vaporizes.

If the temperature at the end 2 of the tube is lowered continuously, condensation of A vapor occurs finally, when the vapor pressure p_A^0 of the pure substance

Fig. 6.34
Dew-point method to determine thermodynamic activities

A at T_2 is equal to the vapor pressure p_A above the alloy A – B at T_1. Knowledge of the p_A^0 – T curve and measurements at T_1 and T_2 permits one to obtain the thermodynamic activity of the more volatile A.

The principle of the dew-point method was proposed by H. Lescoeur (1889). In the mean time many variations of this method became known.

In the effusion method, developed by M. Knudsen (1909), a substance L is located in a container, equipped with a small opening with a very thin edge (see Fig. 6.35). The container is placed in a vacuum. If the surface L of the sample is large compared to the surface of the hole F, then at the given temperature the equilibrium vapor pressure of the substance investigated develops in the container. The vapor effuses through the opening. The number of moles Δn which passes, within a time Δt, through the opening is given by the kinetic theory of gases as:

$$\Delta n = \frac{P \cdot F \cdot \Delta t}{\sqrt{\dfrac{2\pi RT}{M}}} \tag{6.94}$$

P is the total vapor pressure above the sample and M is the molecular weight of the particles in the vapor.

The number of moles Δn of vapor effusing during time Δt, required to calculate the vapor pressure P, can be determined using different methods. In the simplest case the container is weighed before and after the experiment, and the amount Δn determined from the difference in weight. More accurate is a condensation of the effusing gas on a cold plate in front of the effusion opening. In addition to the total vapor pressure, the partial pressures of the components can be obtained by marking the substance with radioactive tracers, by chemical-analytical determination of the concentration of the components, by mass-spectrometric determination of the total mass as well as the mass of each component effusing per unit time. Especially, the use of a mass-spectrometer is of ad-

Fig. 6.35
Knudsen-cell for determination of vapor pressure on the basis of the effusion-method.
L Liquid, D Vapor

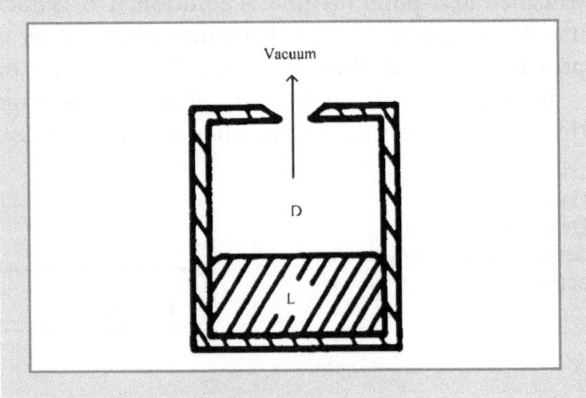

vantage, as a control of the species present in the vapor is immediately possible. In some cases associations occur in the vapor, which influences M in Eq. (6.94), and thus, must be considered. The Knudsen effusion method has proved itself especially for the accurate determination of low vapor pressures.

6.18
Activity Determination from the EMF of Galvanic Cells

In a galvanic cell

$$(-) \,|\, A \,|\, A^{z+} - \text{solution} \,|\, \text{alloy A-B} \,|\, (+) \tag{6.95}$$

whose electrodes consists of an alloy A – B and the less noble component, and whose electrolyte contain ions of the less noble component, the electromotive force (EMF) is directly related to the partial Gibbs energy of mixing of component A in the alloy A – B. With the isothermal and reversible transfer of 1 g-at. A from the electrode |A| to the electrode |A-B|, the partial Gibbs energy of mixing $\overline{\Delta G}_A$ at the composition of the alloy |A-B| is obtained, if no change in composition occurs at the alloy electrode. The condition of constant composition at the transfer is guarantied, because E is measured practically with no current flowing in the cell, which means that no perceivable transfer occurs.

$$\overline{\Delta G}_A = - z \cdot \Phi \cdot E \tag{6.96}$$

Φ is the Faraday constant, and z the valence of the A ions in the electrolyte.

Figure 6.36 shows schematically a simple, often used arrangement of electrodes and the electrolyte. Mostly molten salts are used as electrolyte. It suffices

Fig. 6.36
Determination of thermo-dynamic activities using the EMF-method (schematic). *1, 2* electrodes, *3* electrolyte, *4, 5* wires to contact the electrodes

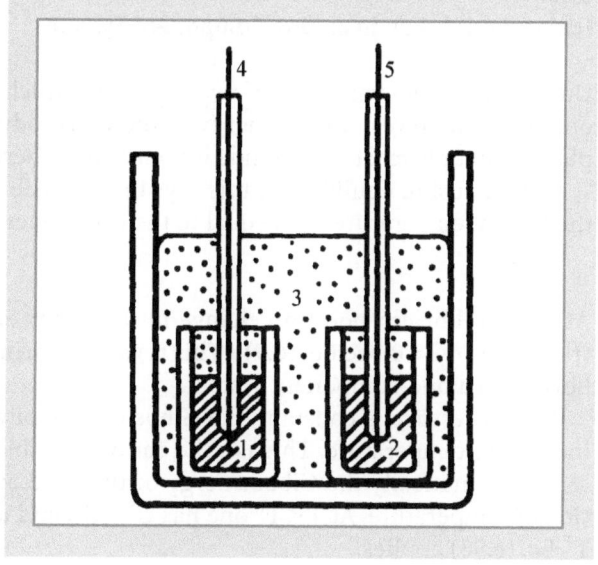

to use a low melting base mixture of inert salts, for example KCl + LiCl, to which a few percent of a salt of the less noble component of the alloy to be investigated are added. Oxygen ions conducting salts, for example solid solution of ThO_2 and Y_2O_3 or ZrO_2 and CaO, are used with great success in the investigation of solid solutions.

The thermodynamic activity of the less noble component can be obtained from the EMF measurements using Eqs. (6.80) and (6.96). Most vapor pressure procedures yield merely the activity of one component. The activity of the other component can be obtained through the Duhem-Margules equation. Based on the relation

$$\Delta G^{mix} = x_A \cdot \overline{\Delta G_A} + x_B \cdot \overline{\Delta G_B} \tag{6.97}$$

the integral Gibbs energy of mixing is also accessible.

The partial enthalpies of mixing can also be obtained from the temperature dependence of the activities, according to Eq. (6.93), which, together with Eq. (6.29), yield the integral quantity, ΔH. Finally, with the help of the Helmholtz-Gibbs equation, the integral entropy of mixing ΔS is obtained. All quantities necessary for the thermodynamic description of a system can be obtained from the knowledge of the thermodynamic activities as functions of temperature and composition.

The value of the experimentally determined temperature dependence of the activity, da_A/dT, naturally has a larger relative error than the activity itself. As a rule a greater accuracy of the integral mixing properties can be expected, when ΔH is measured directly calorimetrically, $\overline{\Delta G}$ obtained from activity measurements, and the entropy of mixing ΔS calculated from these two.

6.19
Fusion Equilibrium in an One-Component System

Under given state variables, of two phases in which a substance can occur, the one with the lowest Gibbs energy is the thermodynamically stable one. Two phases can only exist side by side if their Gibbs energies are equal.

For the fusion equilibrium this says that the Gibbs energy G^S of the solid at the fusion temperature T^F is equal to the Gibbs energy G^L of the liquid:

$$G^S = G^L \tag{6.98}$$

At the equilibrium point no change in the Gibbs energy occurs during phase transition ($\Delta G = 0$, see Eq. (6.9)). This is valid, in principle, for a two-phase equilibrium in a one-component system.

If the pressure is constant, the Gibbs energy of a substance decreases with increasing temperature. This is valid for all possible phases. Figure 6.37 represents schematically the Gibbs energy of the solid and liquid phases as a function of temperature. At $T < T^F$ one has $G^S < G^L$, and at $T > T^F$ one has $G^S > G^L$. At T^F Eq. (6.98) applies.

Fig. 6.37
Gibbs energy G^L of the liquid and Gibbs energy G^S of the solid phase as a function of temperature (schematic) T^F = melting temperature

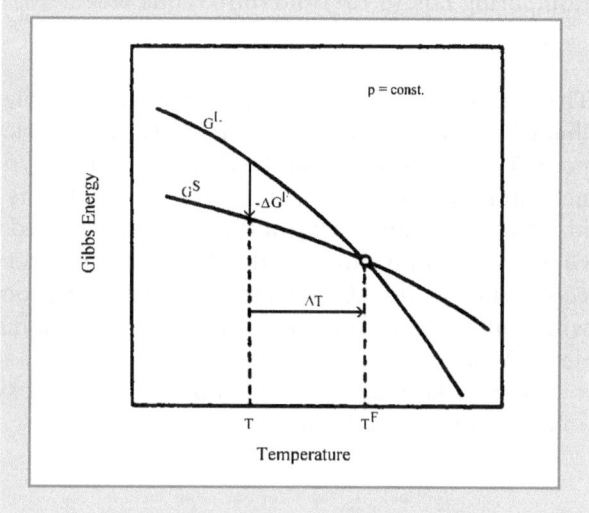

The temperature dependence of both phases is given, according to Eq. (6.91), by

$$\frac{\partial G^S}{\partial T} = -S^S \tag{6.99}$$

$$\frac{\partial G^L}{\partial T} = -S^L \tag{6.100}$$

The entropy S^S of the solid phase is fixed at a temperature T by its heat capacity C_p^S:

$$S_{(T)}^S = \int_0^T \frac{C_p^S}{T} dT \tag{6.101}$$

With the transition of the solid into the liquid phase at the melting point, an increase of the enthalpy of the system occured. This is the latent enthalpy of fusion, ΔH^F. This is conditioned by the bonding and structural differences of the two phases. In an analogous fashion, an increase in entropy of the system occurs during fusion. The entropy of fusion ΔS^F is given by:

$$\Delta S^F = \frac{\Delta H^F}{T^F} \tag{6.102}$$

Thus, the entropy of the liquid phase at temperature T is:

$$S_{(T)}^L = \int_0^{T^F} \frac{C_p^S}{T} dT + \Delta S^F + \int_{T^F}^T \frac{C_p^L}{T} dT \tag{6.103}$$

Comparing Eqs. (6.101) and (6.103) one sees:

$$S^L > S^S \qquad\qquad (6.104)$$

The slope of the G^L-T curve is, at the melting point, fundamentally greater than that of the $G^S - T$ curve. If the liquid is undercooled by ΔT under the melting point T^F, then $G^L > G^S$ and the liquid is thermodynamically unstable. If no kinetic hindrances are present, the liquid transforms spontaneously into the solid. Thus, the Gibbs energy of the system is reduced by $\Delta G^F = G^L - G^S$. ΔG^F is the Gibbs energy of fusion. It is the driving force for crystallization. Because at the equilibrium point $\Delta G^F = 0$, the crystallization proceeds infinitely slowly. Only with finite undercooling can a finite crystallization rate be obtained, which also depends on other factors.

Similar relationships occur in one-component systems in the case of solid-solid two-phase equilibria.

6.20
Fusion Equilibria in Binary Systems

For two-component systems the equilibrium condition (6.98) is only valid in special cases, when the liquid and solid solution phases in equilibrium have the same composition. This is the case, for example, at the melting point maximum of a congruently melting compound, or at the melting point maximum or melting point minimum of solid solutions.

In general, for the fusion equilibrium in a two-component system, the so-called tangent condition applies:

$$\left(\frac{dG^L}{dx_B}\right)_{x_B^L} = \left(\frac{dG^S}{dx_B}\right)_{x_B^S} \qquad\qquad (6.105)$$

This expresses the condition for the minimum Gibbs energy of a mixture of two phases with different compositions.

To illustrate Eq. (6.105): Fig. 6.38 represents the equilibrium between liquid and solid in a binary system with complete solubility at a temperature $T_A > T_1 > T_B$. Because $T_A > T_1$, one has $G_A^L > G_A^S$.

On the other hand, for component B, $G_B^L < G_B^S$ because $T_B < T_1$. The $G^L - x$ curve, starting from G_A^L and G_B^L, and the $G^S - x$ curve, starting from G_A^S and G_B^S, intersect in Q. This point, where $x_{B(Q)}^S = x_{B(Q)}^L$ and $G_Q^L = G_Q^S$, has no special significance. The Gibbs energy of the mixture of these phases with $x_{B(Q)}^S$ and $x_{B(Q)}^L$ can be lowered by the amount δG to G_Q, by moving the composition of the solid phase to x_B^S, and that of the liquid phase to x_B^L. The points x and y at these concentrations on the $G^L - x$ and $G^S - x$ curves, correspond to the contact points of a common tangent. The first derivatives of G^L and G^S with respect to the composition must have the same value at the compositions of the contact points (see

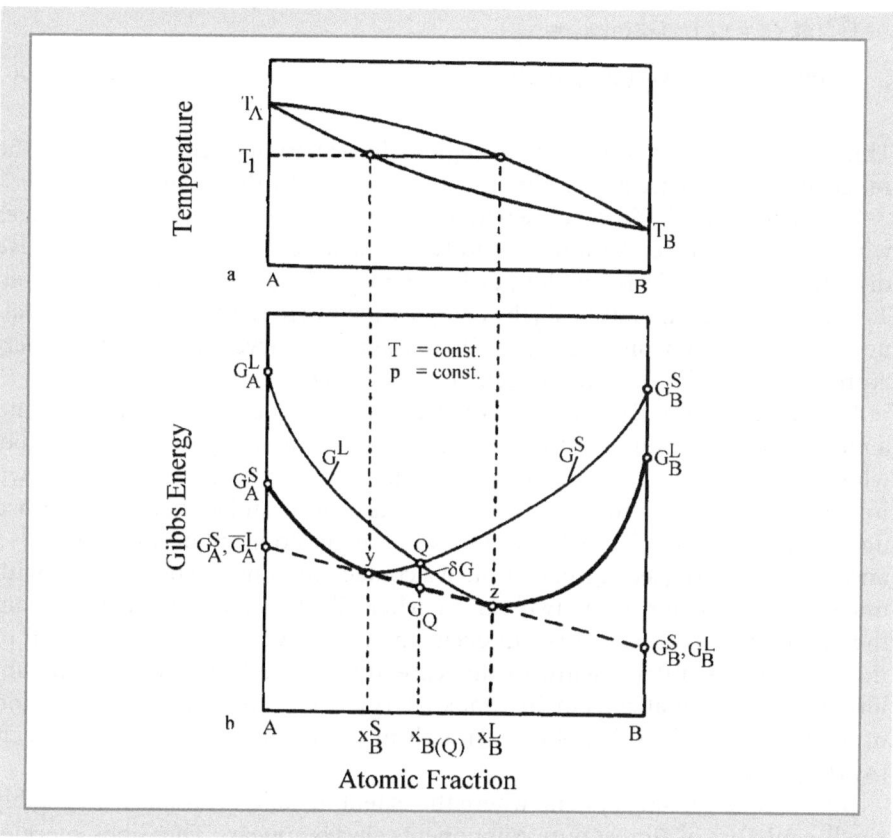

Fig. 6.38 For explanation of the condition of tangent. **a** Phase diagram with uninterrupted formation of solid solutions, **b** Gibbs energy of the liquid and the solid phase as a function of atomic fraction

Eq. (6.105)). The ratio of the amounts of the two phases in the mixture with the overall composition $x_{B(Q)}$ is given by the lever rule. The compositions x_B^S and x_B^L are the solidus resp. liquidus compositions at T_1 (see Fig. 6.38a).

Solutions with $x_B < x_B^S$ are solid, because here $G^L > G^S$ (see Fig. 6.38b). Similarly mixtures with $x_B > x_B^L$ at T_1 are single phase and liquid, because here $G^L < G^S$. In the region $x_B^S < x_B < x_B^L$ a mixture of solid and liquid phases have a lower Gibbs energy than a solid phase or a liquid phase of identical composition. Here solid-liquid equilibria are present (see Fig. 6.38a). The Gibbs energies of all A-B alloys in equilibrium at T_1 lay on the line $G_A^S - y - z - G_B^L$ (drawn heavy in Fig. 6.38b).

The condition of the common tangent for the equilibrium in a two-component system, is, as is immediately apparent from Fig. 6.38b, equivalent to the requirement that the partial Gibbs energies be equal in the solid and liquid state:

$$\overline{G}_A^S(\text{at } x_B = x_B^S) = \overline{G}_A^L(\text{at } x_B = x_B^L)$$

$$\overline{G}_B^S(\text{at } x_B = x_B^S) = \overline{G}_B^L(\text{at } x_B = x_B^L) \qquad (6.106)$$

This is also valid for all two phase regions in two-component systems. In the range of the two phase region the G-x curve is given by the mixing rule.

The position of the G^L-x curve with respect to that of the G^S-x curve, changes with the temperature. For a temperature T_2, for which $T_A > T_2 > T_1$, the relative distance of G_A^S and G_A^L is smaller (at T_A, $G_A^S = G_A^L$). Through it, while going from T_1 to the higher temperature T_2, the composition range is displaced to lower values of x_B. Basically, from a series of $G^L - x$ and $G^S - x$ curves at different temperatures the whole fusion diagram can be constructed.

Figure 6.39 shows $G - x$ diagrams for a system with complete solubility and a melting point minimum. At the melting point minimum, points a, b, c, and d, which at T_1 still belong to two-phase equilibria, coincide at point m, as shown in Fig. 6.39a. Similarly the four contact points a', b', c' and d', belonging to two double tangents (see Fig. 6.39b), unite to a single point m' between the G^L-x and the G^S-x curve (see Fig. 6.39c). Such a contact point m' can be expected with uncomplicated forms of the two curves, when G^L more strongly deviates from the mixing rule than G^S. This can occur in systems where the atomic radii of the components differ greatly. In this case, during the formation of solid solutions a much smaller gain in Gibbs energy occurs than during formation of liquid solutions, so that always, at a given composition and temperature, is $|\Delta G_S| < |\Delta G_L|$.

In a simple eutectic system, where the extent of solid solutions is negligibly small, a mixture of almost pure components always appears. The Gibbs energies of these mixtures are composed of the Gibbs energies of the components G_A^S and G_B^S, according to the lever rule (see Fig. 6.40b).

At a temperature of $T = T_1$ (see Fig. 6.40a) for which $T_A > T_1 > T_B$, the $G - x$ curves are represented schematically in Fig. 6.40b. Because $T_A > T_1$ and $T_B < T_1$, one has $G_A^L > G_A^S$ and $G_B^L < G_B^S$. The solid component B cannot appear at this temperature. From G_A^S a tangent is drawn to the $G_L - x$ curve. The composition of the contact point a' gives the composition of the liquid which is in equilibrium with the solid component at T_1. At the temperature T_2, for which $T_A > T_B > T_2$, solid B can also participate as a partner in a fusion equilibrium. Two regions of solid-liquid equilibria are present (see Fig. 6.40c). At the eutectic temperature the straight line $G_A^S - G_B^S$ itself is the tangent. It touches the $G_L - x$ curve at a point which corresponds to the composition of the eutectic. The phases connected by the tangent are in equilibrium: solid A, liquid with the composition c' and solid B. Below the eutectic temperature all points of the straight line $G_A^S - G_B^S$ lay below the $G_L - x$ curve. No liquid solution can appear. Only a mechanical mixture of A and B is thermodynamically stable in the whole composition range.

Figure 6.41 represents the fusion equilibria and the corresponding energetic relationships for a system with a congruently melting compound. In the solid,

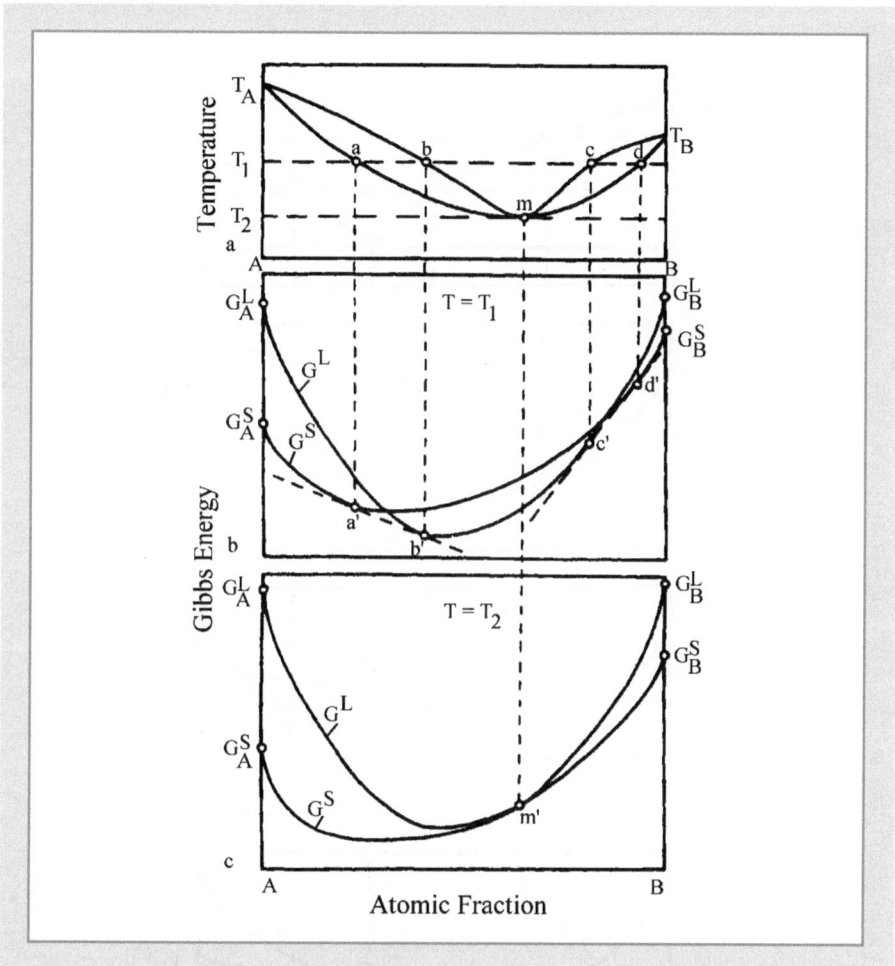

Fig. 6.39 Energetics of phase equilibria in a binary system with uninterrupted formation of solid solution and a melting point minimum

noticeable solid solution formation is present. At the temperature $T_\beta > T_1 > T_B$ no α- or β-solid solutions appear. The $G^\alpha - x$ and $G^\gamma - x$ curves lay above the $G_L - x$ curve. Meanwhile, the line of the Gibbs energy G^β of the congruently melting compound, β, passes through the $G_L - x$ curve at T_1, and at the middle composition ranges below the $G_L - x$ curve. Here, the single phase field of the compound appears between b and c (correspondingly b' and c'). At the temperature which corresponds to the temperature of the melting point maximum, the G^β-x curve touches the G_L-x curve.

Situations analogous to a component without solid solution formation exist for compounds, which exist only at a strictly stoichiometric composition (see

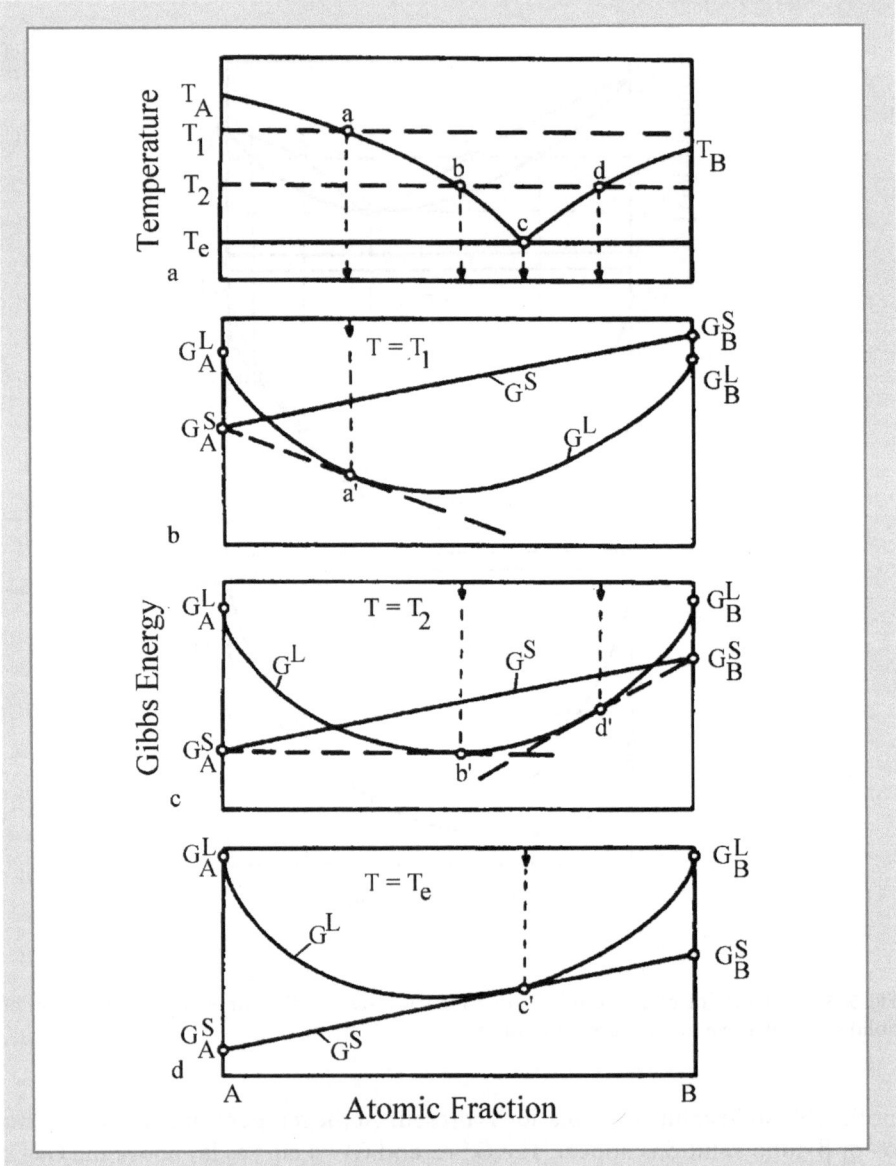

Fig. 6.40 Energetics of phase equilibria in a simple binary eutectic system with negligible small mutual solubility of components in the solid state

Fig. 6.40 for the case of a simple eutectic system). The Gibbs energy at T_2 (see Fig. 6.42) is $G^V (T_2)$. The two branches of the corresponding G^V – x curve degenerate into a vertical line. To determine the composition of the liquids which are in equilibrium with the compound V at T_2, one must lay tangents in Fig. 6.21

Fig. 6.41

Energetics of phase equilibria in a binary system with a congruently melting compound and a markedly formation of solid solution

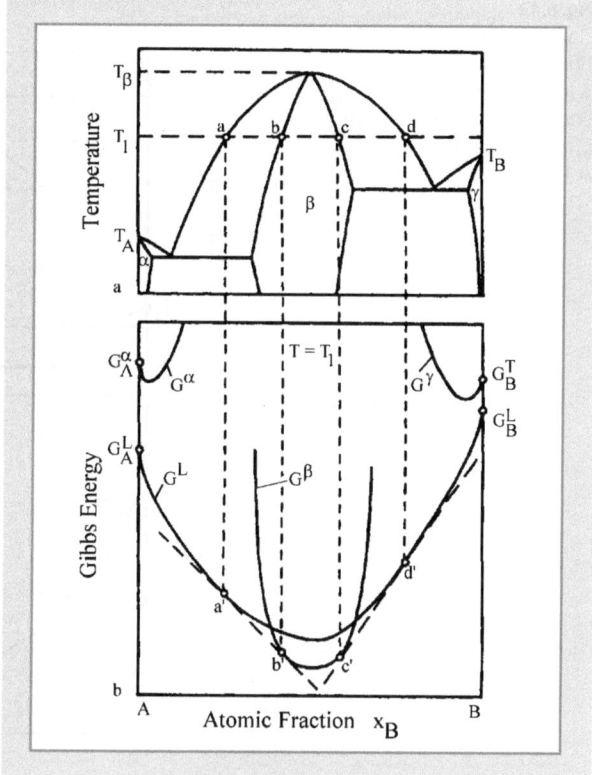

from $G^V (T_2)$ to the G_L-x curve. The contact points a' and b' correspond to the liquidus points a and b in Fig. 6.42a.

At the melting point T_V of the compound, $G^V (T_V)$ lays on the $G_L - x$ curve. At higher temperature ($T_1 > T_V$), $G^V (T_1)$ is naturally higher than G^L for the same composition. Trivially, at x_B^V only the liquid is present.

Fig. 6.42
Energetics of phase equi-
libria in a binary system
with a congruently melting
compound and with negli-
gible small solubility in the
solid state

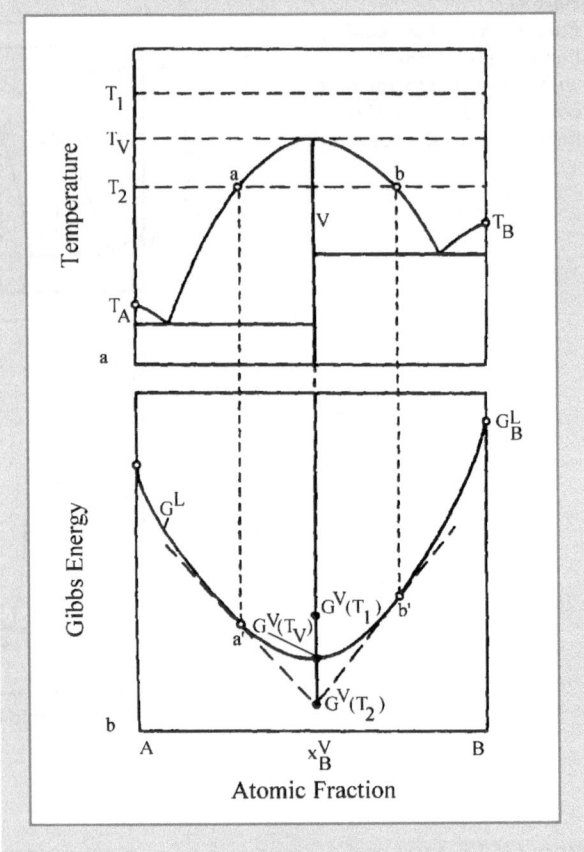

6.21
Equilibrium between a Binary Liquid and the Crystal
of one Component in the Ideal System

To understand the calculation of phase diagrams from thermodynamic data,
first the simplifying assumption of ideal behavior of the solutions is made.

The Gibbs energy G of an ideal solution with the mole fraction x is composed
from the partial Gibbs energies of the components as follows:

$$G = x_A \cdot \overline{G}_A + x_B \cdot \overline{G}_B \qquad (6.107)$$

In addition, for an ideal solution

$$G = x_A \cdot G_A + x_B \cdot G_B - T \cdot \Delta S_i \qquad (6.108)$$

G_A and G_B are the Gibbs energies of the components. With Eq. (6.11), (6.108)
yields

$$G = x_A \cdot G_A + x_B \cdot G_B + x_A \cdot R \cdot T \cdot \ln x_A + x_B \cdot R \cdot T \cdot \ln x_B \tag{6.109}$$

The comparison of Eq. (6.107) with Eq. (6.109) shows, that the partial Gibbs energies of the components are given by:

$$G_{A(i)} = G_A + R \cdot T \cdot \ln x_A$$
$$G_{B(i)} = G_B + R \cdot T \cdot \ln x_B \tag{6.110}$$

In the following the fusion equilibrium will be considered using component A as an example. In general, according to Eq. (6.106):

If a solid and liquid solution coexist at equilibrium, the partial Gibbs energy of component A in them can be expressed as follows:

$$\overline{G}_A^L = G_A^L + R \cdot T \cdot \ln x_A^L \tag{6.111}$$

and

$$\overline{G}_A^S = G_A^S + R \cdot T \cdot \ln x_A^S \tag{6.112}$$

Let us assume for the simplest case of a fusion equilibrium, that the solid phase is practically the pure component A. Thus, $x_A^S = 1$, and $RT \cdot \ln x_A^S = 0$. For this special case $\overline{G}_A^S = G_A^S$. From Eqs. (6.106) and (6.111) follows:

$$G_A^L + R \cdot T \cdot \ln x_A^L = G_A^S \tag{6.113a}$$

Transforming:

$$R \cdot T \cdot \ln x_A^L = G_A^S - G_A^L = -\Delta G_A^F \tag{6.113b}$$

The left side of Eq. (6.113a) contains the state variables x_A^L and T, needed to represent the liquidus line. The right side represents the Gibbs energy of solidification of component A, ΔG_A^F (see Fig. 6.37). For the latter

$$\Delta G_A^F = \Delta H_A^F - T \cdot \Delta S_A^F \tag{6.114}$$

Applying equation (6.102) to component A it yields

$$\Delta G_A^F = \Delta T \cdot \Delta S_A^F \tag{6.115}$$

with

$$\Delta T = T_A - T \tag{6.116}$$

Combining Eqs. (6.113b), (6.102) and (6.115) results in

$$RT \cdot \ln x_A^L = -\frac{\Delta H_A^F}{T_A} \cdot (T_A - T) \tag{6.117}$$

For liquids, which contain only small amounts of B, approximately

$$\ln x_A^L = \ln (1 - x_B^L) \approx -x_B^L \tag{6.118}$$

applies.

Equation (6.118) can therefore be simplified for the case of infinite dilution:

$$x_B^L = \frac{\Delta H_A^F \cdot (T_A - T)}{R \cdot T_A \cdot T} \tag{6.119}$$

This is Raoult's law, valid for the melting point lowering for dilute, ideal solutions.

An analogous relation to Eq. (6.117) applies to component B:

$$RT \cdot \ln x_B^L = -\frac{\Delta H_B^F}{T_B} \cdot (T_B - T) \tag{6.120}$$

If the enthalpies of fusion ΔH_A^F and ΔH_B^F and the temperatures of fusion T_A and T_B are known, the liquidus lines can be calculated with the assumptions; the solution is ideal, the enthalpies of fusion are temperature independent, and solid solution formation is negligible.

As an example, Fig. 6.43 represents the phase diagram cadmium-bismuth. The calculated liquidus points do not deviate significantly from the experimentally determined liquidus lines.

The assumptions for the application of Eqs. (6.117) and (6.118) to the system silver-copper do not fit so well. As is shown in Fig. 6.44, calculations yield liquidus temperatures which are much too low. The eutectic temperature obtained from the calculated liquidus lines is of the order of magnitude 775 K, whereas the experimentally determined one is at 1052 K. Among others, the discrepancy is also caused by the neglect of solid solution formation, which is of noticeable extent in this system.

Fig. 6.43
Phase diagram Cd-Bi; — experimental after M. Hansen and K. Anderko [45], ---- calculated after Eq. (6.117)

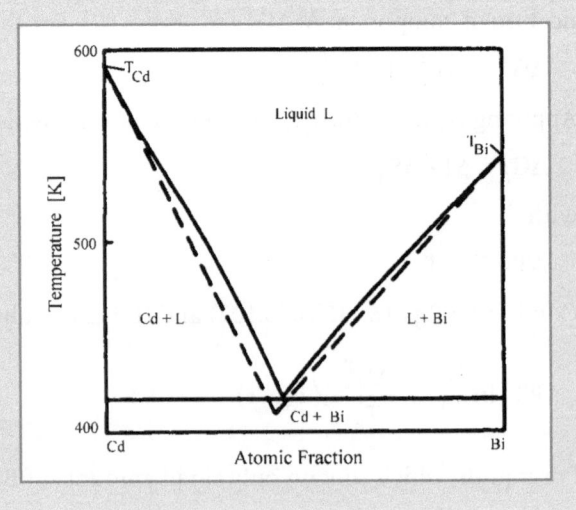

Fig. 6.44

Phase diagram Ag-Cu.
— experimental after
M. Hansen and K. Anderko
[45], --- liquidus calculated
using Eq. (6.117), -.-. liquidus
calculated using Eq. (6.123),
o liquidus points calculated
using Eq. (6.122), but using
thermodynamic activities
instead of atomic fractions

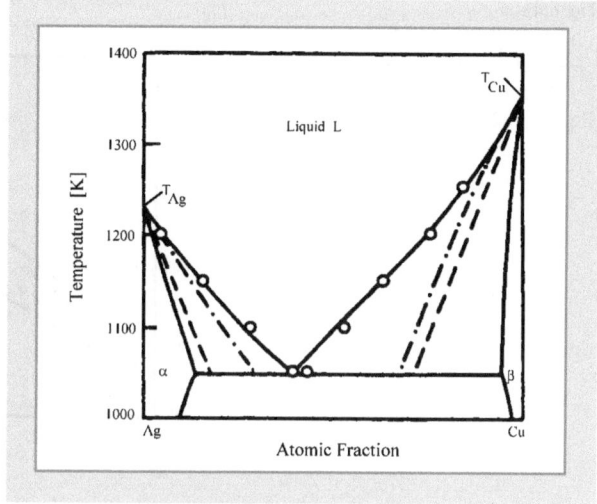

6.22
Equilibrium between a Binary Liquid and a Solid Solution in an Ideal System

For the equilibrium between a liquid and solid solution, combination of Eqs. (6.106), (6.111) and (6.112) yields:

$$G_A^L + R \cdot T \cdot \ln x_A^L = G_A^S + R \cdot T \cdot \ln x_A^S \tag{6.121}$$

Furthermore, with equations (6.102) and (6.115) one has:

$$RT \cdot \ln\left(\frac{x_A^L}{x_A^S}\right) = -\frac{\Delta H_A^F}{T_A} \cdot (T_A - T) \tag{6.122}$$

A direct calculation of the phase diagram with this equation is not possible. In contrast to Eq. (6.117), we have an equation with three unknowns, x_A^L, x_A^S and T. Basically, according to Eq. (6.122) a melting point lowering ($x_A^L < x_A^S$), as well as a melting point increase ($x_A^L > x_A^S$) is possible. With the knowledge of the solidus composition, Eq. (6.122) is a better approximation than (6.117). This is also apparent from Fig. 6.44. Significant discrepancy remains with the true liquidus lines, even after taking into account the solid solubility. This is caused by the assumption of ideal behavior in liquid and solid silver-copper solutions. Introduction of thermodynamic activities (a_A^L, a_A^S) instead of atom fractions (x_A^L, x_A^S) yields an excellent agreement with the liquidus lines determined experimentally.

For the component B, similarly to Eq. (6.122)

Fig. 6.45
Phase diagram Cu-Ni.
— experimental (after [45])
--- Calculated using
Eqs. (6.122) and (6.123)

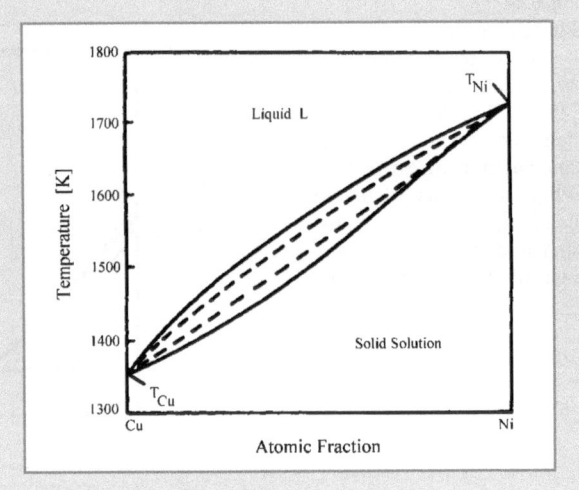

$$RT \cdot \ln\left(\frac{x_B^L}{x_B^S}\right) = -\frac{\Delta H_B^F}{T_B} \cdot (T_B - T) \qquad (6.123)$$

is valid.

Equations (6.122) and (6.123) must be satisfied simultaneously for a system with complete solution formation at a given temperature. By applying both equations, it is thus, possible to calculate the liquidus and solidus composition at a fixed temperature. As an example Fig. 6.45 reproduces the fusion diagram of the system copper-nickel. In this system, the deviations from the true fusion equilibria are relatively minor.

The lense shaped fusion diagram in a binary system with complete solution formation is determined by the temperature dependence of x_A^L and of the ratio x_A^L/x_B^S. As seen in Eqs. (6.122) and (6.123), the enthalpies of fusion ΔH^F, and T^F are the determining factors. The greater the difference $1/T_A - 1/T_B$ is, the greater is the maximum temperature difference between the solidus and liquidus line.

6.23
Fusion Equilibrium in a Regular System

A melting point minimum or a melting point maximum can not occur in a system, in which the solid and liquid solutions are ideal. Application of the regular solution model, however, can describe these equilibria. In this case as with ideal systems, $\Delta S_{(i)}$ the ideal entropy of mixing is used for ΔS. Here the enthalpy of mixing is added as an important quantity. This significantly approaches many systems to the reality, even if the composition dependence of ΔH deviates sometimes greatly from that in real systems. The partial Gibbs energy of component

A of a real liquid solution with the concentration x is given by:

$$\overline{G}_A^L = G_A^L + \overline{\Delta G}_A^L$$
$$= G_A^L + R\,T\ln a_A^L$$
$$= G_A^L + \overline{\Delta H}_A^L - T\,\overline{\Delta S}_A^L \tag{6.124}$$

For a regular solution $\overline{\Delta H}_A^L$ is given by Eq. (6.27), and $\overline{\Delta S}_A^L$ by Eq. (6.35). Equation (6.124) becomes:

$$G_A^L = G_A^L + K^L\,(1 - x_A^L)^2 + RT\ln x_A^L \tag{6.125}$$

The analogous applies to the solid phase.

Based on the equilibrium conditions for the fusion in a two-component system (Eq. (7.106)), one obtains

$$G_A^S + K^S\,(1 - x_A^S)^2 + RT\ln x_A^S = G_A^L + K^L\,(1 - x_A^L)^2 + RT\ln x_A^L \tag{6.126}$$

With

$$G_A^S - G_A^L = -\frac{\Delta H_A^F}{T_A}\cdot(T_A - T) \tag{6.127}$$

(see Eqs. (6.115), (6.102) and (6.113a)) and transformation of Eq. (6.126) yields

$$\frac{\Delta H_A^F}{R}\cdot\left(\frac{1}{T} - \frac{1}{T_A}\right) = \frac{K^S}{RT}\cdot(1 - x_A^S)^2 - \frac{K^L}{RT}\cdot(1 - x_A^L)^2 + \ln\left(\frac{x_A^S}{x_A^L}\right) \tag{6.128}$$

For component B in an analogous fashion

$$\frac{\Delta H_B^F}{R}\cdot\left(\frac{1}{T} - \frac{1}{T_B}\right) = \frac{K^S}{RT}\cdot(1 - x_B^S)^2 - \frac{K^L}{RT}\cdot(1 - x_B^L)^2 + \ln\left(\frac{x_B^S}{x_B^L}\right) \tag{6.129}$$

K^S and K^L are constants, which contain among others, the interaction parameters. If in addition to the values of ΔH^F and T^F, those for K^S and K^L are also given, The phase diagram can be calculated with Eqs. (6.128) and (6.129).

The model of the regular solution is often used to calculate phase equilibria, but is also used in reverse to obtain thermodynamic properties of solutions by evaluating known fusion diagrams. Its advantage is, that the composition dependence of the partial enthalpies of mixing can be expressed by the simple hypothesis contained in Eq. (6.27). Equation (6.27) represents, for many cases, a good approximation. Furthermore the simplification $\Delta S_A = \Delta S_{A(i)}$ is a sufficient approximation. Taking into consideration the significant deviations from the approximations used, yields often to complicated hypotheses, which makes the calculations quite difficult.

6.24
Fusion Equilibrium in a Real System

For a fusion equilibrium in a real system, application of Eqs. (6.106) and (6.124) and an equation analogous to Eq. (6.124) for the solid state yields:

$$G_A^L - G_A^S = \overline{\Delta H_A}^S - T\overline{\Delta S_A}^S - \left(\overline{\Delta H_A}^L - T\overline{\Delta S_A}^L \right) \tag{6.130}$$

The real partial entropy of mixing can be expressed as the sum of the ideal partial entropy of mixing and excess partial entropy of mixing. Considering (6.127) one obtains from Eq. (6.131)

$$\frac{\Delta H_A^F}{T_A} \cdot \left(T_A - T \right) = \overline{\Delta H_A}^S + RT \cdot \ln x_A^S - T \cdot \overline{\Delta S_A}^{S,ex} - \overline{\Delta H_A}^L - RT \cdot \ln x_A^L + T \cdot \overline{\Delta S_A}^{L,ex} \tag{6.131}$$

For component B the analogous applies:

$$\frac{\Delta H_B^F}{T_B} \cdot \left(T_B - T \right) = \overline{\Delta H_B}^S + RT \cdot \ln x_B^S - T \cdot \overline{\Delta S_B}^{S,ex} - \overline{\Delta H_B}^L - RT \cdot \ln x_B^L + T \cdot \overline{\Delta S_B}^{L,ex} \tag{6.132}$$

Knowledge of the enthalpies and temperatures of fusion of the components, and the partial enthalpies and partial entropies of mixing as a function of composition permits the calculation of the liquidus and solidus lines of a real system.

One still has to consider, that Eq. (6.127) only represents the enthalpy of solidification correctly, if the molar heat capacities in the liquid and solid phases are equal within the temperature range considered, when

$$\Delta C_p = C_p^S - C_p^L \tag{6.133}$$

equals zero. Is this not the case, than Eq. (6.127) can still be a good approximation for small deviations (T_A-T) from the equilibrium temperature T_A. The temperature dependence of the enthalpy and entropy of fusion has to be taken into account for large values of ΔC_p, and noticeable differences (T_A-T). In this case, the difference of Gibbs energies between the solid and liquid state at a fixed temperature is given by

$$G_A^S - G_A^L = -\frac{\Delta H_A^F}{T_A} \cdot (T_A - T) + \int_T^{T_A} \Delta C_p \cdot dT - T \int_T^{T_A} \frac{\Delta C_p}{T} \cdot dT \tag{6.134}$$

6.25
Melting Point Minimum

In a real system with complete solubility in the solid phase a melting point minimum or a melting point maximum can occur. Th. Heumann [46] discovered the energetic conditions, which make this possible. Transformation of Eq. (6.131) yields

$$RT\ln\frac{x_A^S}{x_A^L} = \frac{\Delta H_A^F}{T_A} \cdot (T_A - T) - \left(\overline{\Delta H}_A^S - \overline{\Delta H}_A^L\right) + T\left(\overline{\Delta S}_A^{S,ex} - \overline{\Delta S}_A^{L,ex}\right) \qquad (6.135)$$

In many cases the values of the excess entropies of mixing are small. In addition $\overline{\Delta S}_A^{S,ex} \approx \overline{\Delta S}_A^{L,ex}$, so that the difference $\overline{\Delta S}_A^{S,x} - \overline{\Delta S}_A^{L,ex} \approx 0$, and thus, can be neglected in the following discussions.

If the solid and liquid solutions would behave as ideal solutions, we would also have $\overline{\Delta H}_A^S - \overline{\Delta H}_A^L = 0$. No melting point extreme could appear. One would always have $x_A^S > x_A^L$.

A melting point minimum can occur with negligible entropy influence, when $\overline{\Delta H}_A^S > \overline{\Delta H}_A^L$ with endothermic enthalpy of mixing or $\overline{\Delta H}_A^L > \overline{\Delta H}_A^S$ with exothermic enthalpy of mixing.

At the melting point extreme the coexisting phases do not show a difference in composition. With the exception of the general equilibrium condition, Eq. (6.106), the condition applicable to the melting point of pure substances, that the Gibbs energy of the solid and liquid phase are equal, is also valid here $G^L = G^S$.

We have

$$G^L = x_A^L G_A^L + (1 - x_A^L) G_B^L + \Delta G^L \qquad (6.136)$$

and

$$G^S = x_A^S G_A^S + (1 - x_A^S) G_B^S + \Delta G^S \qquad (6.137)$$

(see also Fig. 6.2)

$$\Delta G^L = \Delta H^L - T \Delta S^L \qquad (6.138)$$

and

$$\Delta G^S = \Delta H^S - T \Delta S^S \qquad (6.139)$$

are the Gibbs energies of the liquid resp. solid phase at the melting point extreme. With Eq. (6.127) it follows from Eqs. (6.98), (6.136)–(6.139), and taking into account that $x_A^S = x_A^L = x_A$ one obtains

$$x_A \cdot \frac{\Delta H_A^F}{T_A} \cdot (T_A - T) + (1 - x_A) \cdot \frac{\Delta H_B^F}{T_B} \cdot (T_B - T) =$$
$$= (\Delta H^S - \Delta H^L) - T(\Delta S^{S,ex} - \Delta S^{L,ex}) \qquad (6.140)$$

One can reach this relationship naturally on the basis of the general equilibrium condition, Eq. (6.106). It can be used successfully, as has been shown by Th. Heumann [46], to obtain the difference in integral enthalpy of mixing between the solid and liquid phases coexisting at melting point minimum. The small influence of the excess entropies is here generally neglected.

System	ΔV [cm^3/g-atom]	ΔH^S [J/g-atom]	$\Delta H^S - \Delta H^L$ [J/g-atom]
Cs-K	25.5	644	535
K-Rb	10.7	314	209
Ni-Au	3.8	6870	4803
Cu-Mn	0.3	1930	(1700)

Melting point minima with complete solid solution formation occur in systems where the components show a significant difference in atomic radii. The introduction of an atom too large or too small into a given lattice, requires a lattice deformation, which is connected with an enthalpy input during solution formation. Naturally, this expenditure is larger for the formation of a solid solution, than for the formation of a liquid solution, at a given difference in atomic radii and composition: $\Delta H^S - \Delta H^L > 0$.

Th. Heumann [46] has developed a method to calculate the deformation enthalpies of solid solutions. Important factors which determine the amount of the deformation enthalpy are the compressibility, the thermal expansion, and the difference in atomic volume of the components forming the solid solution. Table 6.7 contains a few systems with a melting point minimum and the corresponding energetic quantities.

6.26
Phase Equilibrium During Demixing

In a binary system in the region of the miscibility gap, two phases (1) and (2) of the same modification, are in equilibrium. The general equilibrium condition apply $\overline{G}_A^{(1)} = \overline{G}_A^{(2)}$ (Eq. (6.106)).

The partial Gibbs energies of phase (1) and (2) can be expressed as the sum of the Gibbs energies of the pure component G_A, and of the partial Gibbs energy $\overline{\Delta G}_A$ in the phase in question:

$$\overline{G}_A^{(1)} = G_A + \overline{\Delta G}_A^{(1)} \tag{6.141}$$

and

$$\overline{G}_A^{(2)} = G_A + \overline{\Delta G}_A^{(2)} \tag{6.142}$$

The Gibbs energy G_A is naturally equal in both phases. The two phase coexisting within the miscibility gap are either both liquid or have the same crystal structure. From Eq. (6.106) with Eqs. (6.141) and (6.142) follows

$$\overline{\Delta G}_A^{(1)} = \overline{\Delta G}_A^{(2)} \tag{6.143}$$

Using Eq. (6.80) one obtains

$$R T \ln a_A^{(1)} = R T \ln a_A^{(2)} \tag{6.144a}$$

$$a_A^{(1)} = a_A^{(2)} \tag{6.144b}$$

The phases which coexist within a miscibility gap have, for the component A in question, the same partial Gibbs energies and thus, same thermodynamic activities. Same applies to component B. Within the miscibility gap the activity isotherm is characterized by a field of constant activity (see Fig. 6.46, isotherms for T_2, T_3 and T_4).

In the case of a miscibility gap the energetic considerations are reduced according to Eq. (6.143) to obtain the Gibbs energy. Naturally, only one $\Delta G - x$ curve has to be considered. The condition given in Eq. (6.143) for the coexistence of two liquids or two solid phases in a miscibility gap, can only be satisfied if the ΔG-x curve has a fold, which permits the construction of a double tangent with two contact points (see Fig. 6.47).

Formation of such a fold can occur at certain temperature, when the enthalpies of mixing are positive. This can be shown easily, based on the regular solution model.

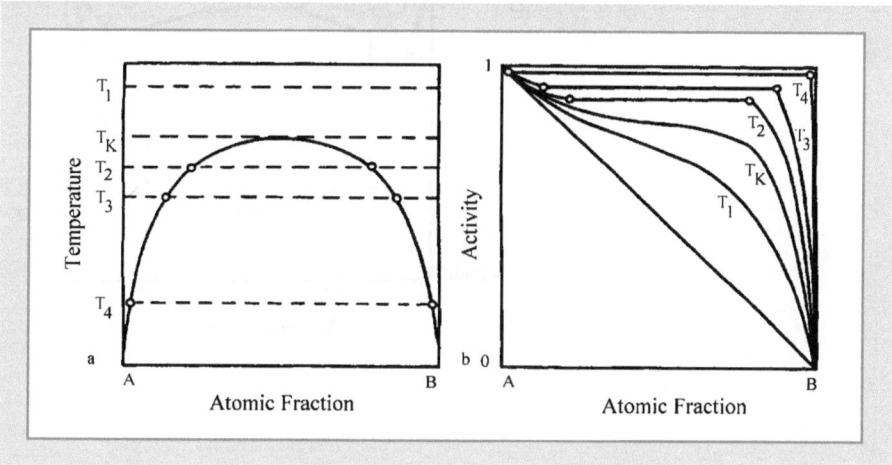

Fig. 6.46 Miscibility gap (a) and corresponding activity isotherms (b)

Fig. 6.47
ΔG – x isotherms in a regular
system K = 12,000 J/g-at.

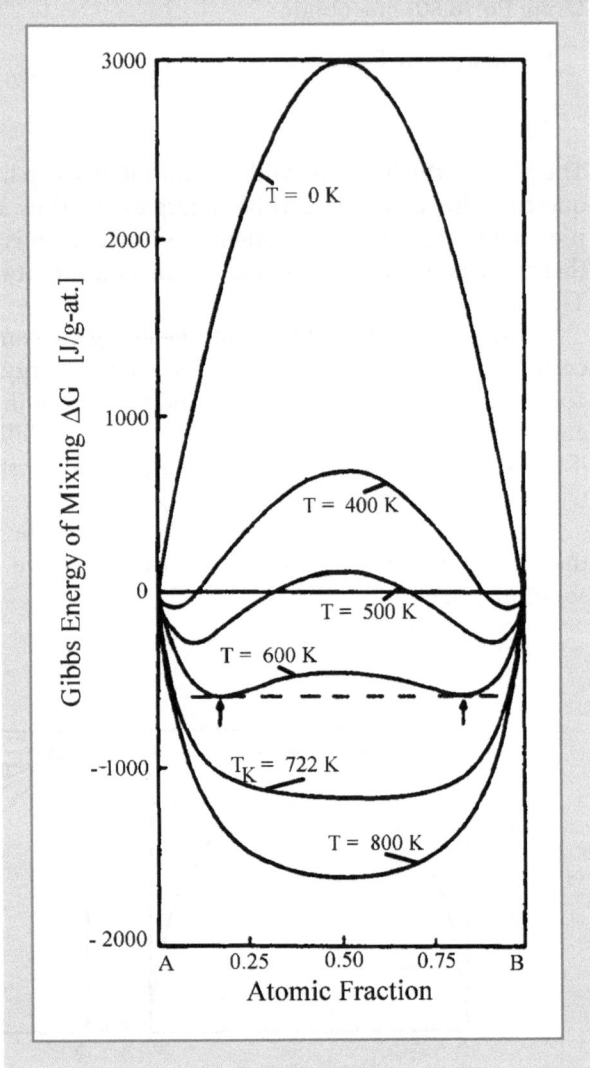

6.27
Calculation of the Miscibility Gap Based on the Regular Solution Model

The Gibbs energy of mixing of a regular solution is, according to Eqs. (6.11) and (6.25):

$$\Delta G = K\, x_A \cdot x_B + R\, T\, (x_A \ln x_A + x_B \ln x_B) \tag{6.145}$$

With a given value of K, ΔG can be obtained as a function of temperature and composition.

As an example, Fig. 6.48 reproduces a few ΔG-isotherms based on a value of $K = 12,000$ J/g-at. In the isotherms at 400, 500 and 600 K a fold is clearly visible. The double tangent drawn to the 600 K isotherm indicates, with the contact points, the composition of the two phases coexisting at this temperature. At the critical point, $T_K = 722$ K, the minima and inflection-points, which appear at lower temperatures in the $\Delta G - x$ curves, coincide at $x = 0.5$. The $\Delta G - x$ curves do not show a turning point above T_K. At these temperatures only single phase regions can exist.

The influence of the entropy of mixing, given by the product $T \cdot \Delta S_{(i)}$, is decreasing, with decreasing temperature. With it, the mutual solubility of the components also decreases. At the absolute zero the entropy term in Eq. (6.145) is zero. At this point the Gibbs energy of mixing is determined solely to the enthalpy of mixing. Because ΔH is positive, basically at $T = 0$ no mutual solubility of the components can exist.

The miscibility gap can be obtained graphically from the $\Delta G - x$ curves, by constructing double tangents. If the regular solution model is valid, it is possible to represent the solubility line by an analytical expression.

In a regular system the ΔS-x and the ΔH-x curve are symmetrical around $x = 0.5$, The critical point lies at $x = 0.5$. Similarly, the contact point of the double tangent are equidistant from the equiatomic composition. In addition the double tangent has a slope zero. Its values at $x_A = 1$ and $x_A = 0$ are equal. This means, that in this case the partial Gibbs energies of the components must be equal

$$\Delta G_A^{(1)} = \Delta G_B^{(1)} = \Delta G_A^{(2)} = \Delta G_B^{(2)} \tag{6.146}$$

Omitting the indication of the phase, it follows for the miscibility gap in a regular system:

$$\Delta G_A = \Delta G_B \tag{6.147}$$

The Helmholtz-Gibbs relation yields

$$\Delta H_A - T \Delta S_A = \Delta H_B - T \Delta S_B \tag{6.148}$$

which, according to Eq. (6.27) resp. (6.28) and using Eq. (6.35) gives:

$$K x_B^2 + R T \ln (1 - x_B) = K (1 - x_B)^2 + R T \ln x_B \tag{6.149}$$

Transformation of Eq. (6.149) gives the following expression for the shape of the miscibility gap in a regular system:

$$RT \ln \frac{1 - x_B}{x_B} = K \cdot (1 - 2x_B) \tag{6.150}$$

The temperature dependence of the solubility line is given solely by the value of the constant K, in the case, that the regular solution is valid.

Equation (6.150) is undefined at the critical point. To obtain a correlation between the constant K and the critical temperature T_K of the miscibility gap, one takes advantage of the fact, that the minima and inflection-points which are present at $T < T_K$ in ΔG-x curves, coincide.

At the critical temperature the double tangents to the minima as well as those to the inflection points have a zero slope. Thus:

$$\frac{d\Delta G}{dx} = 0 \tag{6.151}$$

and

$$\frac{d^2\Delta G}{dx^2} = 0 \tag{6.152}$$

The integral Gibbs energy of mixing of a regular solution is given by:

$$\Delta G = K\,(1 - x_A)^2\,x_A + K\,(1 - x_B)^2\,x_B + R\,T\,[x_A \ln x_A + x_B \ln x_B] \tag{6.153}$$

The second derivative by x_B is:

$$\frac{d^2\Delta G}{dx^2} = -2K + RT\left(\frac{1}{x_B} + \frac{1}{1 - x_B}\right) \tag{6.154}$$

Equation (6.152) is valid for the critical point. Considering further, that the critical composition lies at $x = 0.5$, yield the following expression for the critical temperature:

$$T_K = \frac{K}{2R} \tag{6.155}$$

If R, K and ΔH are not expressed in J/g-at., but in cal/g-at., then Eq. (6.155) yields a simple rule to memorize. One has $R = 1.986$ cal/g-at. (≈ 2 cal/g-at.). From

Fig. 6.48 Miscibility gap and spinodal in a regular system with $K = 12\,000$ J/g-at.

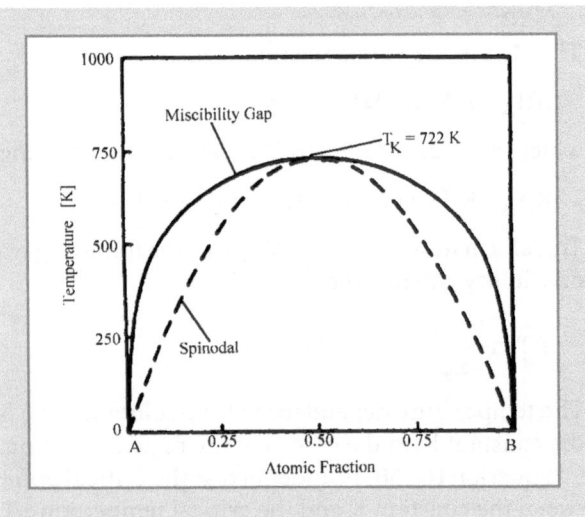

Table 6.8 Experimentally determined critical temperature of the mixing, maximal enthalpy of mixing and maximal excess entropy and further on critical temperature of the mixing calculated by Eq. (6.142)

System	T_K [K] experimentally	ΔH_{max} [cal/g-at.]	ΔS^{ex}_{max} [cal/g-at · K]	T_K [K] calculated Eq. (6.142)
Miscibility gap in the solid state				
Al-Zn	625	840	0.15	735
Au-Ni	1093	1800	0.71	1057
Cr-Mo	1130	1725	0.478	1173
Au-Pt	1525	(1400)	–	–
Miscibility gap in the liquid state				
Ga-Cd	568	640	0	644
Al-In	1125	1400	0.40	1005
Cu-Tl	1520	2050	0.50	1373
Cu-Pb	1253	1600	0.22	1322

Eq. (6.155) follows $T_K \approx K/4$. At $= x\ 0.5$, the enthalpy of mixing is a maximum and, according to Eq. (6.25), $\Delta H_{max} = K/4$.

As an example miscibility gap and spinodal in a regular system is given in Fig. 6.48.

The numerical value of the critical temperature is approximately equal to the maximum enthalpy of mixing ΔH_{max} in cal/g-at. In Table 6.8 the maximum enthalpies of mixing are compared to the critical temperatures of miscibility gaps, in the solid and liquid states. The still present, sometimes significant discrepancies are due, neglecting the experimental errors in the data, to the excess entropy of mixing, which was neglected here. Also, the often significant asymmetric behavior of the miscibility gaps was not considered here, which is due to special bonding conditions.

6.28
Evaluation of Solubility Equilibria

The calculation of a miscibility gap is possible with the knowledge of the required thermodynamic quantities, and it is also feasible to calculate thermodynamic mixing functions from available solubility data. It is not necessary to know the whole miscibility gap up to the critical point. Very often relatively limited solubilities are present, known only in a narrow temperature range. Simplifications can be used in the evaluations, which facilitate the calculation of thermodynamic mixing functions.

In the miscibility gap represented in Fig. 6.46a the solubility at the temperature T_4 on both sides is low. The composition of the A-rich phase (1) is $x_A^{(1)} \cong 1$. Here in good approximation, Raoult's law is valid:

$$a_A^{(1)} \approx x_A^{(1)} \tag{6.156}$$

from this $a_A^{(1)} \cong 1$ follows, and

$$R\,T \ln a_A^{(1)} \cong 0 \tag{6.157}$$

The mole fraction of A in the B-rich phase (2) coexisting at T_4 with phase (1) is very small. It can thus, be assumed that the regular solution model can be applied here as a good approximation. The partial entropy of mixing of component A can be taken as:

$$\Delta S_A = -R \ln x_A^{(2)} \tag{6.158}$$

From Eqs. (6.147), (6.157) and (6.158) and taking into consideration the Helmholtz-Gibbs relationship, it follows that

$$\overline{\Delta G_A} = \overline{\Delta H_A} + R\,T \ln x_A^{(2)} = 0 \tag{6.159}$$

Transforming

$$\ln x_A^{(2)} = -\frac{\overline{\Delta H_A}}{RT} \tag{6.160}$$

or

$$\ln x_A^{(2)} = \exp\left(-\frac{\overline{\Delta H_A}}{RT}\right) \tag{6.161}$$

Equation (6.161) describes the course of the solubility line under the simplifying assumptions of small solubility expressed above.

A little further simplification, which can be applied at noticeable solubilities, is to assume that $a_A^{(1)} \cong x_A^{(1)}$ rather than $RT \cdot \ln x_A^{(1)} = 0$. The numerical value used for $x_A^{(1)}$ is the value of the phase equilibrium:

$$R\,T \ln x_A^{(1)} = R\,T \ln a_A^{(2)} \tag{6.162}$$

At the same time the assumption of regular behavior at the composition of phase (2) can be omitted, especially with not very dilute solutions.

In contrast to the approximation in Eq. (6.158) the excess entropy must be considered. From Eq. (6.162) one obtains

$$R\,T \ln x_A^{(1)} = \overline{\Delta H_A} + R\,T \ln x_A^{(2)} - T\,\overline{\Delta S_A^{ex}} \tag{6.163}$$

Transforming

$$R \cdot \ln \frac{x_A^{(1)}}{x_A^{(2)}} = \frac{\overline{\Delta H_A}}{T} - \overline{\Delta S_A^{ex}} \tag{6.164}$$

or

$$\frac{x_A^{(1)}}{x_A^{(2)}} = \exp\left(\frac{\overline{\Delta H}_A}{RT}\right) \cdot \exp\left(-\frac{\overline{\Delta S}_A^{ex}}{R}\right) \tag{6.165}$$

Equation (6.165) represents the behavior of the solubility line in a real solution with moderate solubility.

The partial functions $\overline{\Delta H}_A$ and $\overline{\Delta S}_A^{ex}$ depend on the composition. To simplify the evaluation, suitable approximations for the composition dependence are recommended. Often symmetrical assumptions, corresponding to the regular solution model or based on it, suffice

$$\overline{\Delta H}_A = K\,(1 - x_A)^2 \tag{6.166}$$

$$\overline{\Delta S}_A^{ex} = L\,(1 - x_A)^2 \tag{6.167}$$

K and L are constants.

Thus, from Eq. (6.164) follows

$$R \ln \frac{x_A^{(1)}}{x_A^{(2)}} = \frac{K \cdot (1 - x_A^{(2)})^2}{T} - L \cdot (1 - x_A^{(2)})^2 \tag{6.168a}$$

or

$$\frac{R \cdot \ln\left(\dfrac{x_A^{(1)}}{x_A^{(2)}}\right)}{(1 - x_A^{(2)})^2} = \frac{K}{T} - L \tag{6.168b}$$

Simplifying

$$\Lambda = \frac{K}{T} - L \tag{6.169}$$

where

$$\Lambda = \frac{R \cdot \ln\left(\dfrac{x_A^{(1)}}{x_A^{(2)}}\right)}{(1 - x_A^{(2)})^2} \tag{6.170}$$

Furthermore, according to the regular solution model

$$K = 4\,\Delta H_{max} = \overline{\Delta H}_A^{\infty} \tag{6.171}$$

$\overline{\Delta H}_A^{\infty}$ is the partial enthalpy of mixing of component A at infinite dilution $(x_A \rightarrow 0)$.

6.29
Evaluation of a Fusion Equilibrium with Small Liquidus Concentrations

Liquidus lines are also solubility lines. A liquidus point corresponds to a liquid, which is saturated in the component, making up the largest part of the corresponding solid. The liquidus lines in the region of very small concentrations of solvent A can be evaluated in similar fashion to the solubility lines in a simple miscibility gap with low solubility.

The fusion relationship valid for a two-component system (Eq. (6.106)) and

$$G_A^L + \Delta G_A^L = G_A^S + \Delta G_A^S \tag{6.172}$$

combined with

$$\Delta G_A^L = \Delta H_A^L - T \cdot \Delta S_A^L \tag{6.173}$$

$$\Delta S_A^L = -R \ln x_A^L + \Delta S_A^{L,ex} \tag{6.174}$$

and

$$G_A^S = R\,T \ln a_A^S \tag{6.175}$$

yields

$$G_A^L + \Delta H_A^L + R\,T \ln x_A^L - T\,\Delta S_A^{L,ex} = G_A^S + R\,T \ln a_A^S \tag{6.176}$$

With Eqs. (6.17) and (6.167) one obtains

$$-\Lambda = \frac{R \ln x_A^L - \dfrac{G_A^S - G_A^L}{T} - R \ln a_A^S}{(1 - x_A^L)^2} = -\frac{K}{T} + L \tag{6.177}$$

$G_A^S - G_A^L$ can be expressed according to Eq. (6.127).

In the case of a simple miscibility gap, $G_A^S - G_A^L = 0$. Equation (6.177) takes the form of Eq. (6.168b).

If the approximations applied are justified in a situation, the plot of Λ versus $1/T$ yields a straight line. As an example, Fig. 6.49 represents binary three phase diagrams, with an extended solubility at the monotectic temperature. Below the monotectic temperature along the liquidus line, crystals of the higher melting component (Al resp. Zn) are in equilibrium with liquid (L_2), which are very rich in the lower melting component.

Figure 6.50 shows the plot of Λ versus $1/T$ for the solid-liquid equilibria (according to Eq. (6.177)), and for the liquid-liquid two-phase equilibria according to Eq. (6.169). In all cases a straight line is obtained, which applies simultaneously to the solid-liquid and liquid-liquid equilibria. The slope of the line gives K, whereas the value of the abscissa yields L. From these quantities, partial enthalpies of mixing and partial excess entropies of mixing can be obtained for the composition ranges, for which the approximations in Eqs. (6.166) and (6.167)

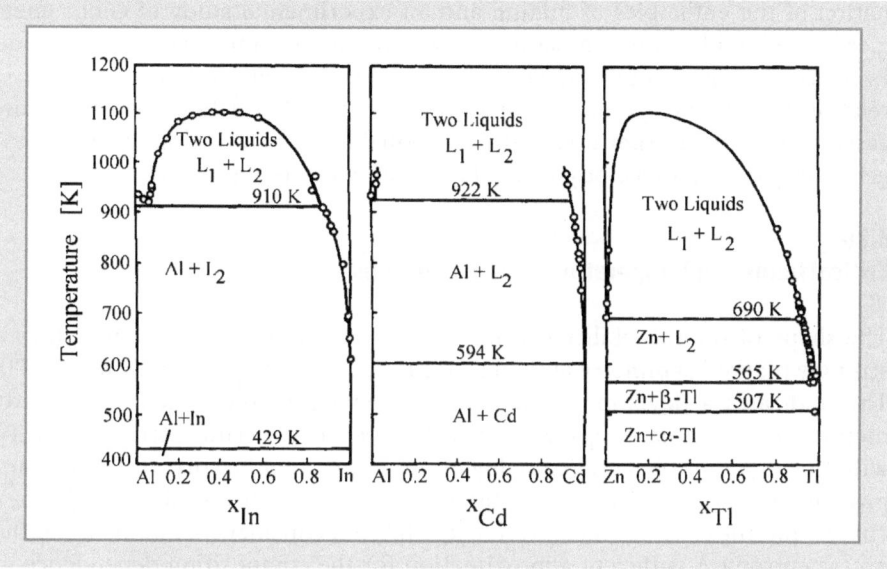

Fig. 6.49 Examples of miscibility gaps in the liquid with small mutual solubility of the components at monotectic temperature (after [47])

Fig. 6.50
Evaluation of solubility data
in the systems Al-In, Al-Cd,
and Zn-Tl with help of the
Eq. (6.168b) and (6.177),
resp., after [47],
o liquid-liquid equilibria,
▲ solid-liquid equilibria

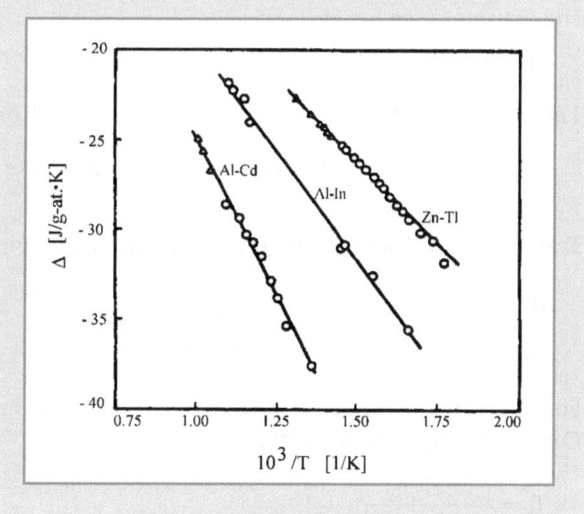

are adequate. Similarly enthalpies of mixing and excess entropies of mixing can be calculated from K and L.

The boiling point of cadmium is at 1040 K. The monotectic temperature in the system aluminum – cadmium is 922 K. In this system, the miscibility gap would only close way above the boiling point of cadmium. The direct determination of the enthalpies of mixing and an experimental study of Gibbs energies in the liquid aluminum-cadmium system is almost impossible to carry out. Evaluations of the solubility data, however, yield information about the energetic relationships of the system. The evaluation of solubility data is useful in many cases, where direct experiments are questionable – due to various causes – and generally to check thermodynamic data obtained by other means.

6.30
Critical Demixing Temperature in Real Solutions

The shape of the miscibility gap in real systems, in general, does not show a strongly marked symmetry at $x = 0.5$ (see for example Fig. 6.49, system Zn – Tl). This indicates a deviation from regular behavior of the system. As already mentioned, in this case Eq. (6.155), which connects the critical temperature T_K with the maximum integral enthalpy of mixing, cannot be correct. A better approximation is obtained, if one retains the symmetry of the $\Delta H - x$ and $\Delta S - x$ curves, and the critical concentration $x_K = 0.5$, but considers the influence of the excess entropy. A sufficient approximation for the composition dependence of ΔS^{ex} is:

$$\Delta S^{ex} = L \, x_A \, (1 - x_A) \tag{6.178}$$

Similarly, for the composition dependence of the integral enthalpy of mixing one assumes (see Eq. (6.25)):

$$\Delta H = K \cdot x_A \, (1 - x_A) \tag{6.179a}$$

The integral Gibbs energy of mixing for the solution is then given by:

$$\Delta G = K \, x_A \, (1 - x_A) + R \, T \, [x_A \ln x_A + (1 - x_A) \ln (1 - x_A)] - T \, L \, x_A \, (1 - x_A) \tag{6.179b}$$

Transformation, and the second derivative by x_A yields:

$$\frac{d^2 \Delta G}{dx_A^2} = -2K + RT \cdot \left[\frac{1}{x_A} + \frac{1}{(1 - x_A)} \right] + 2T \cdot L \tag{6.180}$$

For the critical point of a miscibility gap one has Eq. (6.152). Finally, from Eqs. (6.152) and (6.180) follows

$$T_K = \frac{2 \Delta H_{max}}{R + 2 \Delta S_{max}^{ex}} \tag{6.181}$$

In the case of the regular model ($\Delta S_{max}^{ex} = 0$), a miscibility gap can only occur with positive enthalpies of mixing, according to Eq. (6.155). In real systems, positive T_K values can result with negative ΔH values, only when ΔS_{max}^{ex} is negative, and is sufficiently large. A detailed treatment of models for miscibility gaps with negative enthalpies of mixing follows later (see Sect. 9.3).

Inspection of Table 6.8 shows, that Eq. (6.181) is a better approximation than the thumb rule of Eq. (6.155). The T_K values calculated using Eq. (6.181) agree much better with the critical temperatures determined experimentally, than do the T_K values estimated with Eq. (6.155).

6.31
The Spinodal

The spinodal is defined as the line within the composition range of a miscibility gap, which represents the totality of the T-x points, according to Eq. (6.152). The spinodal is thus, the line, on which all the inflection points of a ΔG-x curve in a T-x plot lie.

For systems, for which the model of the regular solution applies, from Eq. (6.154) if follows in general (and not only in the special case of the critical temperature T_K)

$$-2K + RT \cdot \left[\frac{1}{x} + \frac{1}{1-x} \right] = 0 \tag{6.182}$$

Transformation yields

$$x \cdot (1-x) = \frac{RT}{2K} \tag{6.183}$$

This is the expression for the shape of the spinodal (see Fig. 6.48). The spinodal is important for nuclei formation in precipitation reactions.

6.32
Calculation of a Simple Ordering Reaction in Solid Solutions

Superstructures occur in solid solutions when the bonding energy between dissimilar atoms is larger than the average of the bonding energies between identical atoms. This corresponds to a tendency for compound formation. The probability to find an A – B pair in a solid solution lattice is greater than a random distribution of the atoms on the available lattice sites. The occurrence of an ordered distribution of atoms influences the value of the enthalpy of mixing ΔH, and that of the entropy of mixing ΔS. If it results in a lowering of the Gibbs energy of the system, the ordering can take place spontaneously.

In the presence of a compound forming tendency, each A atom surrounds itself with as many B atoms as possible, and the same with B atoms. With this, the entropy of mixing is lower than by random distribution of atoms.

In the case of complete ordering the entropy of mixing is zero. On the other hand, the enthalpy of mixing has a maximum value. The Gibbs energy, responsible for the ordering reaction, depends on the temperature, according to the Helmholtz-Gibbs equation:

$$\Delta G_{O \to u} = \Delta H_{O \to u} - T \, \Delta S_{O \to u} \tag{6.184}$$

$\Delta H_{O \to u}$ is the enthalpy change during the transition from an ordered distribution of atoms into a random distribution of atoms. As a tendency for compound formation is the condition for ordering (preferential formation of A – B bonds), $\Delta H_{O \to u}$ has a positive sign. $\Delta S_{O \to u}$ is the change in entropy during the transition from an ordered state into a disordered solid solution. During this transition the entropy increases. $\Delta S_{O \to u}$ (entropy of disordering) thus, has a positive value. At a lower temperature, T, the term $T \cdot \Delta S_{O \to u}$ is insignificant. The sign of $\Delta G_{O \to u}$ is determined by the sign of $\Delta H_{O \to u}$; it is positive. A disordering of the superstructure phase does not take place. The ordered structure is stable at low temperatures. With a given value of $\Delta H_{O \to u}$ and sufficiently high temperature $\Delta G_{O \to u}$ is negative according to Eq. (6.145). The superstructure transforms instantaneously into a solid solution with a random distribution of atoms. At high temperatures the disordered state is stable.

The quantitative description of the ordering reaction will be applied using the example of β-brass. The superstructure of β-brass (CuZn) is of the CsCl-type (see Fig. 6.51). Because, in a completely ordered solid solution, all A atoms are on α sites, and B atoms are on β sites, the probability to find an atom on its "correct" site is p = 1. At a complete random distribution half of the α sites are occupied by A atoms, the other half by B atoms. The fraction of A atoms sitting on α sites to the total number of A atoms is p = 1/2.

In addition to the extreme cases of random distribution and complete ordering, intermediate situations can also occur. Let the probability to find an A atom on an α site be p. This is simultaneously the fraction of all A atoms, which sit on their "correct" (α-sites) sites. The fraction of all A atoms, which can be found on β-sites is (1-p). The same applies to B atoms.

At a given temperature a degree of ordering, p, occurs, which minimizes the Gibbs energy of the system. In the following, the effect of the degree of order on

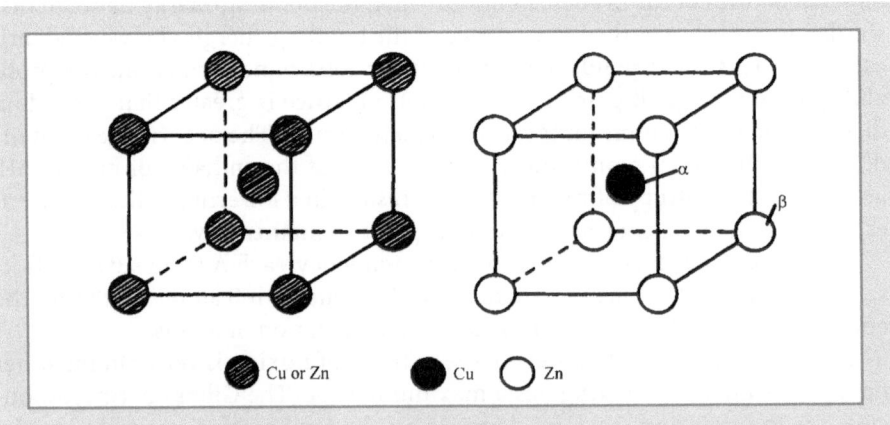

Fig. 6.51 Unit cell of a Cu-Zn alloy at x = 0.5. **a** Statistical distribution of atoms; β-brass, **b** completely ordered distributions of atoms; superstructure of CsCl-type; β'-brass

the enthalpy of mixing and entropy of mixing is considered. Only the influence of the configuration will be considered. Also, the small difference between enthalpy of mixing and internal energy of the solution is neglected.

According to the simple model of Gorsky and that of Bragg and Williams, only interactions between next neighbors will be considered.

The interactions between A-A, B-B and A-B pair of atoms are considered independent from their surroundings.

Let

w_{AA}, w_{BB}, w_{AB} the interaction energy of the pairs
N_{AA}, N_{BB}, N_{AB} the number of the pairs.

The enthalpy of configuration is the sum of all interaction energies:

$$H = N_{AA}\,w_{AA} + N_{BB}\,w_{BB} + N_{AB}\,w_{AB} \tag{6.185}$$

If N is the total of atoms in the crystal, then in the case of AB stoichiometry of the superstructure considered here, $(N/2)A$ atoms and $(N/2)B$ atoms are present. One has:

$(N/2)p$ A-atoms on α sites
$(N/2)(1-p)$ A-atoms on β sites
$(N/2)p$ B-atoms on β sites
$(N/2)(1-p)$ B-atoms on α sites

The coordination number in the CsCl lattice is $z = 8$. In the case of complete order all next neighbors of an A atom are β sites. If no complete order exists, a fraction of these β sites are occupied by A atoms, and the fraction of A atoms on these β sites is $(1-p)$. The number of A neighbors, which each atom has is $8 \cdot (1-p)$. The number of A-A pair bondings are in general:

$$N_{AA} = \frac{N \cdot z}{2} \cdot p(1-p) \tag{6.186}$$

Here it is assumed, that the fraction of not correctly occupied β sites are true on the average for the first coordination sphere of each A atom.

Similarly, the number of the B – B pairs is

$$N_{BB} = \frac{N \cdot z}{2} \cdot p(1-p) \tag{6.187}$$

The number of total pair bondings is $N \cdot z/2$. It must be equal to the sum of all pair bondings:

$$\frac{N \cdot z}{2} = N_{AA} + N_{BB} + N_{AB} \tag{6.188}$$

With Eqs. (6.186) and (6.187) one obtains

$$\frac{N \cdot z}{2} = \frac{N \cdot z}{2} \cdot p(1-p) + \frac{N \cdot z}{2} \cdot p(1-p) + N_{AB} \tag{6.189}$$

Transformation gives the number of A-B pairs

$$N_{AB} = \frac{N \cdot z}{2} - N \cdot z \cdot p \cdot (1-p) \tag{6.190}$$

The configurational enthalpy according to Eq. (6.146) can be expressed by taking into account of Eqs. (6.186), (6.187) and (6.190) as

$$H = \frac{N \cdot z}{2} p(1-p)(w_{AA} + w_{BB}) + \frac{N \cdot z}{2} w_{AB} - N \cdot z \cdot p(1-p) \cdot w_{AB} \tag{6.191}$$

In the case of complete ordering $p = 1$, and Eq. (6.191) simplifies to

$$H_0 = \frac{N \cdot z}{2} \cdot w_{AB} \tag{6.192}$$

Let us consider the transition of a completely ordered superstructure (starting condition with H_0) into a state characterized by the parameter p (final state with H). Subtracting Eq. (6.192) from (6.191) yields the enthalpy of disordering:

$$\Delta H = H - H_0 = -N \cdot z \cdot p(1-p) \cdot \left[w_{AB} - \frac{w_{AA} + w_{BB}}{2} \right] \tag{6.193}$$

with

$$D = \left[w_{AB} - \frac{w_{AA} + w_{BB}}{2} \right] \tag{6.194}$$

one obtains

$$\Delta H = -N z \cdot D \cdot p (1-p) \tag{6.195}$$

The interaction energies w_{AA}, w_{BB} and w_{AB} have a negative sign. Thus, D also has a negative value in the case of compound forming tendency. The maximum value of the energy of disordering is reached, when a completely ordered solid solution is transformed into one with random distribution of atoms with $p = 0.5$. One obtains

$$\Delta H_{max} = -\frac{N \cdot z \cdot D}{4} \tag{6.196}$$

The determination of the entropy change during disordering of an ordered solid solution can be carried out using simple statistical considerations, as is done during calculation of the ideal entropy of mixing. The system with a given degree of order can be considered as a solution consisting of atoms on right and wrong sites. Similarly to the ideal entropy of mixing (see Eq. (6.11))

$$\Delta S_i = -N_L \cdot k \cdot [x_A \ln x_A + (1 - x_A) \ln (1 - x_A)] \tag{6.197a}$$

can be written as

$$\Delta S = -N \cdot k \cdot [p \cdot \ln p + (1-p) \ln (1-p)] \tag{6.197b}$$

$R = N_L \cdot k$ is the general gas constant, where N_L is the Loschmidt number, and k is the Boltzmann constant.

At the formation of a completely ordered state, the entropy of mixing is zero. In general

$$S = R \cdot \ln W \qquad (6.198)$$

where W is the number of ways the atoms can be arranged with a given degree of order. Let us consider a planar mixture of atoms A and B with the AB stoichiometry. With complete order, there are two possibilities:

ABABA BABAB
BABAB ABABA
ABABA BABAB
BABAB ABABA
ABABA BABAB

The two arrangements are identical. One has $W = W_0 = 1$, and thus, $S = S_0 = R \cdot \ln W_0 = 0$.

Let us consider the transition of a totally ordered solid solution into a solid solution not completely ordered. The change in entropy connected with it is

$$\Delta S = S - S_0 = R \ln W - R \ln W_0 = R \ln W \qquad (6.199)$$

S is the entropy of the solid solution with a not completely ordered distribution of atoms, and S_0 with total order. With total order $W_0 = 1$.

The expression $\Delta S = R \ln W$ corresponds to than in Eq. (6.197b). $\Delta S = S - S_0$ is the entropy of disordering.

The Gibbs energy change connected with the disordering is given by

$$\Delta G = G - G_0 \qquad (6.200)$$

where G is the Gibbs energy of the solid solution with incomplete ordering, and G_0 is the Gibbs energy with total order. Combining Eqs. (6.193) and (6.197b) with Eq. (6.200) and applying the Helmholtz-Gibbs equation (Eq. (6.12)) yields

$$\Delta G = -N \cdot z \cdot p \cdot (1-p) \cdot \left[w_{AB} - \frac{w_{AA} + w_{BB}}{2} \right] + $$
$$+ N \cdot k \cdot T \cdot [p \cdot \ln p + (1-p) \cdot \ln(1-p)] \qquad (6.201)$$

or, considering the simplification in Eq. (6.194)

$$\Delta G = - N z D p (1-p) + N k T [p \ln p + (1-p) \ln (1-p)] \qquad (6.202)$$

This equation is formally similar to Eq. (6.145) for the Gibbs energy of mixing of a regular solution.

In a similar way to the determination of the critical demixing temperature at a miscibility gap, the critical ordering temperature T_K is obtained by equating to zero the second derivative of ΔG against x. From Eq. (6.202) it follows after two differentiations

$$2NzD + NkT_K \cdot \frac{1}{p(1-p)} = 0 \tag{6.203}$$

At the critical point $p = 0{,}5$. Thus,

$$T_K = -\frac{zD}{2k} \tag{6.204}$$

Expansion with the Loschmidt number and considering Eq. (6.196) yields

$$T_K = -\frac{2\Delta H_{max}}{R} \tag{6.205}$$

If the value of ΔH_{max} is expressed in cal/g-at., and R in cal/g-at. $\cdot$ K then numerically $T_K \approx \Delta H_{max}$, because $R \approx 1.986 \approx 2$ cal/g-at. $\cdot$ K. ΔH is positive. At the transition order-disorder the system absorbs enthalpy.

The critical temperature for β-brass is $T_K = 733$ K. According to Eq. (6.205) with $x = 0.5$ $\Delta H_{max} = 730$ cal/g-at. Sykes and Wilson found experimentally $\Delta H_{max} = 630$ cal/g-at., which agrees within an order of magnitude with the value of ΔH_{max} calculated from T_K.

The analogy between Eqs. (6.202) and (6.145) permits a similar treatment for demixing and superlattice formation. Instead of the ΔG-x diagram (Fig. 6.44) a ΔG-p diagram is used (see Fig. 6.52), which has the same shape as a ΔG-x diagram. ΔG-p line does not have an inflection point at $T > T_K$. The ΔG–p curve shows a fault at $T < T_K$. In Fig. 6.52 only half of diagram, from $p = 0.5$ to $p = 1$ has physical meaning.

Fig. 6.52
Gibbs energy of disordering
ΔG in dependence of degree
of order p

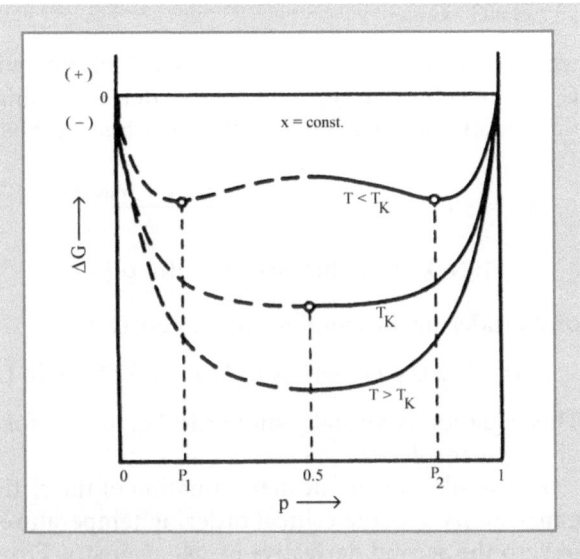

For points with $p < 0.5$ α-sites would be overwhelmingly occupied by B atoms, and vice versa – in contrast to the random distribution where $p = 0.5$. At $p = 0$, all B atoms would be placed on α-sites, and A atoms on β-sites. The half of the ΔG-p diagram from $p = 0$ to $p = 0.5$ is formally identical with the region $p = 0.5$ to $p = 1$. Points which lie to the left and right of $p = 0.5$ correspond to the same state. Therefore a double tangent has no physical meaning. The minima, however, have a physical meaning: They indicate the degree of order for which, at a given temperature, ΔG is minimum. This degree of order occurs in the solid solution. With increasing temperature the minima approach $p = 0.5$. Is $p = 0.5$, a random distribution of atoms is present. Now two minima coincide with the two inflection points situated between them. One has $d\Delta G/dp = 0$, and $d^2\Delta G/dp^2 = 0$. Based on the last condition, the critical ordering temperature T_K was derived in Eq. (6.204).

6.33
Degree of Order in a Superlattice as a Function of Temperature

The degree of order (value of p at the minimum of the ΔG-p curve) as a function of temperature is generally not represented in a p-T diagram, as is the case in an x-T diagram with demixing, but is plotted as an order parameter, s, as a function of the reduced temperature T/T_K.

The order parameter is defined at the AB stoichiometry as

$$s = 2\,p - 1 \ (x = 0.5) \tag{6.206}$$

At random distribution ($p = 0.5$) $s = 0$, and at total order ($p = 1$) $s = 1$.

For any stoichiometry of the superlattice ($x \neq 0$)

$$s = \frac{p_A - x_A}{1 - x_A} = \frac{p_B - x_B}{1 - x_B} \tag{6.207}$$

p_A and p_B are the fraction of α-sites occupied by A atoms, resp. β-sites occupied by B atoms.

In the case AB with the CsCl structure a continuous transition takes place from total order to random distribution with increasing temperature, according to Eq. (6.202) (see Fig. 6.53a). The s-(T/T_K) curve becomes steeper with decreasing order parameter s. As long as the order is slightly disturbed, large amounts of energy has to be spent, to put an atom on the wrong site. With decreasing degree of order the energy required to further destroy the ordering decreases.

For other stoichiometries than AB, s can be calculated in a similar fashion as a function of temperature.

$$T_K = -0.137 \cdot \frac{z \cdot D}{2 \cdot k} \tag{6.208}$$

Thus, the critical temperature for the superstructure AB_3 is lower than that for

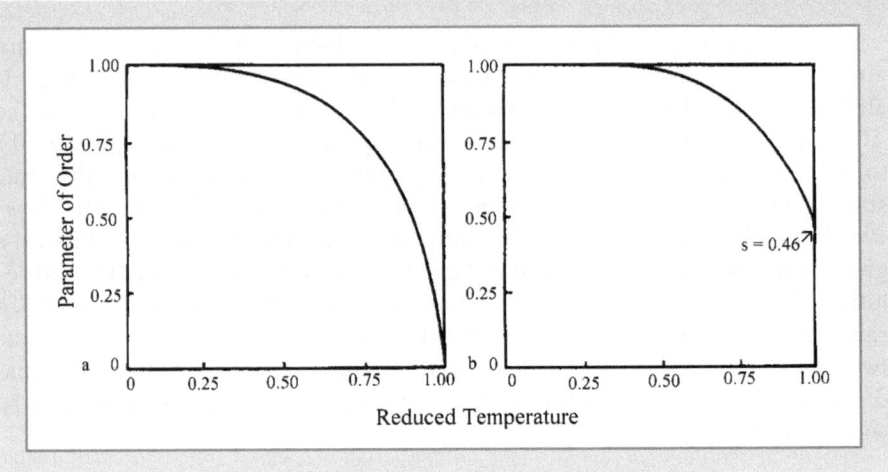

Fig. 6.53 Order parameter s as a function of reduced temperature T/T_K after Bragg and Williams. **a** For the stoichiometry AB, **b** for the stoichiometry AB_3

the superstructure AB with same value of D. To be sure, the different coordination numbers also have to be considered. In a system, which contains the superstructures AB and AB_3, the superstructure AB often has the higher critical temperature. The shape of the s-(T/T_K) curve is also different. The disordering of the AB_3 superstructure with increasing temperature first occurs continuously, until at T_K the value of $s = 0.46$ is reached. Here a discontinuous transition takes place, from the superlattice with $s = 0.46$ to the phase with random distribution of atoms $(s = 0)$.

Based on the character of the s-(T/T_K) curve one can expect, that the disorder reaction of an AB superlattice is not a phase transition of the first order. It is not a transformation point, but a so-called Λ point, analogous to the Curie-point at the transition of a ferromagnetic state into a paramagnetic one.

On the other hand, in AB_3 superstructures one can expect a phase transition of the first order, where at the transition point the heat capacity curve shows a jump, which represents the latent heat of transition. At the temperature T_K two phases are in equilibrium. Above the critical temperature only leftover order can persist (short range order). At T_K two partially ordered phases are in equilibrium. The short range order parameter of the solid solution can be determined by small-angle X-ray diffraction.

The character of the transformation can be determined in principle, by the following properties, which depend on the degree of order (for example electric resistance) as a function of temperature. However, this can result in significant difficulties. An unequivocal decision about the presence of a phase transition of the first order is only possible, if the two coexisting phases can be identified directly (for example with a light- or electron-microscope). This must

Fig. 6.54
Schematic presentation of
regions of phases at order/
disorder-transformations,
which occur as transitions of
I. order

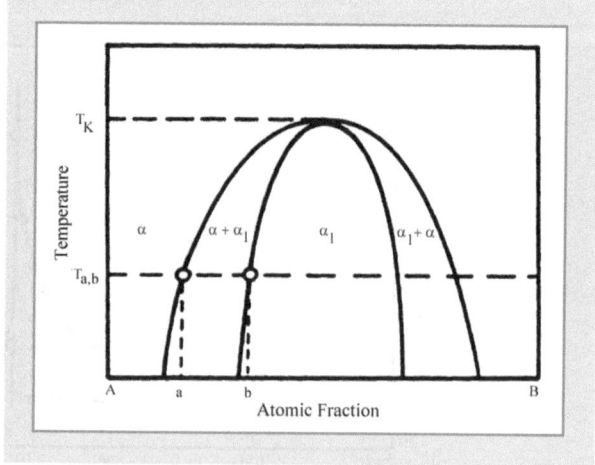

not necessarily be done at the critical temperature T_K, where the two phases in equilibrium have the same composition. That is shown in Fig. 6.54, where the phase equilibria for the order-disorder transitions are represented schematically, which proceed according to a phase transition of the first order (see for example the system Au–Cu in Fig. 4.50). It is easier to prove experimentally the existence of two phases in equilibrium α and α_1 having compositions a and b at $T_{a,b}$, than at T_K.

At a stoichiometry AB the order-disorder transition can also be a phase transition of the first order. In this case the description for the energetic situation of Bragg and Williams does not suffice. Other models have been developed (for example by Bethe and by Kirkwood), which, however, require a greater mathematical expenditure.

6.34
Comments on the Character of Phase Transformations

Transformations which represent the transition of one state of aggregation into another, or a transition from a distinct crystallographic modification of a substance into another, are called transformations of the first order. At the transformation point two phases are in equilibrium. The enthalpy-temperature curve shows a jump at the transformation temperature. The jump represents the enthalpy of transformation. The heat capacity $C_p = (\Delta H/\Delta T)_p$ has a discontinuity at the transformation point.

On the other hand, in the consideration of the order-disorder transformation in β-brass, the order parameter moved continuously from the fully ordered superlattice to the solid solution with random distribution of atoms (see Fig. 6.53a). No jump occurs in the H–T curve at the ordering temperature T_K: this function is also continuous at T_K. To be sure, the C_p–T curve indicates a

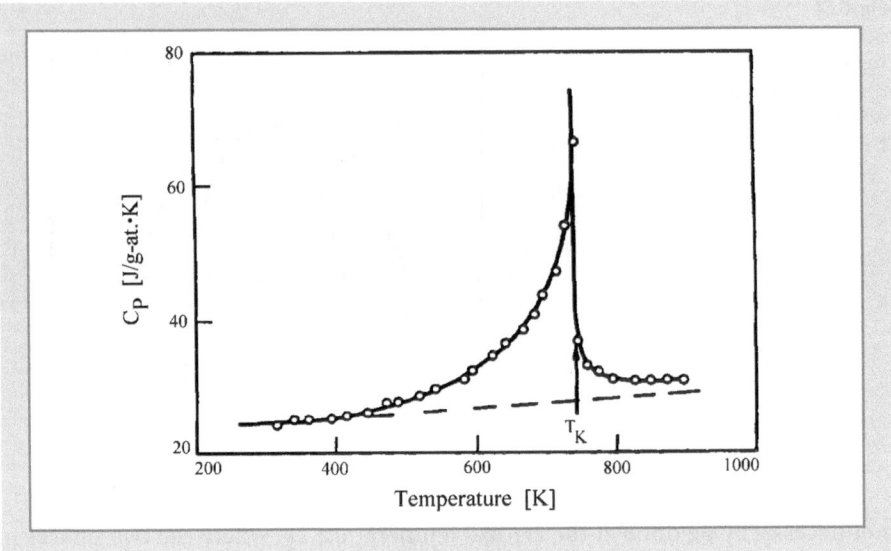

Fig. 6.55 Molar heat C_p of β-brass as a function of temperature (after [48])

discontinuity. Such processes are called transformation of the second order, because only the second derivative of the enthalpy versus temperature is discontinuous.

Figure 6.55 represents the heat capacity of β-brass as a function of temperature. In the neighborhood of the critical temperature T_K the C_p – T curve has the shape of the capital Greek letter Λ. Such a transition point is called a Λ-point. With increasing temperature in approaching T_K more and more enthalpy is needed to remove the ordered distribution of atoms. Slightly above T_K the heat capacity is not equal to the heat capacity one would expect from the extrapolation of the data way below T_K. The increase above the extrapolated value is caused by an atomic short range order, which is present at temperatures slightly above T_K, and which disappears slowly with increasing temperature.

To the transformations of second order also belong so-called magnetic transformations (ferromagnetic state → paramagnetic state). The critical temperature in this case is called the Curie-point. Iron shows such a transition with the Curie-point at 1042 K. Two phases are not in a heterogeneous equilibrium at the critical temperature of such transitions. Here only one phase exists.

6.35
Thermodynamic Properties of Ternary Alloys

The relationships valid for binary solid solutions can be transferred in an analogous fashion into ternary solutions. Sometimes, for an approximate description of the thermodynamic properties of poly-component solutions, the regular solution model is applied. It must be remembered, that in this case, a random distribution of atoms is assumed, and only the interaction between two next neighbors considered.

Accordingly, in a ternary solution made up of component A,B and C the bonding energies between the pairs A – A, B – B, C – C, A – B, A – C, and B – C are considered. Counting these pairs in a random solution with the mole fractions x_A, x_B, and x_C yields for the integral enthalpy of mixing

$$\Delta H = K_{AB}\, x_A\, x_B + K_{AC}\, x_A\, x_C + K_{BC}\, x_B\, x_C \tag{6.209}$$

The constants K_{AB}, K_{AC} and K_{BC} contain the contributions from the difference in interaction energy between pair of atoms, the coordination number z, and the Loschmidt number N_L. In particular one obtains

$$K_{AB} = N_L \cdot \frac{z}{2} \cdot (2w_{AB} - w_{AA} - w_{BB}) \tag{6.210}$$

$$K_{AC} = N_L \cdot \frac{z}{2} \cdot (2w_{AC} - w_{AA} - w_{CC}) \tag{6.211}$$

$$K_{BC} = N_L \cdot \frac{z}{2} \cdot (2w_{BC} - w_{BB} - w_{CC}) \tag{6.212}$$

In general, for a system with n components one obtains

$$\Delta H = \sum K_{ij} \cdot x_i \cdot x_j; \quad (i,j = 1,\ldots n; \quad i \neq j) \tag{6.213}$$

with

$$K_{ij} = \frac{N_L \cdot z}{2} \cdot (2w_{ij} - w_{ii} - w_{jj}) \tag{6.214}$$

The entropy of a poly-component solution with n components and random distribution of atoms is given based on statistical considerations:

$$\Delta S_n = -R \cdot \sum x_i \cdot \ln x_i; \quad (i = 1 \ldots n) \tag{6.215}$$

Thus, for ternary alloys we have

$$\Delta S = -R \cdot [x_A \cdot \ln x_A + x_B \cdot \ln x_B + x_C \cdot \ln x_C] \tag{6.216}$$

Excess functions are defined for ternary solutions as was done for binary systems.

The experimental determination of thermodynamic mixing functions of ternary solutions can be carried out in analogy to binary solutions. For example the integral enthalpies of mixing can be obtained directly using a calorimeter. With activity measurements the necessity presents itself often, to obtain the integral Gibbs energy of mixing from the partial Gibbs energy of only one component. This calculation can be carried out using a formalism developed by Darken. It is represented in Eq. (6.217) for the case of the excess Gibbs energy of mixing:

$$\Delta G^{ex} = (1-x_A) \cdot \int_1^{x_A} \frac{\overline{\Delta G_A}^{ex}}{(1-x_A)^2} \cdot dx_A \Bigg|_{\left(\frac{x_B}{x_C}=const.\right)} -$$

$$x_B \cdot \int_1^0 \frac{\overline{\Delta G_A}^{ex}}{(1-x_A)^2} \cdot dx_A \Bigg|_{x_C=0} -$$

$$x_C \cdot \int_1^0 \frac{\overline{\Delta G_A}^{ex}}{(1-x_A)^2} \cdot dx_A \Bigg|_{x_B=0} \tag{6.217}$$

The integration is naturally only possible, if for $\Delta G_A^{ex}/(1-x_A)^2$ at $x_A \to 0$ a finite value is present, for the section $x_B/x_C = const.$ for the bounding binary system A–B ($x_C=0$) and for the bounding binary system A–C ($x_B=0$). It can be shown, that these limiting values are really finite. No finite limiting value appears, when instead of $\overline{\Delta G}_A^{ex}$ the quantities $\overline{\Delta G}_{A(i)}$ or $\overline{\Delta G}_A$ are used. Finite limiting values exist also for the quantities $\overline{\Delta H}_A/(1-x_A)^2$ and $\overline{\Delta S}_A^{ex}/(1-x_A)^2$.

In comparison to binary systems, the number of ternary systems, which have been thoroughly investigated from a thermodynamic point of view is relatively small. It suggests itself, to obtain the thermodynamic properties of solutions with three components, one starts from the corresponding properties of the binary systems making up the ternary system. For this purpose several empirical methods have been developed.

In Eq. (6.217) the first term on the right side contains only data of the ternary solid solution, whereas the second and third terms relate to the binary systems A–B and A–C. This fact is used as a starting point by Kohler [49] and Toop [50] to express the thermodynamic functions of ternary solutions by corresponding functions of the constituting binary solutions. The relationship given by Kohler is:

$$\Delta G^{ex} = (1-x_A)^2 \cdot [\Delta G_{BC}^{ex}]_{\underset{x_C}{x_B}} + (1-x_B)^2 \cdot [\Delta G_{AC}^{ex}]_{\underset{x_C}{x_A}}$$

$$+ (1-x_C)^2 \cdot [\Delta G_{AB}^{ex}]_{\underset{x_B}{x_A}} \tag{6.218}$$

Figure 6.56 gives an overview about the positions of the integral excess Gibbs energy of the binary bounding systems

$[\Delta G_{BC}^{ex}]_{xB/xC},$
$[\Delta G_{AC}^{ex}]_{xA/xC}$

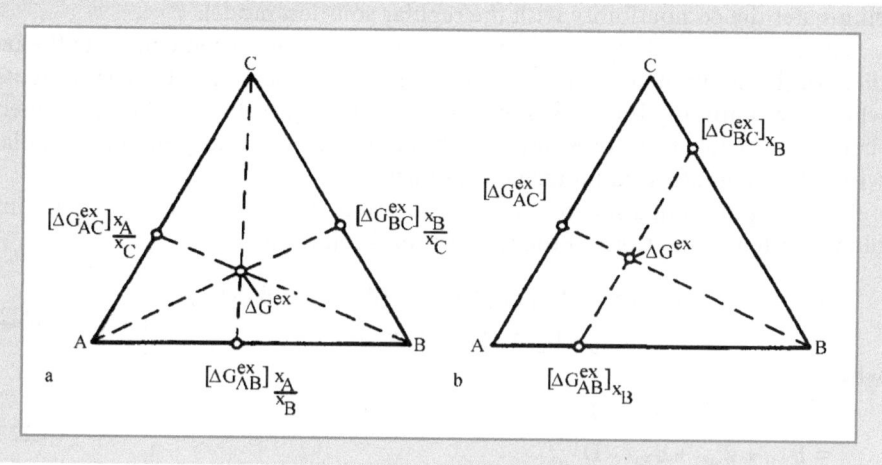

Fig. 6.56 Presentation of the concentrations of $\Delta G_{ij}{}^{ex}$ values of constitutioning bounding systems for calculation of the ΔG^{ex} values of a ternary solution. **a** After Kohler equation (6.218), **b, c** after Toop equation (6.220) respectively Bonnier equation (6.221)

$$[\Delta G_{AB}^{ex}]_{x_A/x_B}$$

which are necessary according to Eq. (6.218) to obtain the excess Gibbs energy ΔG^{ex} of the ternary solution.

This relationship was transferred by Kehianian to multi-component systems:

$$\Delta G^{ex} = \sum (x_i - x_j)^2 \cdot [\Delta G_{ij}^{ex}]_{\frac{x_i}{x_j}} \tag{6.219}$$

Toop proposes the following relationship:

$$\Delta G^{ex} = \left[\frac{x_A}{1-x_B} \cdot \Delta G_{AB}^{ex} + \frac{x_C}{1-x_B} \cdot \Delta G_{BC}^{ex}\right]_{x_B} + (1-x_B)^2 \cdot [\Delta G_{AC}^{ex}]_{\frac{x_A}{x_C}} \tag{6.220}$$

The functions of the binary bounding systems needed to determine the ΔG^{ex} values of the ternary solution according to the above equation are shown in Fig. 6.56b.

The equations of Kohler and Toop are strictly valid for regular solutions. They can be used as an approximation, if deviations from regular behavior are present. For exactly regular solutions Eqs. (6.218) and (6.220) are identical.

The relationship developed by Bonnier [51] is similar to the Toop equation:

$$\Delta G^{ex} = \left[\frac{x_A}{1-x_B} \cdot \Delta G_{AB}^{ex} + \frac{x_C}{1-x_B} \cdot \Delta G_{BC}^{ex}\right]_{x_B} + (1-x_B) \cdot [\Delta G_{AC}^{ex}]_{\frac{x_A}{x_C}} \tag{6.221}$$

The difference compared to Eq. (6.220) is in the coefficient of the last term. This eliminates the compatibility with the regular solution model.

The equations of Toop and Bonnier are not invariant when changing the indices of the components. This has especially been observed in ternary systems, where the bounding binary systems behave differently. It must be mentioned, that both the equations of Kohler and Bonnier are pure interpolation formulas without taking into account ternary interactions.

The Margules equation contains, in addition to functions of the bounding binary systems, functions of the ternary solid solution:

$$\Delta G^{ex} = x_A x_B \cdot [A_{AB} x_A + B_{AB} x_B] + x_A x_C \cdot [A_{AC} x_A + B_{AC} x_C] + $$
$$+ x_B x_C \cdot [A_{BC} x_B + B_{BC} x_C] + x_A x_B x_C \cdot C \qquad (6.222)$$

where

$$C = A_{AB} + A_{AC} + A_{BC} - D$$
$$= B_{AB} + B_{AC} + B_{BC} - D' \qquad (6.223)$$

The coefficients A_{ij} and B_{ij} are the partial excess Gibbs energies of dilute binary solutions. The coefficients D and D' must be obtained from experimental data of the ternary solution. They take care of ternary interactions. Equation (6.222) is suitable for the interpolation of data in ternary systems. For organic systems it could be shown that the term in Eq. (6.222), which takes into account the ternary interaction ($x_A x_B x_C \cdot C$), has little influence on ΔG^{ex}.

The equation of Krupkowski can similarly be used for interpolation in the ternary:

$$\Delta G^{ex} = \frac{R \cdot \alpha_{AB}}{T^{(k_{AB}-1)} \cdot (m_{AB} - 1)} \cdot x_B \cdot [1 - (1 - x_A)^{m_{AB}-1}]$$

$$+ \frac{R \cdot \alpha_{BC}}{T^{(k_{BC}-1)} \cdot (m_{BC} - 1)} \cdot x_C \cdot [1 - (1 - x_B)^{m_{BC}-1}]$$

$$+ \frac{R \cdot \alpha_{AC}}{T^{(k_{AC}-1)} \cdot (m_{AC} - 1)} \cdot x_A \cdot [1 - (1 - x_C)^{m_{AC}-1}] \qquad (6.224)$$

The quantities a_{ij}, k_{ij} and m_{ij} are constants. They can be obtained from experimental data of the bounding binary systems. The partial excess Gibbs energy of component i of a binary system is connected with these constants according:

$$\overline{\Delta G_i^{ex}} = \frac{\alpha_{ij}}{T^{(k_{ij}-1)}} (1 - x_i)^{m_{ij}} \qquad (6.225)$$

The somewhat complicated equation of Krupkowski is only applicable, when all bounding binary systems show the same sign in the deviation from ideality.

Figure 6.57 compares, in two ternary systems, the $\overline{\Delta G^{ex}}$ values obtained experimentally with the calculated ones from the excess Gibbs energies of the bounding binary systems, using the methods of Kohler and Bonnier. It is obvious, that the calculations using Eqs. (6.218), (6.220) and (6.221) yield values

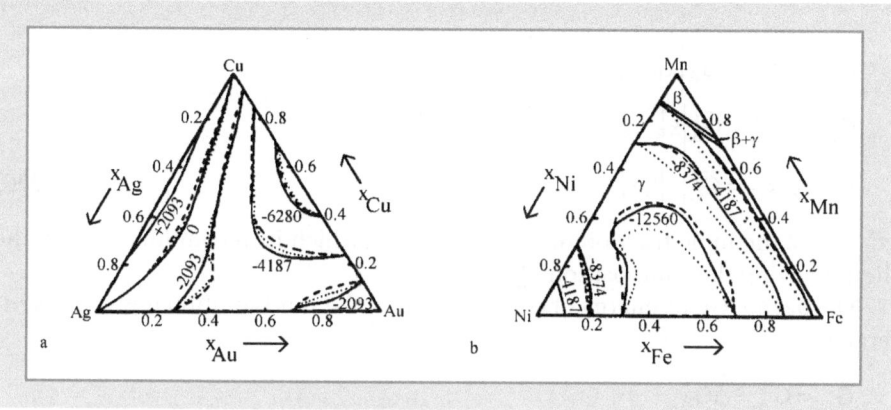

Fig. 6.57 Lines of constant ΔG^{ex} values of ternary solutions after Spencer, Hayes and Kubaschweski [52]. **a** For liquid Ag-Au-Cu alloys at 1350 K, **b** for solid Ni-Fe-Mn alloys at 1232 K; experimental, ... after Kohler, Eq. (6.218), ---- after Bonnier, Eq. (6.221)

of ΔG^{ex}, which are sufficiently accurate for many purposes.

It is useful, to carry out a few measurements of the excess Gibbs energies in ternary solutions, to increase the reliability of these ΔG^{ex} values, and use them for a better fit. In this case Eqs. (6.222) and (6.224) can be useful. Lück [53] studied the limits of applicability of the Kohler, Toop, Bonnier and Krupkowski (Muggianu [54]) extrapolation methods.

6.36
Calculation of Fusion Equilibria in Ternary Systems

Though a large number of ternary phase diagrams have been investigated, the knowledge of phase equilibria here is much more incomplete than in the case of binary systems. This applies to the number of ternary combinations investigated, as well as to the completeness of the investigated systems. In many cases only narrow, especially interesting, or experimentally easily accessible composition and temperature ranges were investigated. Because of the large number of measurements needed it will not be possible, in the near future to investigate experimentally all, industrially interesting, ternary systems. The outlook is even bleaker, to obtain this for systems with more than three components. The calculation of phase equilibria suggests itself.

In the following the solid-liquid equilibria in a ternary system are considered, where complete solubility exists in the solid and liquid phases. In systems with more than two components, as in binary systems, the condition, that at equilibrium the partial Gibbs energies of the individual components in the coexisting phases are equal, is valid. For the ternary system A – B – C in analogy to Eq. (6.106) one has:

$$\overline{G}_A^S\big|x_B^S, x_C^S = \overline{G}_A^L\big|x_B^L, x_C^L$$

$$\overline{G}_B^S\big|x_B^S, x_C^S = \overline{G}_B^L\big|x_B^L, x_C^L$$

$$\overline{G}_C^S\big|x_B^S, x_C^S = \overline{G}_C^L\big|x_B^L, x_C^L \qquad (6.226)$$

x_B^S, x_C^S are the mole fraction of the solid phase, which is in equilibrium with the liquid having the composition x_B^L, x_C^L.

The partial Gibbs energy of component A in the solid solution is given in general by:

$$\overline{G}_A^S = G_A^S + \overline{\Delta G}_A^S$$
$$= G_A^S + \overline{\Delta H}_A^S - T\overline{\Delta S}_A^S \qquad (6.227)$$

Similar relationship applies to the partial Gibbs energy of component A in the liquid.

Provided that the partial quantities in the ternary are known for the liquid and solid solutions, the phase equilibria can be obtained using Eqs. (6.226) and (6.227), as is done for a binary system.

Figure 6.58 represents schematically the Gibbs energy surfaces of three phases a, b and t in a ternary system. Instead of a common tangent to the ΔG-x curve in the binary case, here a common tangent plane must be placed to the ΔG-x_B-x_C surfaces, to obtain the composition of the coexisting phases.

If no directly measured thermodynamic data are available for a ternary system, for which phase equilibria have to be obtained, it is possible to calculate rough estimates of the phase equilibria based on suitable models. In the simplest case, the regular solution model can be applied.

The partial enthalpies of mixing of the three components in a ternary regular solution can be expresses as follows as a function of composition:

$$\overline{\Delta H}_A = K_{AB}\, x_B^2 + K_{AC}\, x_C^2 + \Delta K\, x_B\, x_C \qquad (6.228)$$

$$\overline{\Delta H}_B = K_{AB}\,(1-x_B)^2 + K_{AC}\, x_C^2 - \Delta K\, x_C\,(1-x_B) \qquad (6.229)$$

$$\overline{\Delta H}_C = K_{AB}\, x_B^2 + K_{AC}\,(1-x_C)^2 - \Delta K\, x_B\,(1-x_C) \qquad (6.230)$$

where

$$\Delta K = K_{AB} + K_{AC} - K_{BC} \qquad (6.231)$$

The constants K_{AB}, K_{AC} and K_{BC} have the same meaning as in Eq. (6.209).

The ideal partial entropies of mixing of the components can be obtained immediately from Eq. (6.216). One has:

$$\overline{\Delta S}_{A(i)} = -R \ln x_A \qquad (6.232)$$

Fig. 6.58
Explanation of the equilibrium between three phases a, b, and c in a ternary system. Hatched: common tangential plane $[A'\,B'\,(BC)'\,(AC)']$

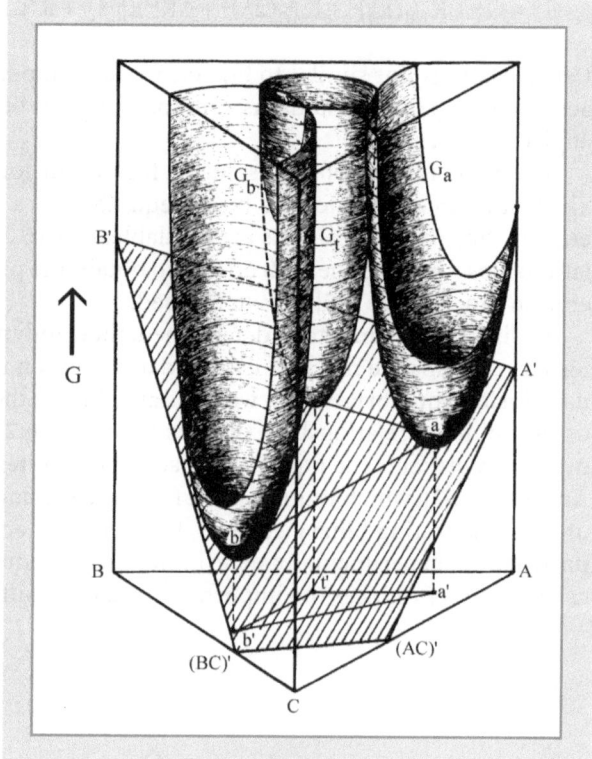

$$\overline{\Delta S}_{B(i)} = -R \ln x_B \tag{6.233}$$

$$\overline{\Delta S}_{C(i)} = -R \ln x_C \tag{6.234}$$

For component A in the case of equilibrium one obtains with Eqs. (6.227), (6.228) and (6.232)

$$G_A^S + K_{AB}^S (x_B^S)^2 + K_{AC}^S (x_C^S)^2 + \Delta K^S x_B^S x_C^S + R\,T \ln x_A^S =$$
$$G_A^L + K_{AB}^L (x_B^L)^2 + K_{AC}^L (x_C^L)^2 + \Delta K^L x_B^L x_C^L + R\,T \ln x_A^L \tag{6.235}$$

T is the equilibrium temperature. Analogous equations apply to components B and C.

Finally, one can use Eq. (6.127) for the difference of the partial Gibbs energy of the pure component in the liquid and solid state.

In the case, that no mentionable mutual solubility exits in the solid between the components, and thus, $\overline{\Delta G}_A^S$ is negligibly small, Eq. (6.227) is simplified to:

$$\overline{G}_A^S = G_A^S \tag{6.236}$$

and with Eq. (6.235) to

$$G_A^S = G_A^L + K_{AB}^L (x_B^L)^2 + K_{AC}^L (x_C^L)^2 + \Delta K^L x_B^L x_C^L + R\,T \ln x_A^L \qquad (6.237)$$

Using relationships similar to Eq. (6.237) for component B or C, and taking into account of Eq. (6.127), liquidus compositions can be calculated in a simple way at given temperature.

In the general case of solid solution formation, four unknowns, x_A^S, x_B^S, x_A^L, and x_B^L must be calculated for the fusion equilibria at a given temperature. However, only three basic equations are available: (Eq. (6.235)) and two analogous relationships for components B and C. To obtain the phase equilibria, a suitable iteration process (see [55]) has to be used.

As direct, experimentally determined, thermodynamic mixing quantities for ternary systems are mostly missing, data are often used for the calculation ternary fusion equilibria, which were obtained from the binary bounding systems, using empirical relationships (for example Eqs. (6.218), (6.220) or (6.221)). Figure 6.59 represents the liquidus surface of the system Cd–Sn–Zn according to Ansara [56, 57], which was obtained from the thermodynamic mixing functions of the constituting binary systems, based on the equation of Kohler. For comparison, the experimentally determined fusion equilibria are also shown. The calculation reproduces the important lines with sufficient accuracy.

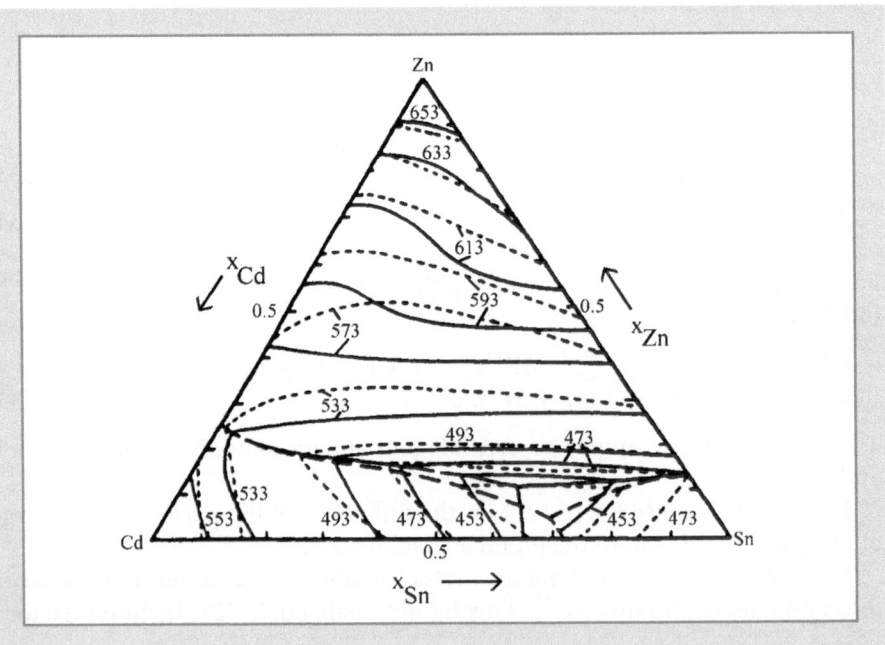

Fig. 6.59 Liquidus sphere of the system Cd-Sn-Zn (after [56, 57]) Temperature in K. experimental, … calculated with equation after Kohler (6.218) using thermodynamic values of the binary bounding systems

References

Citations

[1a] R. Hultgren, P.D. Desai, D.T. Hawkins, M. Gleiser, K.K. Kelley and D.D. Wagman, "Selected Values of Thermodynamic Properties of the Elements", Amer. Soc. for Metals, Metals Park, Ohio (1973)

[1b] R. Hultgren, P.D. Desai, D.T. Hawkins, M. Gleiser and K.K. Kelley, „Selected Values of the Thermodynamic Properties of Binary Alloys", Amer. Soc. for Metals, Metals Park, Ohio (1973)

[2] F.E. Wittig and G. Keil, Z. Metallkde., 54, (1963) 576

[3] O. Alpaut and Th. Heumann, Acta Met Vol. 13, 543-8 (1965)

[4] B. Predel, "Thermodynamik und Aufbau von Legierungen – einige neuere Aspekte" Vortrag 232, Rheinisch-Westfälische Akademie der Wissenschaften, Westdeutscher Verlag (1972)

[5] B. Predel, „Experimentelle Untersuchungen zur Thermodynamik der Legierungsbildung", Forschungsbericht No 2558, Westdeutscher Verlag, Opladen (1976)

[6] F. Sommer Journal of Non-Crystalline Solids, 117/118 (1990) 505–512

[7] F. Sommer, Z. Metallkde 73 (1982) 77

[8] F. Sommer, Z. Metallkde 73 (1982) 72

[9] H.-G. Krull, R.N. Singh and F. Sommer, Z. Metallkde 91 (2000) 356

[10] F. Sommer, M. Keita, H.-G. Krull and B. Predel, J. Less Common Metals 137 (1988) 267

[11] F. Sommer and H.-G. Krull, Z. Metallkde 83 (1992) 533

[12] F. Sommer, Ber. Bunsenges. Phys. Chem. 87 (1983) 749

[13] K. Nagarajan and F. Sommer, J. Less-Common Met. 146 (1989) 89

[14] T.B. Massalski, H. Okamoto, P.R. Subramian and L. Kacprzak: "Binary alloy phase diagrams" 2nd Ed., ASM, Materials Park, Ohio

[15] R. Blachnik and B Gather, Z. Metallkde 74 (1983) 172

[16] B. Predel and G. Oehme, Z. Metallkde 70 (1979) 450

[17] M. Hoch, and I. Arpshofen, Z.Metallkde 75 (1984) 23

[18] M. Hoch, Calphad 11 (1987) 219

[19] E.A. Guggenheim, Mixtures, Clarendon Press, Oxford (1952)

[20] L. Pauling, The Nature of the Chemical Bond, Third Edition, Cornell University Press, Ithaca, New York (1960)

[21] C. Wagner, Thermodynamic of Alloys, Addison-Wesley Publishing Co., Reading MA (1952)

[22] W. Kauzmann, Chem. Rev., 43 (1948) 219

[23] M. Hoch, High Temperatures-High Pressures 24 (1992) 87

[24] M. Hoch, Met. Trans. 23B (1993) 309

[25] M. Hoch, Z. Metallkde 83 (1992) 820

[26] M. Hoch, AiChE Symposium Series, 73 (170), pp. 36-8 (1977); W. R. Schmeal, Ed.

[27] J.-F. Babelot and M. Hoch, High Temperature – High Pressures. 21, 79 (1989) 79

[28] S. Hassam, M. Gaune-Escard, J.P. Bros and M. Hoch, Met. Trans., 19A (1988) 2075

[29] W. Ptak, Z. Moser and W. Zakulski, Archives of Metallurgy 33, (1988) 331–354

[30] M. Hoch and Z. Moser, Archives of Metallurgy 37 (1992) 283–296

[31] M. Gambino, J.P. Bros and M.Hoch, Thermochimica Acta, 314 (1998) 247–254

[32] M. Hoch, J. of Alloys and Compounds 220, (1995) 27–31

[33] V.N. Lozovskiy and N.F. Politova, Russ Met. (3) (1977) 28

[34] O. Kubaschewski, C.B. Alcock and P.J. Spencer, Materials Thermochemistry, Sixth Edition, Pergamon Press, NY, 1992

[35] M. Hoch, J. of Phase Equilibria 14 (1993) 296–302

[36] F.E. Rosztoczy and D. Cubicciotti, J. Phys. Chem., 69 (1965) 1687

[37] M. Hoch, J. of Phase Equilibria 14 (1993) 710–17

[38] E.M. Levin, C.R. Robbins and H.F. McMurdie, Phase Diagrams for Ceramists, Vol. I, American Ceramic Society, Columbus Ohio, 1964

[39] M. Hoch, CALPHAD 18 (1994) 409-428

[40] M. Hoch, CALPHAD 20 (1996) 363-377

[41] K.S. Pitzer and L. Brewer, revised edition of "Thermodynamics" by G.N. Lewis and M. Randall, McGraw Hill (1961)

[42] R.A. Robinson and R.H. Stokes, Electrolyte Solutions, 2nd ed. revised Butterworth Scientific Publications, London, 1968

[43] H.S. Harned and B.B. Owen, The Physical Chemistry of Electrolyte Solutions, ACS, Reinhold, New York, 1958

[44] W.F. Linke, Solubilities, Inorganic and Metal-Organic Compounds, 4th ed., Am. Chem. Soc. Washington DC. (1965)

[45] M. Hansen and K. Anderko, "Constitution of Binary Alloys", McGraw-Hill Book Comp., New York (1958)

[46] T. Heumann, Z. Elektrochemie, 57, (1953) 724

[47] B. Predel, Z. Metallkde., 56, (1965) 791

[48] H. Moser, Zeitschrift Physik, 37, (1936) 737

[49] F. Kohler, Monatshefte Chem., 91, (1960) 738

[50] G.W. Toop, Trans. AIME, 233, (1965) 850

[51] E. Bonnier and R. Caboz, C. R. Acad. Sci., 250, (1960) 527

[52] P.J. Spencer, F.H. Hayes and O. Kubaschewski, Rev. Chim. Minerale, 9, (1972) 13

[53] R. Lück, U. Gerling and B. Predel, Z. Metallkde 77 (1985) 442–448

[54] Y.N. Muggianu, M. Gambino and J.P. Bros, J. Chim. Phys. 72 (1975) 83.

[55] L. Kaufman and H. Bernstein, "Computer Calculations of Phase Diagrams", Academic Press, New York (1970)

[56] I. Ansara, P. Desré and E. Bonnier, C. R. Acad. Sci., 270, (1970) 1098

[57] I. Ansara, Proceedings of a Symposium held at Brunel University and the National Physical Laboratory, 1971, National Physical Laboratory, London, Her Majesty's Stationery Office (1972)

[58] B. Predel and J. Piel, Z. Metallkde 66 (1975) 33

General References

B.L. Averbach, "Solid Solution Formation", in "Energetics in Metallurgical Phenomena", Vol. II, Editor: W. M. Mueller, Gordon and Breach Science Publishers, New York (1965)

L.S. Darken, J. Amer. Chem. Soc., 72, (1950) 2909

P. Dörner, L.J. Gaukler, H. Krieg, H.L. Lukas, G. Petzow and J. Weiss, Calphad 3, (1979) 241

J.D. Fast, „Entropie", Philips Technische Bibliothek, Eindhoven (1960)

R. Haase, „Thermodynamik der Mischphasen", Springer-Verlag, Berlin (1956)

R. Haase und R. Schönert, "Solid-Liquid Equilibrium", in: "The International Encyclopedia of Physical Chemistry and Chemical Physics", Topic 13, Mixtures, Solutions, Chemical and Phase Equilibria, Vol. I, Editor: M. L. McGloshan, Pergamon Press, Oxford (1969)

J. Lumsden, "Thermochemics of Molten Salt Mixtures", Academic Press, London (1960)

J. Nyvlt, "Solid-Liquid Phase Equilibria", Elsevier Scientific Publishing Comp., Amsterdam (1977) Fifth Edition, Pergamon Press, London (1979)

B. Predel, Z. Metallkde., 49, (1958) 226

A. Prince, "Alloy Phase Equilibria", Elsevier Publishing Comp., Amsterdam (1966)

H. Schmalzried and A. Navrotsky, "Festkörperthermodynamik", Verlag Chemie, Weinheim (1975)

W. Schottky, „Thermodynamik", Springer-Verlag, Berlin (1973)

R.A. Swalin, „Thermodynamics of Solids", J. Wiley and Sons, New York (1972)

C. Wagner, "Thermodynamics of Alloys", Addison-Wesley, London (1952)

CHAPTER 7

Nucleation During Phase Transitions

7.1
General

Experience has shown that, during cooling of a liquid, crystallization does not start immediately when the solidification temperature T_g is reached. The new phase is formed only after a more or less severe undercooling ΔT below the equilibrium temperature. This phenomenon is especially noticeable with transformation or precipitation reactions in solids. It is based on the fact, that in a situation, void of energy influences, casual fluctuations of the atomic structure of the starting phase cause the appearance of a very small volume fraction of an atomic arrangement, which is equal, or at least similar to that in the end phase. These new phases, called nuclei, can grow into macroscopic dimensions. This kind of new phase formation is called a nucleation-growth process. Only when all phases belonging to a heterogeneous equilibrium are formed, can the laws of phase equilibria be valid.

Nuclei formation can be influenced by a series of energetic-structural factors. The homogeneous nucleation, caused by fluctuation only, without other influences, are the exception. As a rule, nucleation is accelerated by "catalysts". The understanding of the catalytic action on so-called heterogeneous nucleation is very important for industry. During solidification of metals, emphasis is placed on easy nucleation. The crystallization can then start at a lot of places in a large volume of liquid. For each nucleus, a crystal is formed. With large number of nuclei many crystals are formed which, during their growth, encounter each other soon. The crystallographic orientation of the many small crystals is, in general, random. For example, this leads to quasi-isotropic mechanical properties of the solid, which are often desired.

7.2
Homogeneous Nucleation Without Change of Composition

The characteristic features of homogeneous nucleation – formation of nuclei, able to grow in a homogeneous phase by fluctuation of the atomic structure – have been developed by Volmer and Weber [1] and also by Becker and Döring

[2]. We consider first the case, where the new phase has the same composition as the starting phase. The driving force of the phase transformation is strictly given by the change in the state of aggregation, or the change in the crystallographic structure.

The formation of a small volume fraction, which has the structure of the end phase, is connected to a decrease of the Gibbs energy of the system by ΔG_V. Simultaneously, a phase boundary appears between the nucleus and the starting phase. The nucleation is thus, connected with an increase in the boundary Gibbs energy by ΔG_σ. The total change in Gibbs energy ΔG of the system connected to the nucleation is given by

$$\Delta G = \Delta G_V + \Delta G_\sigma \tag{7.1}$$

In the simplest case one can assume that the nucleus is spherical. The change of the Gibbs energy during nucleation is then related to the radius r of the nucleus:

$$\Delta G = 4/3 \ \pi \ r^3 \ \Delta G_C + 4 \ \pi \ r^2 \ \sigma \tag{7.2}$$

ΔG_C is the change in Gibbs energy per unit volume for the formation of the new phase and σ is the grain boundary energy.

A graphical representation of the Gibbs energy balance is given in Fig. 7.1. ΔG_V has a negative, ΔG_σ a positive sign. The sum of the two values, ΔG, shows with increasing radius r of the nuclei, first a positive sign. The formation of a nucleus able to grow, by random fluctuation of the structure, is first associated with an increase of the Gibbs energy of the system.

Fig. 7.1
Balance of the Gibbs energy concerning homogenous formation of nuclei without change of concentration

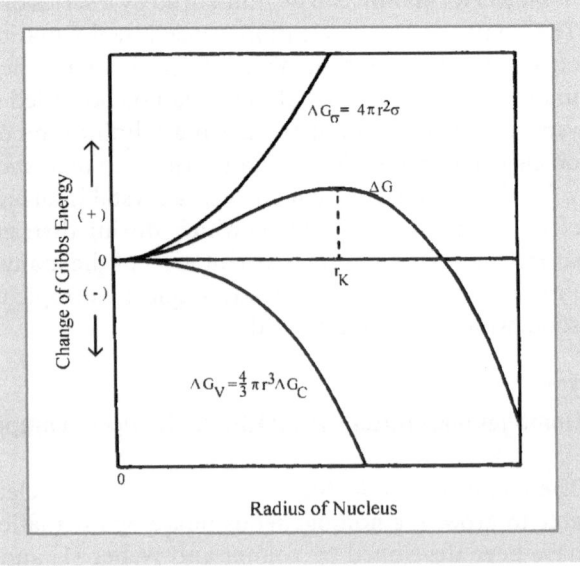

Fig. 7.2
Schematic presentation of
the dependence of ΔG_C on
the supercooling ΔT

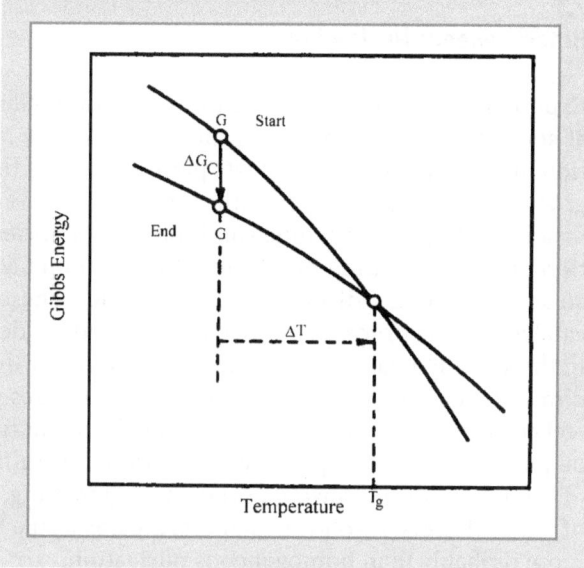

After the nucleus reached a critical radius r_K, each addition of atoms to increase the region, which has the structure of the end phase, is connected with a decrease of the Gibbs energy of the system. Such formations can grow spontaneously. At $r < r_K$ the probability is greater towards shrinking than towards growing, at $r > r_K$ the inverse is true. At $r = r_K$ the ΔG-r curve has a maximum:

$$\frac{d\Delta G}{dr} = 0 \qquad (7.3)$$

The radius of the critical nucleus is obtained using Eq. (7.2)

$$r_K = -2\frac{\sigma}{\Delta G_C} \qquad (7.4)$$

The critical radius is determined by the ratio of specific grain boundary energy and ΔG_C. Both quantities are temperature dependent: As a rule σ changes much less with temperature than ΔG_C. Thus, σ can be considered approximately as temperature independent. The dependence of ΔG_C on T is shown in Fig. 7.2.

Below T_g, the equilibrium temperature, ΔG_C, is negative. Its value increases with increased undercooling ΔT. Thus, according to Eq. (7.4) the critical radius r_K decreases with increasing ΔT, and the probability, through statistical fluctuations, to obtain a nucleus able to grow, increases. At T_g naturally $\Delta G_C = 0$, and thus, $r_K = \infty$. At the equilibrium temperature the probability for nucleation is zero.

7.3
Heterogeneous Nucleation

Experience has shown that the formation of nuclei does not occur at any, but at some very specific places in the volume of the starting phase. At solidification of a liquid, nucleation starts preferentially at the boundary to the container walls, at oxide particles, or other solid bodies in contact with the liquid ready to solidify. In polycrystalline solid substances nucleation will, during precipitation or transformation processes, often be found at grain boundaries or other structural defects – dislocations or vacancies. These places serve as "nucleation catalysts". This process eliminates the structural defects, having higher energy in the starting phase, in the nucleated volume, or the grain boundary energy is diminished at the point where the nucleation takes place. Thus, the catalytic effect consists of a greater lowering of the Gibbs energy of the system, than would be possible with homogeneous nucleation. It is difficult to avoid the formation of grain boundaries towards solids during melting, and to avoid the formation of lattice defects in solid substances, thus, as a rule heterogeneous nucleation is more probable than homogeneous nucleation.

A lowering of the grain boundary energy between the starting phase and a foreign body is only possible, if the foreign body in better wetted by the new, forming phase than by the starting phase. In the simplest case the nucleus does not have a spherical form, but the shape of a calotte, as shown in Fig. 7.3. A

Fig. 7.3
Explaining the principle of
heterogenous nucleation

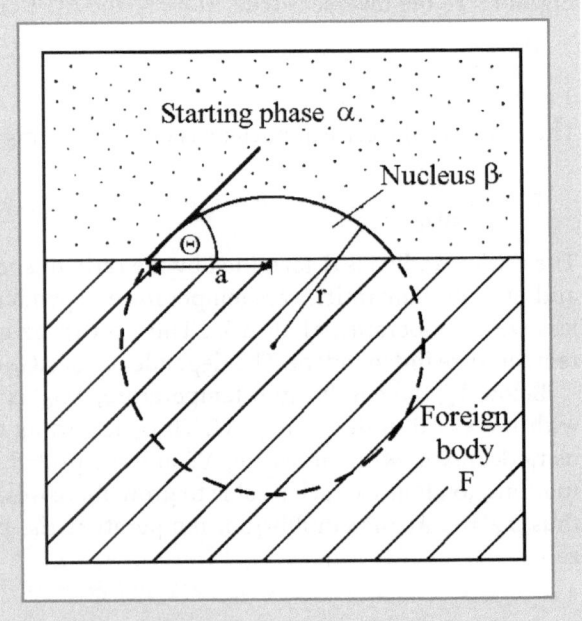

measure of the wetability is the angle of contact, Θ. With increasing wetability Θ decreases. The contact angle is fixed by the equilibrium relationships between the effective surface energies:

$$\sigma_{\alpha F} = \sigma_{\beta F} + \sigma_{\alpha\beta} \cos \Theta \tag{7.5}$$

where

$\sigma_{\alpha F}$ = the specific surface energy of the surface between starting matrix α and foreign body F,

$\sigma_{\beta F}$ = the specific surface energy of the surface between nucleus β and foreign body F,

$\sigma_{\alpha\beta}$ = the specific surface energy of the surface between starting matrix α and nucleus β.

The size of the nucleus can be expressed by the radius of the base of the calotte, and the contact angle Θ, as shown in Fig. 7.3. In analogy to the spherical nuclei during homogeneous nucleation, one can also calculate a Gibbs energy balance, based on the volume and surface of the calotte as well as the quantities ΔG_C and $\sigma_{\alpha\beta}$. This leads to the relationship

$$a_K = r_K \sin \Theta \tag{7.6}$$

where a_K is the radius of the base of the calotte, and r_K the corresponding radius of the sphere for the critical nucleus. In the case of homogeneous nucleation a complete sphere with radius r_K must be formed as the critical nucleus. In the presence of a foreign body in the starting material a part of this sphere can already represent a viable nucleus. The size of this calotte shaped part of a sphere depends on the calotte angle Θ. This is given by the ratio of the effective surface energies. With complete wetting of the foreign body by the newly forming phase ($\Theta \rightarrow 0$) a very small calotte volume forms a variable nucleus. Only a few atoms are needed, to form this nucleus. The probability for nucleus formation is correspondingly greater, than without this boundary. Based on the above considerations, grain refining substances are selected, and added to the melts, to obtain a fine crystalline material on an industrial scale.

The preferential nucleation on grain boundaries, as it is often observed during transformation or precipitation reactions in polycrystalline solids, follows analogous principles (see Fig. 7.4). At the place of the nucleus, the boundary between the crystals, the grain boundary is eliminated and the grain boundary Gibbs energy contained in the grain boundary liberated. It enters into the balance of the Gibbs energy as a gain due to nucleation. Two new boundaries are created, with input of energy: crystal 1 – nucleus, and crystal 2 – nucleus. The shape of the nucleus corresponds to the calottes joined at the base. The angle Θ of the double calotte is composed by the interaction of the boundary energies of the boundaries crystal 1 – crystal 2, crystal 1 – nucleus, and crystal 2 – nucleus.

Fig. 7.4
Nucleation at grain
boundaries

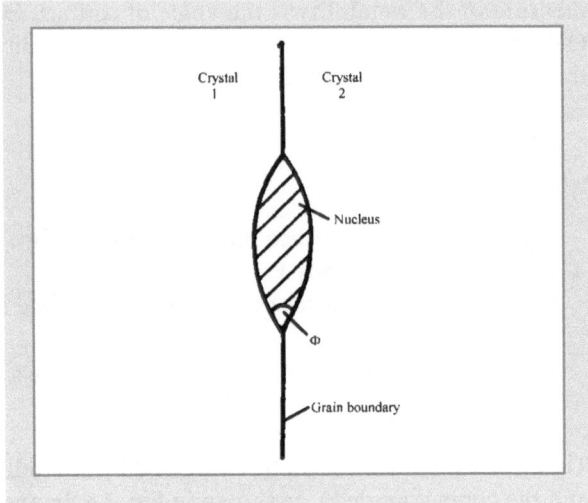

Fig. 7.5
Example of a preferential
precipitation of a new phase
at a phase boundary K
(W. Gust und H. Kuhlmann).
Precipitation of an intermediate phase δ from supersaturated Cu-In solid, solution
containing 7.5 at.-% In after
43 h at 738 K

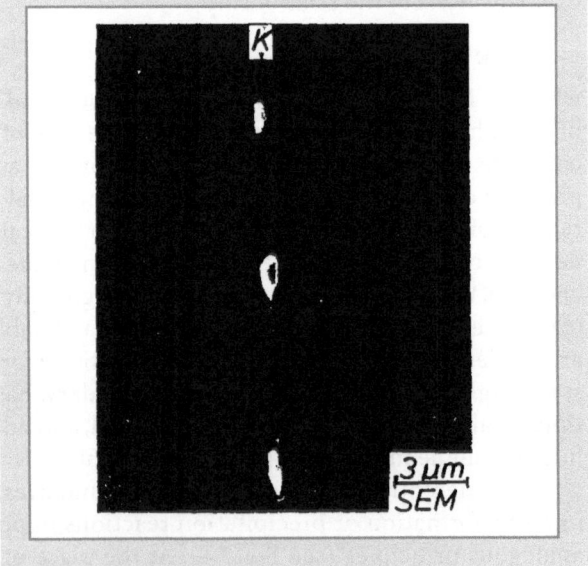

As an example Fig. 7.5 reproduces a preferential nucleation at the grain boundaries of the precipitation of an intermetallic phase (δ) from an supersaturated Cu-In (7.5 at.-% In) solid solution.

Lattice defects are also places in the starting matrix of increased energy content for preferential nucleation. If a nucleus is formed at this place, the elastic energy in the lattice deformation is removed. This provides, in principle, a similar gain in Gibbs energy as in the case of nucleation at grain boundaries, and which is available for a certain part of the energy needed for the construction of the new boundaries nucleus-matrix. Plastic deformation can increase the lattice defects in a substance. This can be used to increase the nuclei in a transformation ready matrix.

It has to be remembered, that at the formation of a nucleus in a solid starting matrix – differently from nucleation in a melt – the noticeable change in volume connected with the formation of a new phase can lead to lattice deformations in the starting matrix and the nuclei. This increases the Gibbs energy during nucleation. This increase is different for different shapes of the nuclei. The shape of a sphere is no longer the optimal form for nuclei, but as a rule it is an ellipsoid, the great and small axis depending on the elastic properties of the systems. Finally the grain boundary energy at the nucleus-starting material boundary in a crystalline matrix depends on the crystallographic orientation of the two neighboring phases, which provides preferential nucleation and growth in preferential directions.

7.4
Homogeneous Nucleation with Change in Composition

In multi-component systems the phases in equilibrium can have different compositions. Therefore, at nucleation, the change in Gibbs energy due to the change in composition has to be taken into account. Naturally, the other quantities, which contribute to the energy balance of nucleation, when a transformation from liquid to the solid state, or a transformation of a solid into a different crystallographic modification occurs, remain, when solidification or change in structure and simultaneous composition change occur during nucleation.

To simplify the discussion, the nucleation in a multi-component system will be considered in a miscibility gap. In this case the precipitated phase has the same structure as the starting phase.

The $G-x$ curve shows a fold in the region of the miscibility gap. The equilibrium compositions are given naturally by the contact points of the double tangent. Becker [3] has developed the ideas about the nucleation of the new phase by undercooling of the starting matrix below the equilibrium temperature given by the miscibility gap. Let:

N = number of atoms in the total considered volume
n = number of atoms in the nucleus
x_0 = composition of the starting phase
α = composition of the matrix after nucleus formation
β = composition of the nucleus
Ω = average volume of an atom

$G(x_0), G(\alpha), G(\beta)$ = Gibbs energies per unit volume at the corresponding compositions (at given state variables temperature T, and pressure p).

If a region is formed out of n atoms, which has the composition of the new phase β, the change in Gibbs energy of the system ΔG_V is given by:

$$\Delta G_V = \Omega \, [n \, G(\beta) + (N - n) \, G(\alpha) - N \cdot G(x_0)] \qquad (7.7)$$

If $\alpha \approx x_0$, thus, only small supersaturation exists, Eq. (7.7) can be written as

$$\Delta G_V = n \cdot \Omega \cdot \left[G(\beta) - G(x_0) - (\beta - \alpha) \cdot \left(\frac{\partial G}{\partial x} \right)_{x_0} \right] \qquad (7.8)$$

During the formation of a nucleus with n atoms, a new interface (between nucleus and matrix) with the size F is formed simultaneously. If s is the number of atoms per unit surface, then the Gibbs energy, ΔG_σ, needed to form this interface is

$$\Delta G_\sigma = s \, F \, z \, K \, (\beta - x_0)^2 \qquad (7.9)$$

Equation (7.9) is based on the rules of the regular solution model; K is the interaction parameter and z the coordination number.

The total change in Gibbs energy at nucleation (with $\alpha \approx x_0$) is obtained from Eqs. (7.8) and (7.9):

$$\Delta G = -n \cdot \Omega \cdot \left[G(x_0) - G(\beta) + (\beta - x_0) \cdot \left(\frac{\partial G}{\partial x} \right)_{x_0} + s \cdot F \cdot z \cdot K \cdot (\beta - x_0)^2 \right] \qquad (7.10)$$

The gain in Gibbs energy per unit volume of the precipitated phase is

$$\Delta G' = - \left[G(x_0) - G(\beta) + (\beta - x_0) \cdot \left(\frac{\partial G}{\partial x} \right)_{x_0} \right] \qquad (7.11)$$

This quantity is the line $B_1 B_2$ in Fig. 7.6.

When the phase capable to precipitate during cooling has reached a point on the miscibility gap, we have $\alpha = x_0$. In this case the Gibbs energy of the system cannot be reduced by nucleation, $\Delta G' = 0$: nucleation cannot take place. If, however, during further cooling, the starting phase reaches the region surrounded by the miscibility gap, the inflection point S on the A side of the G–x curve, whose fold increases with decreasing temperature, approaches the composition x_0. Correspondingly, the slope of the tangent to the G-x curve at point x_0 increases. With increasing approach of the inflection point S to x_0 the value of $\Delta G'$ increases, and is maximum when x_0 lays at the inflection point of the G–x curve.

Analogous relationships also apply, when a phase with a different crystal structure is formed from that of the starting matrix. Figure 7.7 shows the Gibbs energies at the pearlite formation in the system iron-carbon as a function of composition, for a temperature which is below the eutectoid temperature T_e. At equilibrium, phases with the compositions x_α and x_{Fe_3C} coexist, determined by the common tangent, a, to the G-x curve of the ferrite and the G–x curve of the

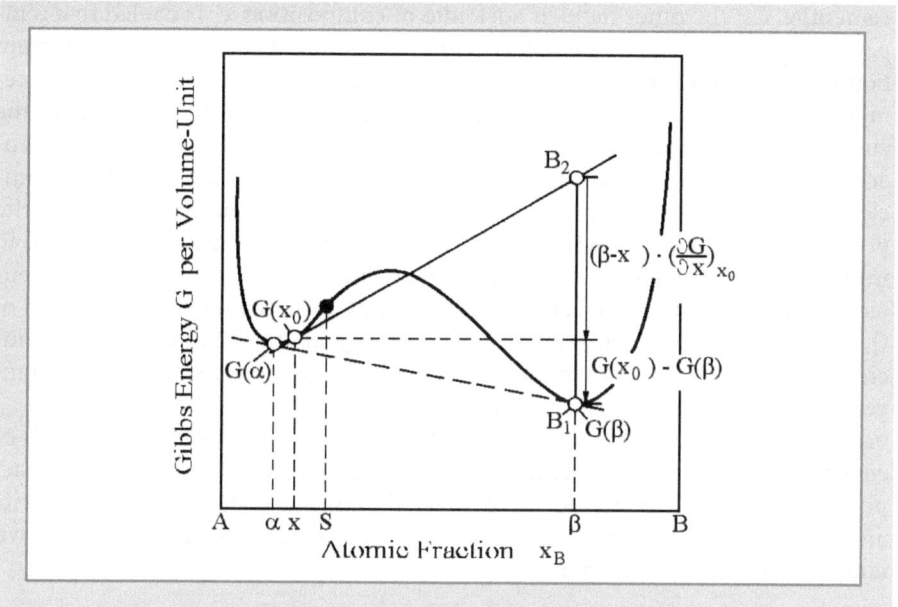

Fig. 7.6 Principle of homogenous nucleation followed by change of concentration

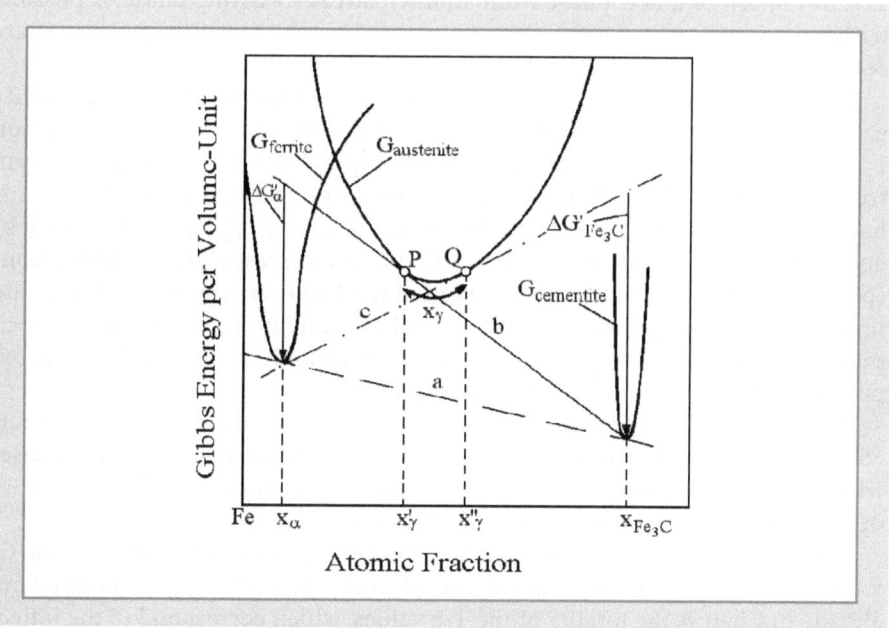

Fig. 7.7 Schematic presentation of the change in free enthalpy by nucleation of pearlite in the system Fe-C

cementite. On the other hand, if austenite of composition x'_γ is cooled to a temperature $T < T_e$, ferrite nucleation occurs exclusively. The nuclei form at grain boundaries. The change in Gibbs energies at the formation of ferrite resp. cementite can be determined graphically by the tangent, b, to the G-x curve of the austenite at the composition x'_γ (contact point P). The decrease in the Gibbs energy at the formation of ferrite nuclei is large, at the formation of cementite nuclei almost zero. No cementite nuclei can form. In the measure that the ferrite precipitations grow, the austenite matrix next to them becomes impoverished in iron. With increasing local carbon content (x'_γ moves in direction of x''_γ) the contact point P of the tangent, b, to the G-x curve of austenite moves in direction of Q, and $\Delta G'_\alpha$ decreases. On the other hand $\Delta G'_{Fe_3C}$ assumes finite values, and increases continuously. If the contact point of the tangent has reached Q (now tangent c), the value of $\Delta G'_\alpha$ has fallen to almost zero, whereas the value of $\Delta G'_{Fe_3C}$ became significant. Now nucleation of cementite follows. During the growth of cementite nuclei the austenite matrix close to it is enriched in iron. The contact point Q moves in direction of P, and so on. By this method, nucleation of ferrite and nucleation of cementite change periodically, which leads to the well known lamellar structure of pearlite.

7.5
Spinodal Decomposition

The formation of a new phase from another, already existing phase, is possible without nucleation, if certain energetic conditions are met. The conditions are seen in Fig. 7.8.

Figure 7.8 shows a G-x line, with a fold, as is typical inside of a miscibility gap. If, in a thought experiment, the single phase solution with the composition x_B^1 is transformed into two phases, whose compositions are not much different from the original phase, then, according to the lever rule, one of the phases must have a composition richer in B, x_B^2, the other poorer in B, x_B^3. Naturally, the Gibbs energies of these phases lie on the G-x line, at the corresponding compositions (points A and B). As can be seen in Fig. 7.8 the Gibbs energy of this two phase mixture is greater than that of the single phase starting solution. Such a precipitation reaction cannot start spontaneously, because it is connected with an expenditure of Gibbs energy ($\Delta G^{2,3}$ is positive).

The situation is different, if one starts with the supersaturated (or undercooled) solution with the composition x_B^4. Here the formation of two phases with the compositions x_B^5 and x_B^6 (points C and D) is connected with the decrease of the Gibbs energy of the system ($\Delta G^{5,6}$ is negative).

The two regions considered here differ in the sign of the curvature of the G-x line. At the inflection point the sign of the curvature changes. As pointed out already in Chap. 6, the totality of the T-x values, which correspond to the inflection points in the G-x lines is the spinodal line, or spinodal. The composition x_B^1 lies outside the spinodal, the composition x_B^4 inside.

Fig. 7.8
Explanation of the energetic
principle of the spinodal
demixing

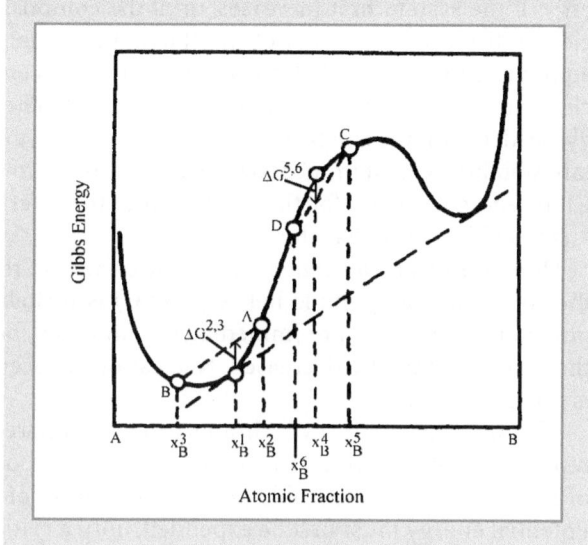

Fig. 7.9
Miscibility gap M'-M'' and
spinodal S'-S'' in the system
with demixing tendency

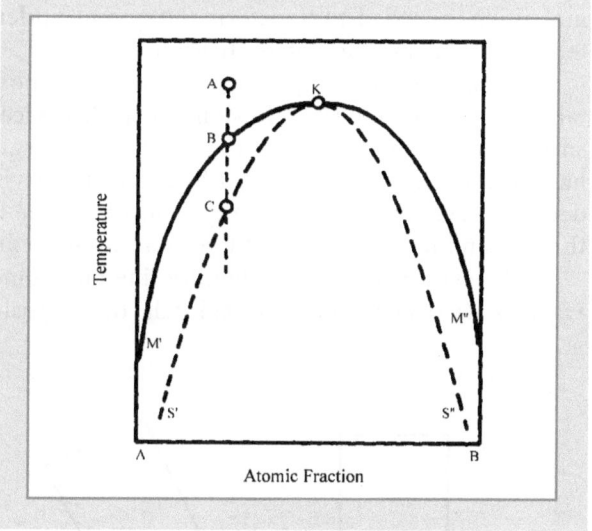

Figure 7.9 reproduces schematically a miscibility gap (solubility line M'-M'')
and the corresponding spinodal S'-S''. Single phase solutions at temperatures
above the miscibility gap (for example at A) are stable, single phase solutions
in the region between the miscibility gap and the spinodal are metastable, single phase solutions within the spinodal are unstable. Cooling of a single phase
solution, A, below point B, makes it metastable. Precipitation of a second phase
through a nucleation-growth process can take place. At nucleation the Gibbs en-

ergy of the system first increases, until the composition of the nucleus is sufficiently displaced from that of the starting material towards that of the second equilibrium phase, so that during further displacement and growth of the nucleus the Gibbs energy of the systems decreases. The expenditure of Gibbs energy, needed with the small changes in composition, gives the starting phase a certain stability against immediate decomposition into the equilibrium phases. The Gibbs energy, needed for the formation of the interface between the two newly formed phases, also has the same effect.

On the contrary, a single phase solution within the T-x range surrounded by the spinodal (for example below point C) is unstable. Thermally caused composition fluctuations are immediately reinforced, because the Gibbs energy of the system is thereby decreased. Nucleation, as occurs with metastable phases, is not necessary.

The different stability conditions can be compared with a mechanical system represented in Fig. 7.10. If a brick, in a metastable position, is tilted over the edge of its base towards the stable position, its centre of gravity must first be lifted. Potential energy must first be expended, until a critical tilt angle is reached, and when it is passed, the system goes spontaneously into the stable situation under decrease of its potential energy. With an unstable starting position, only minimal tilting is needed, to transform the system, under immediate decrease of potential energy, into the stable situation.

At the spinodal decomposition statistical composition fluctuations are spontaneously reinforced. The diffusion, which always leads, outside of the T-x region surrounded by the spinodal, to a reduction of the given concentration gradient, has the opposite effect (up-hill diffusion). The driving force in both cases is the decrease in Gibbs energy of the system. Within the spinodal this can only occur through an increase of the random concentration gradients.

At the critical point, K, solubility line and spinodal touch each other (see Fig. 7.9). Here the region of the spinodal decomposition is immediately reached

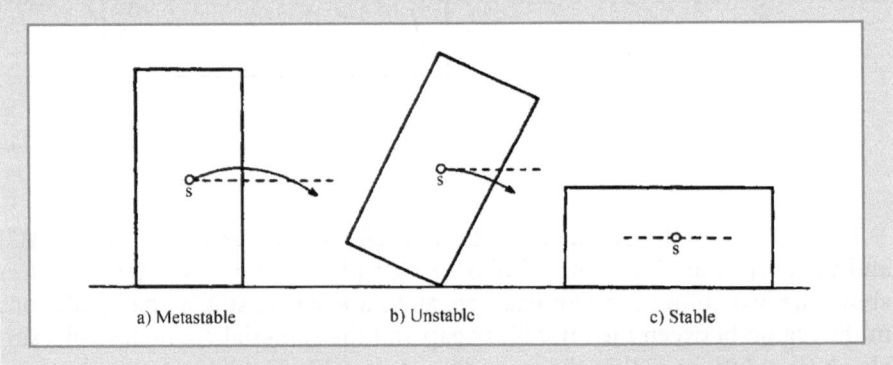

Fig. 7.10 Mechanical analagon to explain the situation of stability in a system with demixing tendency. S = centre of gravity

by cooling of a liquid or solid single phase solution, without passing through a metastable intermediate region. The spinodal region can also be reached at other compositions in the case of solid solutions; however, as a rule, the cooling rate must be high enough to be able to pass the metastable region without nucleation starting. The nucleation is, in certain systems, greatly retarded, due among others, to a low diffusion rate. The spinodal decomposition yields, in general, a very finely dispersed structure.

References

Citations
[1] M. Volmer and A. Weber, Z. Physikal. Chem., 119, (1926) 277
[2] R. Becker und W. Döring, Annalen der Physik, 24, (1935) 719
[3] R. Becker, Z. Metallkde., 29, (1937) 245, Annalen der Physik, 32, (1938) 128

General References
M.B. Bever, "Nucleation Processes", in: "Energetics in Metallurgical Phenomena", Editor: W.M. Mueller, Gordon and Breach Science Publ., New York (1965)
B. Chalmers, "Principles of Solidification", J. Wiley and Sons, New York (1964)
G. Matz, "Kristallisation", Springer-Verlag, Berlin (1969)
W.C. Winegard, "An Introduction to the Solidification of Metals", The Institute of Metals, London (1964)
A.C. Zettlemoyer (Editor), "Nucleation", Marcel Dekker Inc., New York (1969)

CHAPTER 8

Metastable Phases

8.1
Energetics of the Nucleation of Metastable Crystalline Phases

The situation that, during nucleation of a new phase, the sum of the Gibbs energy terms, ΔG, contains terms with different signs, creates the possibility of the formation of metastable phases. This possibility will be explained based on Fig. 8.1.

In Fig. 8.1 the Gibbs energies of the starting phase α and two further phases β and γ, at a given temperature are represented. At equilibrium, with the overall composition of the starting phase $x_B^{\alpha,0} > x_B^{\alpha,(\beta)}$, a two phase mixture, $\alpha + \beta$, is present. The possibility of nucleation of the β phase is given in principle: $\Delta G'_\beta$, the gain in Gibbs energy during nucleation is considerable.

Fig. 8.1
Explanation of the energetics
of nucleation of a metastable
phase. α starting phase,
β generated stable phase,
γ metastable phase

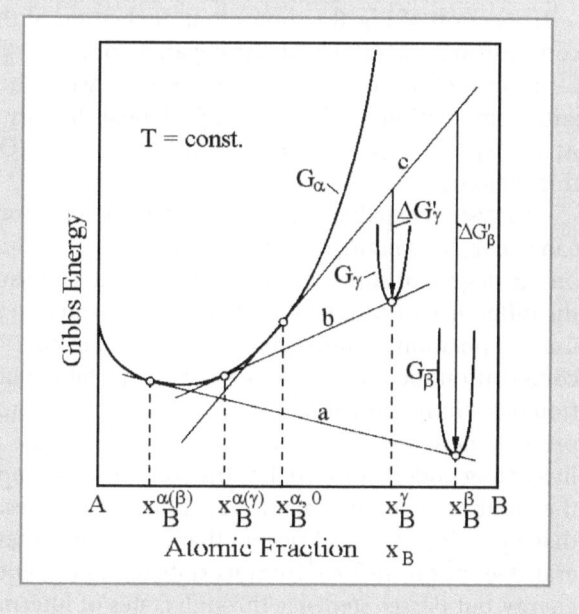

Phase γ cannot coexist with α in a stable equilibrium, because the Gibbs energies of the mechanical mixture, given by the common tangent b drawn through phases α and γ to the G – x line, are higher, than that of the equilibrium mixture α and β (tangent a). There is a driving force for the nucleation of γ, but the gain in $\Delta G'_\gamma$ in the sum of the Gibbs energies is significantly lower than for the nucleation of β. For the probability of nucleation, $\Delta G'_\sigma$, the Gibbs energy expended for the formation of an interface, is also of significance. The interface energy depends on the structural difference of the neighboring phases, and with two solid phases also on the orientation difference at the common interface, and of the position of the interface. The lattice distortions connected with the nucleation can also involve a large expenditure of Gibbs energy. Thus, it is possible, that due to a close structural relationship between α and γ, expenditure of interfacial and distortion energy is much smaller, so that in spite of $\Delta G'_\beta > \Delta G'_\gamma$, the nucleation of β begins earlier than that of the thermodynamically stable β phase.

At temperatures way below the melting point, where the diffusion rate is relatively low, the thus, formed metastable phases can exist for a long time, without transforming into stable phases. As an example, we give the metastable, intermetallic phase Fe_3C (cementite), which forms either during solidification of industrial Fe-C melts with high enough carbon content, or by precipitation from a supersaturated austenite during a solid-solid reaction. Only during a long storage at elevated temperatures does a decomposition into graphite and a Fe-C solid solution take place. It must be mentioned, that martensite, formed by rapid cooling of austenite in a diffusionless process, is metastable. In the hardening process of steel with heat treatment, the metastable phases are technically more important than the equilibrium phases.

Inspection of Fig. 8.1 shows that nucleation of the γ-phase can only occur at compositions $x_B > x_B^{\alpha,(\gamma)}$. $\Delta G'_\gamma$ is negative for $x <> x_B^{\alpha,(\gamma)}$. For the nucleation of the β-phase another driving force is present: $\Delta G'_\beta$ is negative for $x_B < x_B^{\alpha,(\beta)}$. Alloys with compositions $x_B^{\alpha,(\beta)} < x_B < x_B^{\alpha,(\gamma)}$ must directly nucleate the stable β-phase. At $x_B < x_B^{\alpha,(\beta)}$ the β-phase cannot nucleate either. Only the α-phase is stable in this region.

In a system capable of reactions, in principle more than one metastable phase can occur. How many really can appear at fixed state variables T and x, depends on the degree of supersaturation. The greater the supersaturation, the greater is the difference in composition between the starting phase $x_B^{\alpha,0}$, and the equilibrium composition of the α-phase, $x_B^{\alpha,(\beta)}$ (see Fig. 8.1). With increasing difference in composition the number of possible metastable phases γ, δ, ε ..., whose nucleation has a finite probability because a negative volume term $\Delta G'_\gamma$, $\Delta G'_\delta$, $\Delta G'_\varepsilon$ ... is present, increases. This is the energetic base of the rule found empirically: With increasing undercooling below the equilibrium temperature of the starting phase, the number of possible metastable phases increases. The "Ostwald step rule" says that a physico-chemical system that can occur in several energetic states, in general, does not transform from its state of highest energy into that with the lowest energy, but passes stepwise through states of intermediate energy.

8.2
Guinier-Preston Zones in Al-Cu Alloys

By special heat treatment, in Al-rich Al-Cu alloys, metastable phases can occur. Figure 8.2 (a) shows the stable phase equilibria of the Al-part of the Al-Cu phase diagram. Figure 8.2 (b) indicates the heat treatment: A sample from the region of α-solid solution (a) is quenched to almost room temperature (b). At this temperature no precipitation of a second phase takes place.

Raising the temperature up to about 200 °C (473 K) initiates diffusion of the atoms. At first, the stable phase Al_2Cu does not precipitate, but some Cu-atoms, smaller than Al-atoms form two-dimensional copper clusters in the shape indicated in Fig. 8.3.

These two-dimensional Cu-clusters with a diameter of up to about 500 nm are called Guinier-Preston zone I (GPI). They are effective obstacles for the gliding of dislocations. If they are small enough, they can be cut during moving of these two-dimensional lattice defects, but only by higher driving forces than would be necessary for the motion of dislocations in the same solid solution (see Fig. 8.4). That results in a hardening of the alloy.

As indicated in Fig. 8.5 the Cu-content of the GPI zones in metastable equilibrium with the stable tetragonal Al_2Cu phase is higher than the solubility of Cu in the stable equilibrium with Al_2Cu.

In due time, nucleation of other kinds of precipitation, GPII, Θ and Θ′ phases take place with increasing diameters of particles. With greater particle size

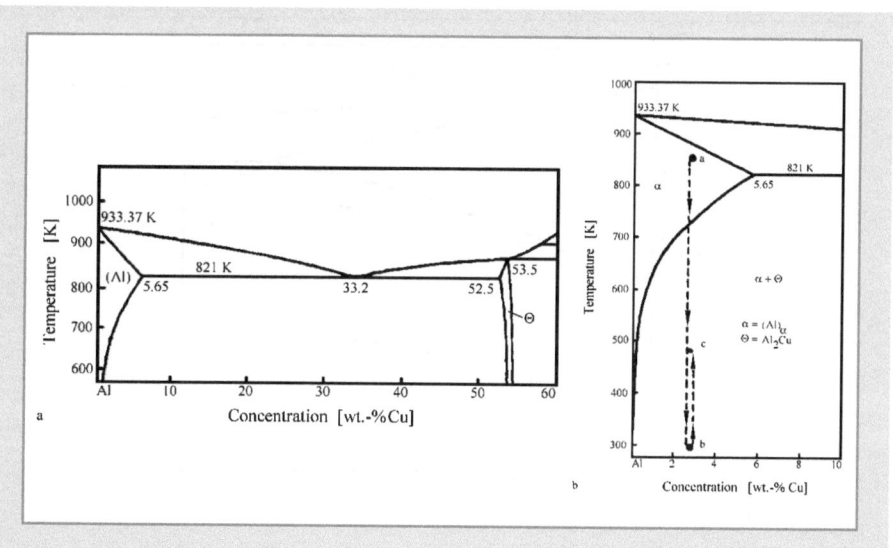

Fig. 8.2 a Part of the phase diagram Al-Cu, **b** Al-rich part of the Al-Cu phase diagram to explain the heat treatment for producing GP-zones

Fig. 8.3
Section across a Guinier-
Preston zone (GPI) parallel
to (200)-plane (after Gerold)

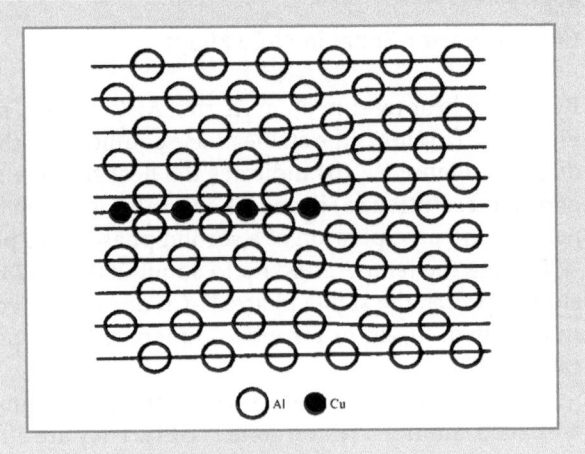

Fig. 8.4
Cutting of precipitated particles by moving dislocations
(after Hornbogen)

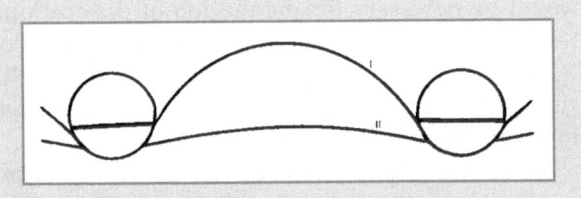

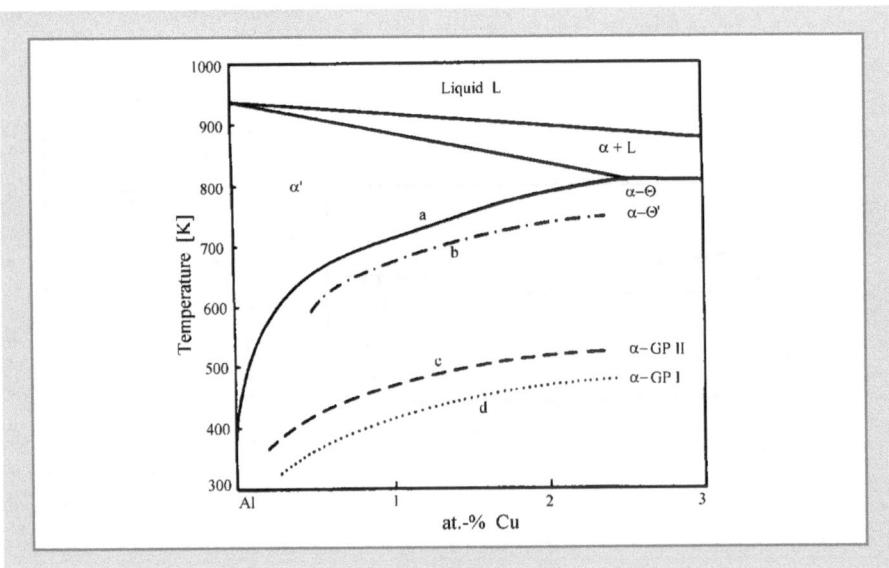

Fig. 8.5 Al-rich part of the phase diagram Al-Cu after[1] with solubility limits for equilibria α-θ (a), for α-θ' (b), α-GP II (c) and α-GP I (d) partially schematic

the force for driving the dislocations become bigger. If the diameter of the zones is small enough and the atomic arrangement is coherent with those of the matrix, the dislocation still cuts the obstacles as seen in Fig. 8.6. But if there is no coherence due to growing obstacles and due to change of this atomic arrangement of both phases there is no more cutting, but by moving the dislocation the precipitates became surrounded by dislocations like those shown in Fig. 8.6 (Orowan mechanism). During this Orowan process there is no more hardening, but a decrease of the strength. There is less force necessary for the dislocation to be driven through the field of precipitates. Due to these facts, by annealling and

Fig. 8.6
Orowan mechanism of motion of a dislocation forming a ring around precipitation particles

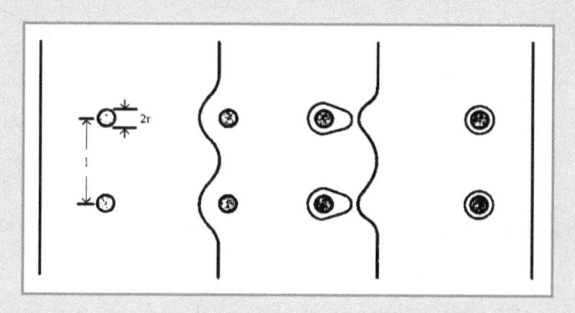

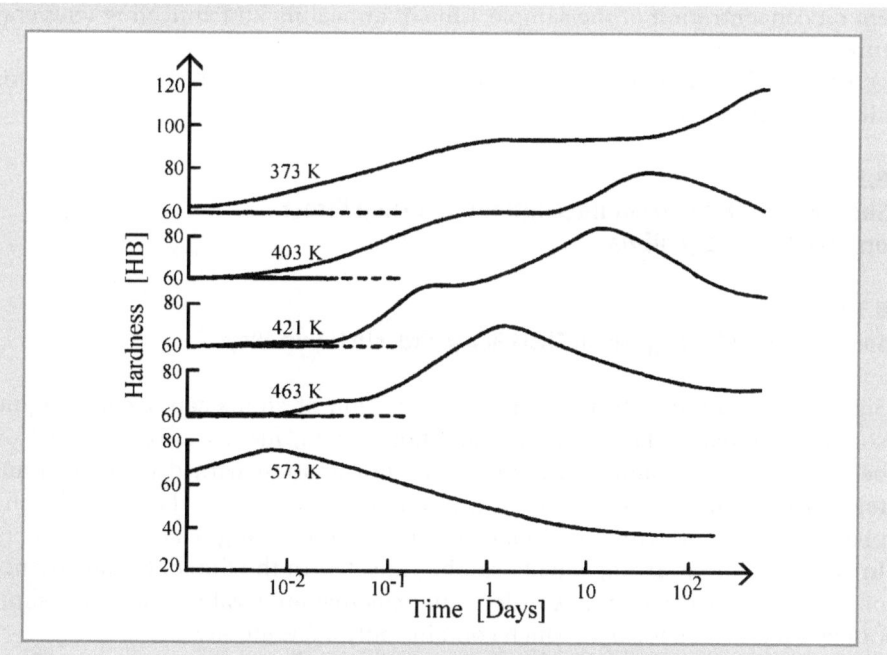

Fig. 8.7 Hardness as a function of time and annealing temperature

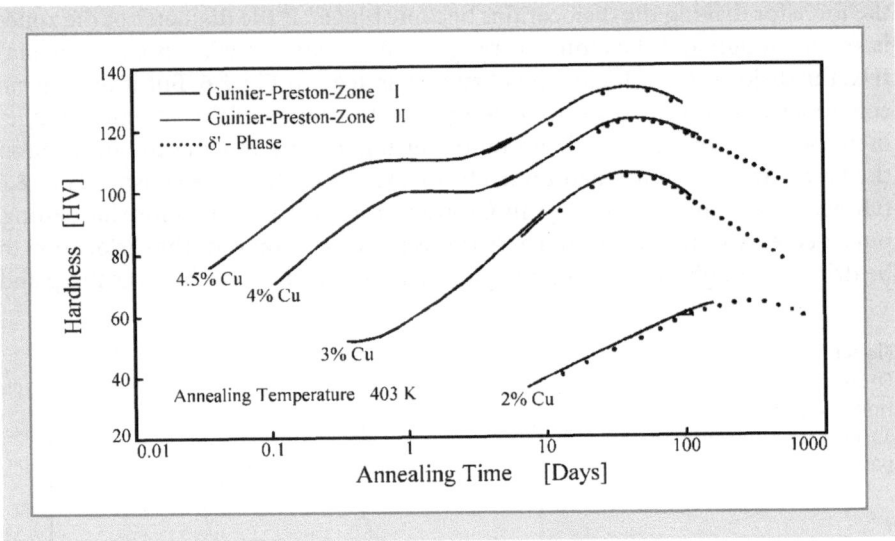

Fig. 8.8 Hardness obtained by annealing at 403 K as a function of time and composition

therefore by influence of kinetical and thermodynamical reasons special hardening process as take place during annealing. This hardening process is dependent on concentration of the sample, time of annealing and annealing temperature, as can be seen in Figs. 8.6, 8.7 and 8.8, respectively.

This hardening is used in producing aluminum alloys of high strength for aircraft components.

8.3
Short Range Order in Liquid Solutions and Their Effect on Solid Liquid Equilibria

8.3.1
Some Empirical Findings about Short Range Order in Liquid Alloys

Significant deviations from random distribution of atoms can occur in liquid solutions. Of particular importance are those where, due to a strong interaction between dissimilar atoms, a preferential coordination with dissimilar atoms takes place. This is often found in the composition ranges, where in the solid phase a compound with individual structural and bonding relationships exists. In general, similar arrangements can be expected in the liquid. In the absence of a crystal lattice, long range order is in principle impossible. Thus, only short range order can be formed, which contains only a few atoms.

Besides, regions can exist, which present a random, or at least not completely ordered distribution of atoms. The degree of order or the fraction of coordination with dissimilar atoms of the total solution is maximum at compositions, where the bonding conditions for the formation of a compound are optimal. This composition often corresponds to the stoichiometry of a compound which exists in the solid state. This correlation is not forcing, because, in the presence of a certain bonding tendency, the formation of certain atomic arrangements in the liquid and solid are subject to different energetic and geometric factors. The effect of a strong coordination with dissimilar atoms will be discussed with examples of binary metallic systems.

The presence of coordination with dissimilar atoms in metallic liquids is observable in structure sensitive properties. Figure 8.9 shows the viscosity of indium-tin alloys at different temperatures as a function of composition. Two

Fig. 8.9
a Viscosity of liquid In-Sn alloys as a function of mole fraction and of temperature, respectively [2] **b** In-Sn phase diagram after [3]

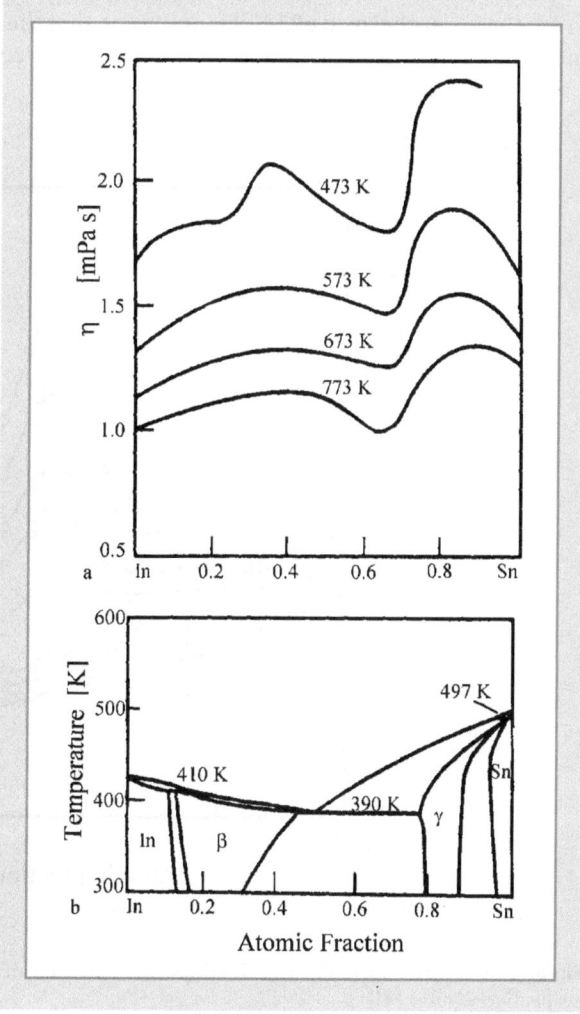

intermetallic compounds are present (β and γ). An increase in the viscosity as a function of composition, above the normal value, takes place in the region where these intermetallic phases occur. At lower temperatures these maxima are strong, at higher temperature they become weaker. With increasing temperature the structural and binding factors lose their effect in the region where the β- resp. γ-phase exists. Short range order increases with decreasing temperature.

The deviation from random distribution of atoms also shows its effect in thermodynamic quantities. Figure 8.10 represents the heat capacity of liquid cadmium-antimony alloys as a function of temperature. At high temperatures the Neumann-Kopp rule, about the additivity of heat capacities is satisfied. The heat capacities show an almost linear variation with composition. At temperatures below 803 K at a composition of $x_{Cd} = 0.6$, however, an increase above this linear behavior is visible, which becomes stronger with decreasing temperature. From the C_p isotherm at 693 K it is apparent, that the maximum is at the stoichiometry Cd_4Sb_3. It is remarkable, that during slow cooling of cadmium-antimo-

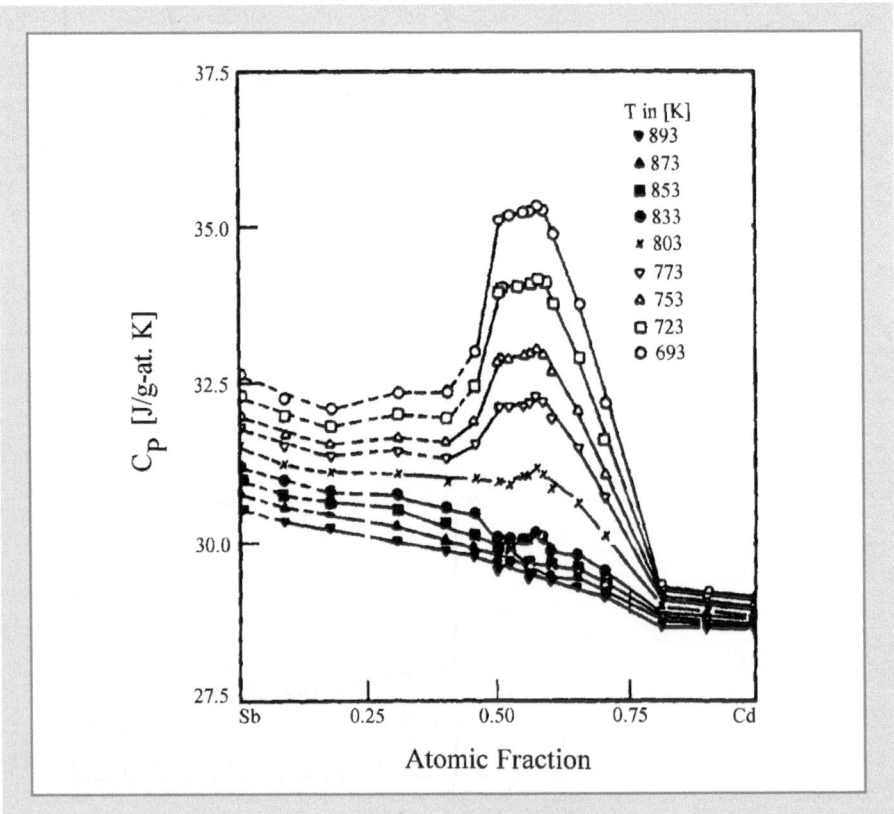

Fig. 8.10 Molar heat of liquid Cd-Sb-alloys as a function of concentration and temperature, respectively (after [4])

ny liquids of intermediate composition, a metastable intermetallic phase with this composition crystallizes. The crystallization does not lead immediately to the stable phase CdSb. Evidently the liquid contains regions of short range order, where the atoms are at least similarly arranged as in the metastable phase Cd_4Sb_3. With increasing temperature the short range order decreases. The decrease of this short range order is connected to an expenditure of energy. This causes the significant increase of the heat capacities above the values expected from the Neumann-Kopp rule. Only after increasing the temperature above 840 K does the fraction of dissimilar coordination become negligible, and the Neumann-Kopp rule is obeyed.

Finally, X-ray and neutron diffraction experiments on liquids with compound formation tendencies, show that short range order with dissimilar coordination is present. One has also to mention, that the existence of a pronounced short range order can be observed at certain composition ranges of liquid alloys in the composition dependence of partial and integral mixing functions. As an example, the enthalpy of mixing will be considered.

The interaction between dissimilar atoms is stronger in the short range order regions, than where a random distribution of atoms is present. The experimentally determined enthalpy of mixing is made up of the two parts ΔH_{stat} and ΔH_{ord} from the two regions:

$$\Delta H = \Delta H_{stat} + \Delta H_{ord} \tag{8.1}$$

In the composition regions where a strong short range order is present, the ΔH-x curve deviates from the shape of the enthalpy of mixing – composition curve which can be expected from a random distribution of atoms (regular solution model). From this deviation the presence of short range ordered liquid regions can be deduced, as well as the amount of short range order.

The short range order influence on the thermodynamic mixing functions also extends to fusion equilibria. Figure 8.11 reproduces the gold-silicon phase diagram. It is characterized by simple solidification equilibria. During cooling, practically pure gold resp. practically pure silicon precipitates. The eutectic point lies much below the melting point of the lower melting component (gold). Stable intermetallic compounds do not appear, although one could expect them based on the rule of Hume-Rothery. Alloying the fourvalent silicon with the monovalent gold, Hume-Rothery phases could be assumed to exist at valence electron concentrations of 1.5 (β-phase), 1.66 (γ-phase) and 1.75 (ε-phase). The absence of stable Hume-Rothery phases can be ascribed to the crystal structure of silicon and to the relatively small enthalpy of formation of electron compounds from the elements. For example the enthalpy of formation of the β-Hume-Rothery phase in the Au-Zn system is $\Delta H_\beta = -26$ kJ/g-at. The element with the higher valence has a hexagonal-closed-packed structure when pure. A similar value could be expected at the formation of the β-Hume-Rothery phase in the Au-Si system, if silicon would have a closed-packed metallic structure. At this reaction one starts with the stable modification of silicon (diamond structure), thus, the

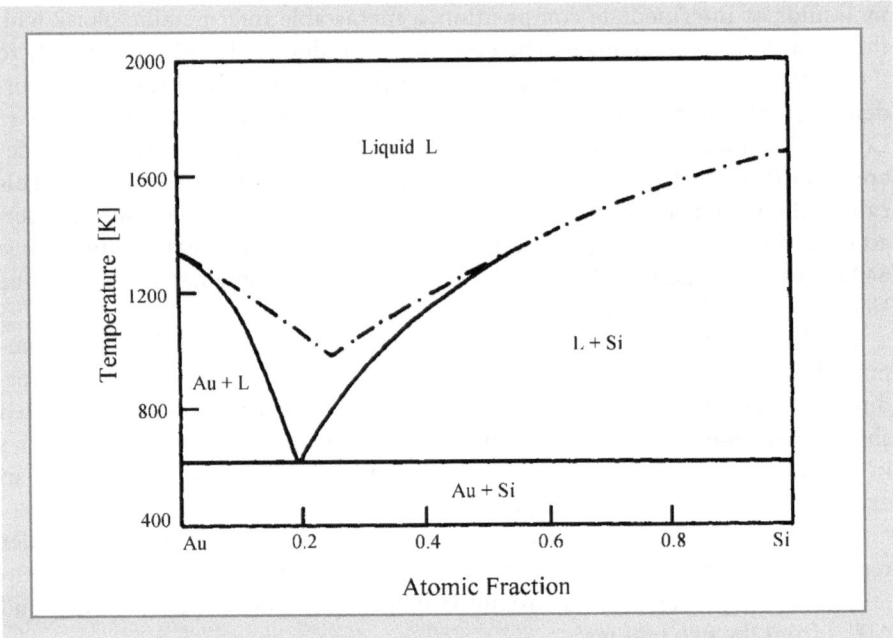

Fig. 8.11 Phase diagram Au-Si after [5, 6]. ──── real liquidus, ----- liquidus calculated supposing a statistical distribution of atoms in the liquid

enthalpy for the transformation of silicon with the diamond structure into a hypothetical silicon modification with the densest packing enters into the enthalpy balance for the total reaction. This enthalpy of transformation is $\Delta H_U > +60 kJ/g$-at., and has, as a consequence, that the Hume-Rothery phases are less stable than a mixture of gold and silicon solid solutions (with very small concentration of the minority component in the solid solution).

In the liquid, silicon has a higher density than in the diamond structure. The coordination number of liquid silicon is significantly higher than four. In the presence of the required valence electron concentrations, ordering can occur in the liquid gold-silicon system, which shows coordination relationships corresponding to those in Hume-Rothery phases, without the long range order possible in lattices. This is shown by a thermodynamic analysis of the liquidus lines of the phase diagram.

Figure 8.12 represents the ΔG^{ex}–x curve, as calculated from the liquidus lines. The compositions, where the electronic conditions are optimal for the formation of Hume-Rothery phases are also shown. Here a short range order maximum can be expected. At a given temperature it becomes smaller as one deviates more and more from the Hume-Rothery composition. At small gold contents the short range order part of the solution is certainly negligible. An almost random solution is present, and the regular solution model is applicable.

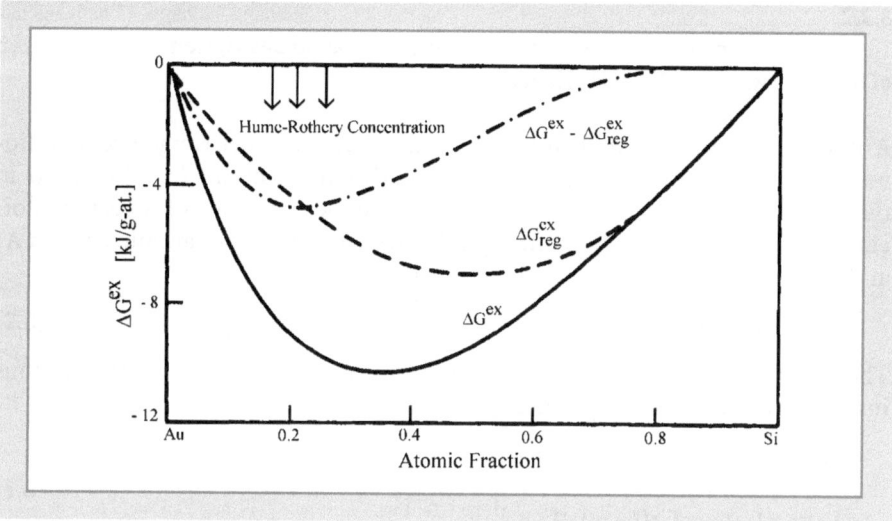

Fig. 8.12 Excess Gibbs energy of mixing ΔG^{ex} calculated from liquidus in the Au–Si system. ΔG^{ex}_{reg} is the integral excess Gibbs energy to be expected, if the distribution of atoms is random.

A parabola, fitted according to Eq. (6.25) ($\Delta H_{reg} = \Delta G^{ex}_{reg} = K \cdot x_A \cdot x_B$) to the ΔG-x curve at $x_{Au} \to 0$ give values of ΔG^{ex}_{reg}, the excess Gibbs energy, which would be present in the absence of any short range order in the Au-Si liquid. The difference $\Delta G^{ex} - \Delta G^{ex}_{reg} = \Delta G^{ex}_{ord}$ can be accounted for by the existence of short range order. Effectively, the maximum of the order effect is to be found in the composition regions, where the conditions for the formation of Hume-Rothery phases are satisfied.

In the hypothetical case, when – under otherwise similar conditions – no deviation from random distribution of atoms occurs, the liquidus lines can be calculated from the ΔG^{ex}_{reg}-x curves. The results are shown in Fig. 8.11. The liquidus points are much higher than in the real system around $x_{Si} \approx 0.2$. The eutectic temperature is now 980 K, whereas experimentally it was found to be at 610 K. The very low lying eutectic point in the gold-silicon system can obviously be accounted for by a strong short range order in the liquid. Low lying eutectic points can in general be interpreted on the presence of an ordered distribution of atoms in the liquid. The region of existence of the liquid is expanded at the expense of the solid phases towards lower temperatures. In systems with metastable intermetallic compounds deviations from random distribution of atoms can also have a significant effect on fusion equilibria. This will be described in detail based on a quantitative formulation of the short range order.

8.3.2
The Model of Homogeneous Equilibria for the Quantitative Description of the Short Range Order in Liquid Alloys

A number of experimental results indicate, that the short range order in liquid alloys with strong interactions between dissimilar atoms (A, B) consists in the formation of "molecules". These molecule like particles of a certain stoichiometry A_iB_j are in a dynamic equilibrium with the monoatomic species A_1 and B_1:

$$i\,A_1 + j\,B_1 \leftrightarrow A_iB_j \tag{8.2}$$

The mass action law controls this reaction. If for simplicity, concentrations instead of activities are used then:

$$\frac{n_{A_iB_j}[1-(i+j-1)\cdot n_{A_iB_j}]^{(i+j-1)}}{(1-n_B-i\cdot n_{A_iB_j})^i \cdot (n_B-j\cdot n_{A_iB_j})^j} = K_{A_iB_j} \tag{8.3}$$

$K_{A_iB_j}$ is the mass action law constant. n_{A_1}, n_{B_1} and $n_{A_iB_j}$ are the number of moles of the corresponding species, n is the total number of moles of the liquid alloy and n_B is the total number of moles of component B.

The fraction of A_iB_j particles of the total number of particles in the liquid depends on the overall composition and temperature. This naturally affects the thermodynamic properties of the liquid alloys. First, the interaction between the particles (A_i, B_j and A_iB_j) has to be considered. If for simplicity, one assumes that these particles are distributed randomly, then the three different interactions between the dissimilar species contribute the following amounts to ΔH, the integral enthalpy of mixing, according to the regular solution model:

$$\frac{n_{A_1}n_{B_1}}{n}\cdot C^{Reg}_{A_1,B_1} \qquad \text{Interaction between } A_i \text{ and } B_j$$

$$\frac{n_{A_1}n_{A_iB_j}}{n}\cdot C^{Reg}_{A_1,A_iB_j} \qquad \text{Interaction between } A_i \text{ and } A_iB_j$$

$$\frac{n_{B_1}n_{A_iB_j}}{n}\cdot C^{Reg}_{B_1,A_iB_j} \qquad \text{Interaction between } B_i \text{ and } A_iB_j$$

Finally, the contribution to ΔH has to be considered, due to the interaction of the atoms within the molecular species A_iB_j. It is proportional to the number of moles $n_{A_1B_j}$ of the A_iB_j molecules:

$$n_{A_iB_j}\cdot C^{Mol}_{A_iB_j} \qquad \text{Interaction between A and B atoms in } A_iB_j$$

$C^{reg}_{A_1,B_1}$, $C^{reg}_{A_1,A_iB_j}$ and $C^{reg}_{B_1,A_iB_j}$ are the interaction parameters. $C^{mol}_{A_i,B_j}$ is the molar enthalpy of formation of the liquid from the pure components, when the liquid consists only of A_iB_j species (molar enthalpy of formation of molecular species).

If only one type of "molecules" are present in the liquid, than the integral enthalpy of mixing of the liquid solution is given by:

$$\Delta H = \frac{n_{A_1} n_{B_1}}{n} \cdot C_{A_1, B_1}^{Reg} + \frac{n_{A_1} n_{A_i B_j}}{n} \cdot C_{A_1, A_i B_j}^{Reg} + \frac{n_{B_1} n_{A_i B_j}}{n} \cdot C_{B_1, A_i B_j}^{Reg} + \\ + n_{A_i B_j} \cdot C_{A_i B_j}^{Mol}$$

(8.4)

In analogy to systems, where more than one intermetallic compound can occur, several molecular species, each with its own stoichiometry, are possible. Correspondingly more terms are needed to describe the composition dependence of the integral enthalpy of mixing. The individual terms can have a more or less important effect on ΔH, depending on the particular property of the system. For example, the enthalpy of mixing at 1600 K in the system Pd-Si can be represented with sufficient accuracy, as shown in Fig. 8.13, by the equation

$$\Delta H = \frac{n_{Pd_1} n_{Pd_2 Si}}{n} \cdot C_{Pd_1, Pd_2 Si}^{Reg} + n_{Pd_2 Si} \cdot C_{Pd_2 Si}^{Mol}$$

(8.5)

It must be mentioned, that for many systems only two terms, the first and last in Eq. (8.4) are of importance.

In the example of liquid Pd-Si alloys practically only one molecular species, Pd_2Si appears, though in the solid state several intermetallic compounds are present. In addition, the interactions between Si and Pd_2Si and Si and Pd are so

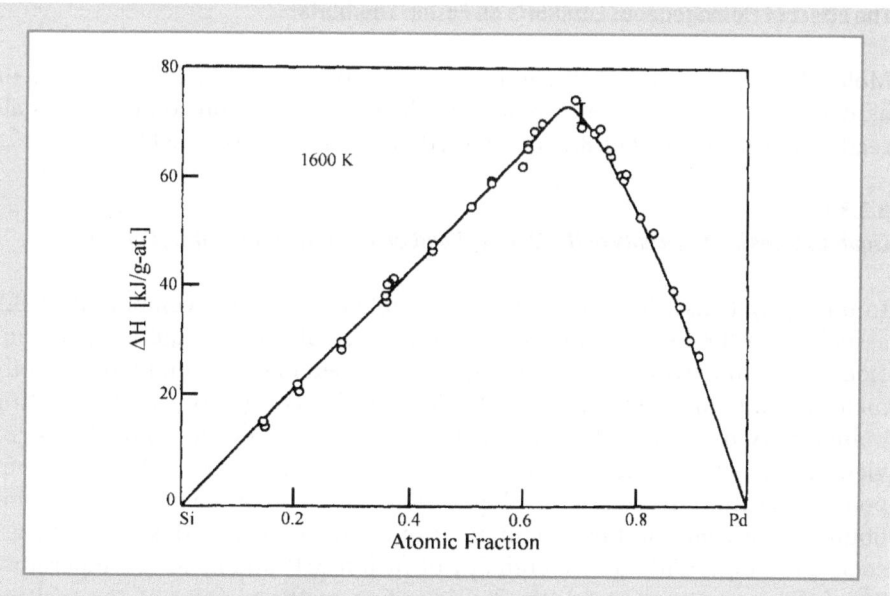

Fig. 8.13 Enthalpy of mixing of liquid Pd-Si-alloys at 1600 K after [7]. O experimental, I limit of error, — calculated after Eq. (8.5)

small that they have no effect. For this example the constants have the following numerical values [7]:

$$C^{Reg}_{Pd_1,Pd_2Si} = -75\,kJ\,/\,g\text{-}atom$$

$$C^{Mol}_{Pd_2Si} = -221\,kJ\,/\,g\text{-}atom$$

$$K_{Pd_2Si} = 1364$$

The fraction of the atoms bound in the Pd_2Si species is naturally maximum, when the liquid has the composition of Pd_2Si. The order of magnitude is 90 %.

The entropy of mixing of a binary liquid solution which contains A_iB_j particles is given by:

$$\Delta S = -R\,[n_{A_1} \ln x_{A_1} + n_{B_1} \ln x_{B_1} + n_{A_iB_j} \ln x_{A_iB_j}] + n_{A_iB_j} S_{A_iB_j} \qquad (8.6)$$

$S_{A_iB_j}$ is the molar entropy of formation of a liquid consisting exclusively from A_iB_j particles. ΔG can be evaluated using Eqs. (8.4) and (8.6), and thus, the influence of the homogeneous equilibria present in the liquid onto the fusion equilibria can also be, in principle, evaluated. The relationships presented here have been tested on a large number of binary metallic systems. They represent correctly the composition and temperature dependence of the thermodynamic mixing properties.

8.3.3
The Effect of Homogeneous Equilibria on Fusion Equilibria

Molecular type species of dissimilar atoms (A_iB_j) are also called associates or associations. The effects of associates in the liquid on fusion equilibria was already mentioned (see Fig. 8.11). Two further cases are considered here.

8.3.3.1
Liquidus Lines in the Vicinity of the Melting Point of a Congruently Melting Compound

In binary systems which present a very stable intermetallic compound (A_iB_j), associates of the same composition appear, as a rule, in the liquid. Let us consider a system in which this compound has its own melting point (congruently melting compound). The shape of the liquidus lines originating at this melting point are fixed, among others, by the fraction of the associates in the liquid, or viewed differently, by the amount of dissociation, which occurs when this compound goes from the solid to the liquid state. The ratio of the number of atoms bound in associates to the total number of atoms in the liquid is called the associate fraction. With known enthalpy of fusion ΔH^F and melting temperature T^F of the compound (A_iB_j), the liquidus line in the vicinity of the melting point of the compound can be calculated, if the associate fraction is given. One obtains [8]

$$\frac{\Delta H^F}{T^F} \cdot \Delta T = \pm \int \Delta x_B \frac{\partial(\overline{\Delta G_A^L} - \overline{\Delta G_B^L})}{\partial x_B} \cdot dx_B \qquad (8.7)$$

Δx_B is the deviation of the overall composition x_B of the alloy from the composition of the intermetallic phase, $x_B^V = j/(i+j)$. The partial Gibbs energies in Eq. (8.7) can be expressed explicitly with good approximation as a function of Δx_B and the associate fraction [8].

As an example, Fig. 8.14 represents the course of the liquidus line in the vicinity of the melting point of an intermetallic compound with the composition A_2B, as a function of the assumed associate fraction in the liquid. The lines were calculated with Eq. (8.7) with preset values of the enthalpy of fusion, the temperature of fusion and the associate fraction. They possess a hyperbolic shape, and with a strong association the branches approach the two asymptotes. If the degree of dissociation of the compound approaches zero when the com-

Fig. 8.14
Liquidus temperature near the melting point of a congruently melting compound A_2B from concentration depending from portion of A_2B associates in the liquid after [8]. Δx_B = difference between atomic fraction of the component B in the liquid at given liquidus temperature and the atomic fraction x_B of the associate A_2B.
------- asymptotes for the case of complete association (100 %)

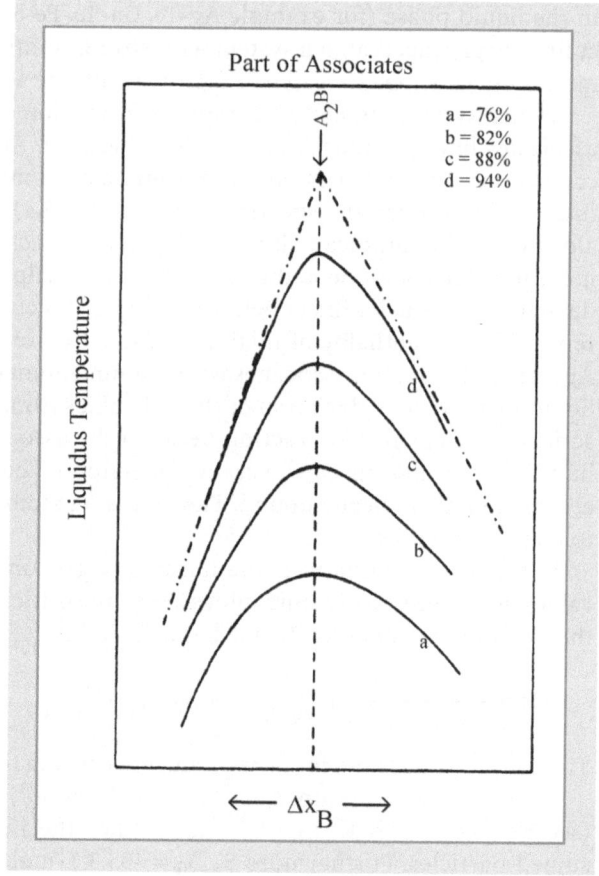

pound transforms into the liquid state, that is the associate fraction is close to 100 %, the horizontal tangent disappears at the melting point. The liquidus lines meet under a finite angle at the melting point. However, this situation cannot be attained completely, because, based on the mass action law in Eq. (8.3), $K_{A_i\text{-}B_j}$ would move towards ∞ and thus, the Gibbs energy of reaction $\Delta G_{A_iB_j}$ towards ∞. Naturally, in the case that ΔH^F and T^F of the compound are known, the associate fraction in the liquid can be obtained from the concentration dependence of the liquidus temperature.

8.3.3.2
Miscibility Gap in a Binary Liquid Alloy with a Strong Compound Forming Tendency

According to the regular solution model, in principle, a miscibility gap can only occur when the enthalpy of mixing of the solution has a positive sign (demixing tendency of the solution). A series of binary alloys which show, in the solid state, high melting intermetallic compounds, and whose enthalpies of mixing in the liquid phase have large negative values, can, however, show a miscibility gap in the liquid phase (for example Ag-Te, Ga-Te, Fe-S and so on). This behavior, at first surprising, that in a system with strong compound forming tendencies a miscibility gap appears, can be explained with the associate model.

In such systems an almost complete association of the atoms is present. The associates have a strong ionic binding character. At complete association one can expect a linear dependence of the enthalpy of mixing on composition, if no interaction between the species occurs. In Eq. (8.4) then, only the fourth term alone would be significant. It is striking, that in such systems, on the side of the more metallic component, the values of the enthalpy of mixing show a positive deviation from such a linear behavior of the ΔH-x curve. As an example, Fig. 8.15 reproduces the enthalpy of mixing as a function of composition in the system Ag–Te. In this system associates with the stoichiometry Ag_2Te are present in the liquid. This positive deviation of the ΔH values from linear behavior can be described by a positive interaction between the associates and the excess metallic atoms. This finding can be easily understood. The atoms in the Ag_2Te associates are mainly ionically bound. The excess Ag atoms, on the other hand, aspire to metallic bonding.

To evaluate quantitatively the miscibility gap, one can start with the known values of ΔH and ΔS of liquid alloys above the critical demixing temperature. In the case of the system Ag-Te one has at 1378 K:

$$\Delta H = \frac{n_{Ag_1} \cdot n_{Ag_2Te}}{n} \cdot C^{reg}_{Ag_1,Ag_2Te} + n_{Ag_2Te} \cdot C^{mol}_{Ag_2Te} \tag{8.8}$$

The constants have the following values according to [9].

$C^{reg}_{Ag_1,Ag_2Te} = 22.5$ kJ·mol^{-1}; $C^{mol}_{Ag_2Te} = -62$ kJ·mol^{-1}, where the mass action law (see Eq. (8.6)) with $K_{Ag_2Te} = 10^5$ determines the number of moles of the assumed particles. Furthermore $S_{Ag_2Te} = -33.3$ J·mol^{-1}·K^{-1} (see Eq. (8.6)). Based

Fig. 8.15
Enthalpy of mixing and
entropy of mixing of liquid
Ag-Te-alloys at T = 1378 K
(after [9]).
oooo experimental ΔH-
values, —— ΔH calculated
by help of Eq. (8.8)
------- entropy of mixing
calculated by help of
Eq. (8.6) with
$S_{Ag2Te} = -33.3$ J mol^{-1} K^{-1}
------- linear dependence of
ΔH-x , as it is to be expected,
if complete association and
no interaction between the
species Ag_1 and Ag_2Te exists

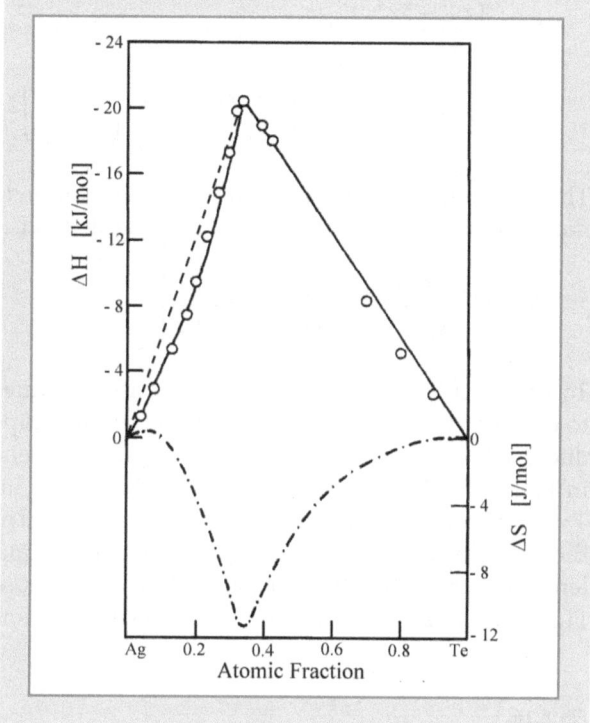

on Eqs. (8.6) and (8.8) the Gibbs energy of mixing of the liquid alloy as a function of composition can be calculated.

For an accurate evaluation of the number of moles, and thus, also of $\Delta G(x, T)$, it is necessary to bring the model for the Gibbs energy into a consistent relationship with the mass action law, because, in the model treated up to now, a mass action law with an ideally associated alloys was used, whereas in the description of ΔH (see Eq. (8.8)) interactions between the free silver atoms with the associates were considered. The activity coefficients, which up to now in first approximation were taken as one in the mass action law (see Eq. (8.3)), must be evaluated.

Very simple relationships result for the liquid Ag-Te alloys, as an almost complete association is present (very large value of K_{Ag_2Te}). In the composition range of interest $0 < x_{Te} < 0.333$, the total number of moles, which corresponds here to the total mole fraction x_{Te}, is equal to the moles of the Ag_2Te associates ($n_{Ag_2Te} = n_{Te} = x_{Te}$).

To calculate the miscibility gap, the common tangent to the ΔG-x curve can be drawn (see schematic Fig. (8.16a)). One obtains for ΔG, based on Eqs. (8.6) and (8.8), the known constants and with $n_{Ag_2Te} = n_{Te}$ for $0 < x_{Te} < 0.333$:

$$\Delta G = \frac{(x_{Te} - 3x_{Te}^2)}{1 - 2x_{Te}} \cdot C_{Ag_1,Ag_2 Te}^{Reg} + x_{Te} \cdot C_{Ag_2 Te}^{Mol} +$$

$$+ RT\left((1 - 3x_{Te})\ln\frac{1 - 3x_{Te}}{1 - 2x_{Te}} + x_{Te}\ln\frac{x_{Te}}{1 - 2x_{Te}} \right) - x_{Te} T S_{Ag_2 Te}. \qquad (8.9)$$

The equation applicable generally for the common tangent to the ΔG-x curve at the points x_α and x_ε and $T = \text{const}$ (see Fig. 8.16a) is:

$$\frac{\Delta G(x_\varepsilon) - \Delta G(x_\alpha)}{x_\varepsilon - x_\alpha} = \frac{\partial \Delta G}{\partial x}\bigg|_{x_\varepsilon} = \frac{\partial \Delta G}{\partial x}\bigg|_{x_\alpha} \qquad (8.10)$$

To fix the shape of a miscibility gap using a numerical evaluation, it is better to consider the Gibbs energy as a function of composition. Figure 8.16b reproduces a $\partial\Delta G/\partial x$ - x curve, which corresponds schematically to the ΔG - x curve in Fig. 8.16a. The position of the miscibility gap in the phase diagram – with $T = \text{const}$. and the position of x_α and x_ε – is determined by the condition that the surfaces $\alpha\beta\gamma$ and $\gamma\delta\varepsilon$ in Fig. 8.16b must be equal. This condition is equivalent to the condition, that the hatched surface enclosed by the curve $\alpha\beta\gamma\delta\varepsilon\alpha$ (see Fig. 8.16b) must be zero. Thus, the following relationship must be valid:

Fig. 8.16
Determination of the points of a miscibility gap, if the ΔG-x curve is known. **a** ΔG-x curve for liquid alloys, showing immiscibility in the concentration range between α and ε, **b** first derivation to concentration of the ΔG-x function shown in **a** (F. Sommer and R. Lück)

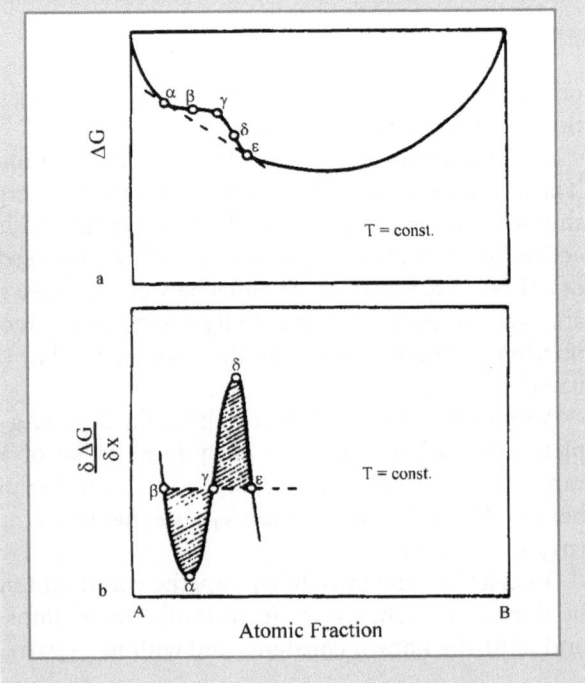

$$\int_{x_\alpha}^{x_\varepsilon} \left[\frac{\partial \Delta G}{\partial x} - \frac{\partial \Delta G}{\partial x}\Big|_{x_\alpha} \right] = dx = 0 \tag{8.11}$$

If Eq. (8.10), setting the conditions for the common tangent, is introduced into Eq. (8.11), one sees that Eq. (8.11) is satisfied. The determination of the miscibility gap using a known $\partial\Delta G/\partial_x$ –x curve is equivalent to the drawing of a common tangent to the ΔG – x curve.

This method of determining the solubility equilibria is useless in the special case of the critical point of the miscibility gap, because points α, β, γ, δ, and ε collapse into one point (see Fig. 8.16a, b). However, the critical composition x_K and the critical temperature T_K can be obtained by setting the second and third derivative of the Gibbs energy of mixing versus the composition equal to zero. The two equations, setting the conditions, result:

$$RT(1-2x_{Te})^2 - 2C^{reg}_{Ag_1,Ag_2Te} \cdot (x_{Te} - 3x^2_{Te}) = 0 \tag{8.12}$$

$$2RT(1-2x_{Te}) + C^{reg}_{Ag_1,Ag_2Te}(1-6x_{Te}) = 0 \tag{8.13}$$

From Eqs. (8.12) and (8.13) one obtains $x_{Te(K)} = 0.25$ and $T_K = 1353$ K.

Figure 8.17 reproduces the silver rich side of the Ag-Te phase diagram. The existence of a miscibility gap is only indicated by the monotectic reaction at

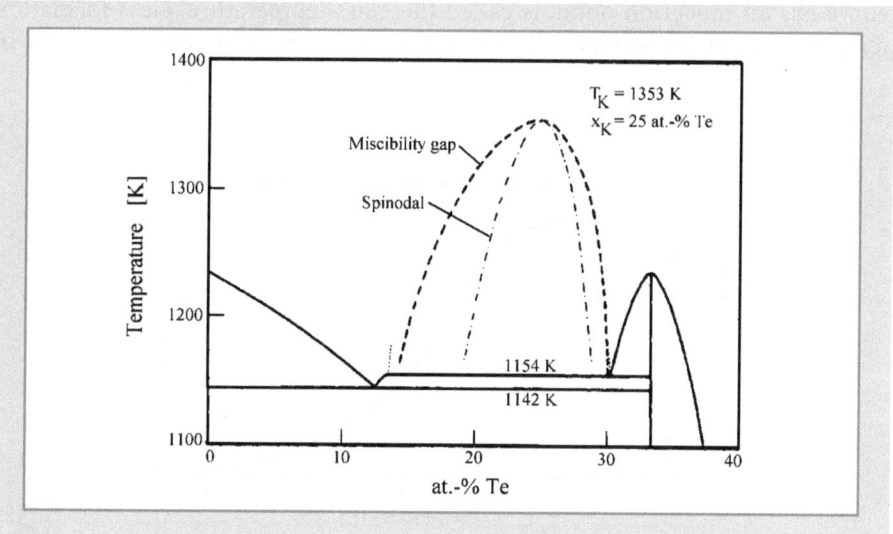

Fig. 8.17 Ag-rich part of the Ag-Te phase diagram (after [9]). ------ miscibility gap, --·--·- spinodal in liquid state calculated on the basis of the associate model

1148 K. The miscibility gap calculated with Eqs. (8.9) and (8.11), and the spinodal obtained using Eq. (8.12) complement well the up to now known phase equilibria, obtained experimentally.

8.4
Metallic Glasses

Short range order of the atomic structure is even more pronounced in a series of non-metallic liquids, compared to liquid metallic solutions with compound forming tendencies. Examples are the liquid silicates. Here, a three-dimensional network of SiO_4-tetrahedrons exists. This indicates a high viscosity. Commercial glasses, at the pouring temperature, have a viscosity of 10 Pa·s. As a comparison, the viscosity of liquid metals slightly above the melting point is of the order of magnitude 10^{-3} Pa·s. The speed of nuclei formation in a highly viscous, undercooled silicate melt is generally very low, because the probability is very small, that the atoms can rearrange themselves rapidly enough to nuclei which correspond to the stoichiometry and structure of the crystalline body. Even during slow cooling of these silicate melts, that is a long time spent in the undercooled region, practically no nucleation occurs.

With decreasing temperature the viscosity increases rapidly. As a rule the probability of nucleation decreases further. The viscosity can increase by many powers of ten (see Fig. 8.18). The consistent of a solid is reached at about 10^{12} Pa·s. Noncrystalline silicates can have a viscosity of 10^{18} Pa·s at room temperature. As an example, Fig. 8.18 represents the temperature dependence of the viscosity of an alkali-calcia glass. The temperature, T^G, at which the log η–T curve has an inflection point, is called the glass temperature, glass formation temperature or transformation temperature. At this point not only the course of

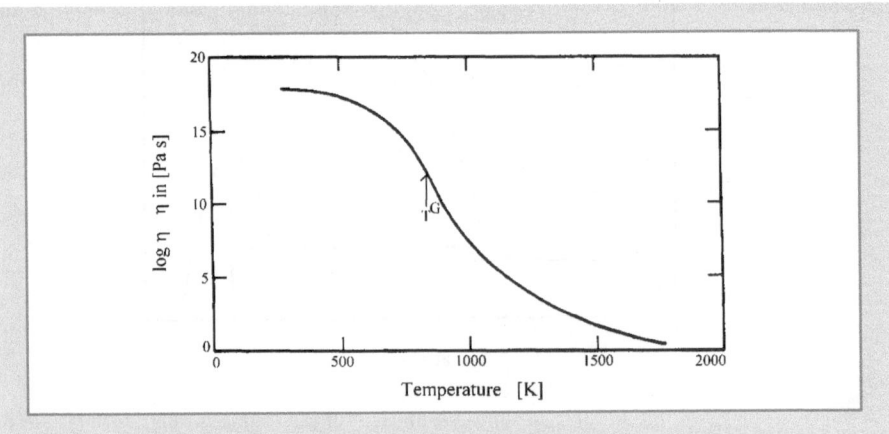

Fig. 8.18 Logarithm of viscosity of a soda-lime-glass as a function of temperature (after [10])

the viscosity changes with temperature (in the vicinity of T^G the change in viscosity is maximum), but a pronounced change in other properties (heat capacity, density, etc.) occurs. This does not occur with a jump, as in transformations of the first order, but in a temperature interval around T^G, which is not well defined. Above T^G an undercooled liquid is present. The non-crystalline substance existing below T^G is called glass. The viscosity at the inflection point of the log η-T curve is of the order of magnitude 10^{12} Pa·s for all glasses.

Above T^G the mobility of the structural elements of the melt is high enough, so that during cooling the structural arrangement for the minimum in the Gibbs energy is rapidly reached at every temperature. Below T^G, where the viscosity $\eta > 10^{12}$ Pa·s, this possibility does not exist any more due to the low mobility. The atomic arrangement existing at T^G is frozen. An essential reason for the change in thermodynamic properties at T^G consists in the fact, that below this temperature only one single configuration of the atoms occurs, whereas above T^G the building blocks of the undercooled melt can be arranged in a large number of different ways. Whereas the undercooled liquid is in a metastable equilibrium relative to the stable crystalline situation, in which internal equilibrium can occur, the glass is in a metastable equilibrium which is not in internal equilibrium.

The glass forming temperature T^G depends on the cooling rate. T^G is higher with rapid cooling rates than with slow ones. In each case a different structure is frozen in. Transformations can occur to a limited extent even below T^G, to optimize the structural and bonding relationships (relaxation processes).

Figure 8.19 represents schematically the heat content of a glass forming substance as a function of temperature. The heat content of the glass is higher than the values obtained by extrapolation of the data of the metastable liquid (above T^G).

Fig. 8.19
Schematic presentation of the heat content of a silicate as a function of temperature. *a* stable melt, *b, d* supercooled melt, *c* glass, *e* crystalline solid, ΔH^F enthalpy of melting

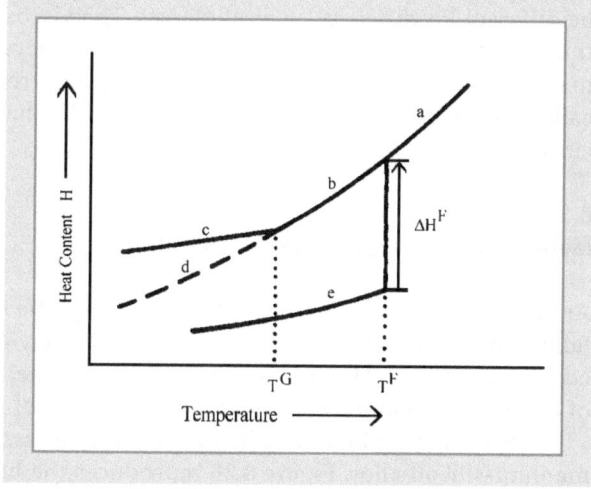

The glass forming temperature T^G, appearing at a given cooling rate, is fixed by many factors, whose type and importance can be very different from case to case. Thus, glass formation can occur in very different types of substances. It is often limited to a narrow composition range in multi-component solutions. One must mention that glasses, – in addition to the already mentioned silicates – can be formed in borates and phosphates.

Also, simple salts such as $ZnCl_2$, can solidify as glasses. For example in the system KNO_3-$Ca(NO_3)_2$ glass formation is found in the region between 40 and 60 wt.-% $Ca(NO_3)_2$, where the eutectic composition of this solution is. In addition many organic substances, for example alcohols, with long molecular chains, and many other can form glasses. Glasses are found even in metallic systems, which will be discussed later.

Common to all glass forming systems is a strong interaction between dissimilar atoms, resp. a blocking structure of the liquid. Easy nucleation is thus, hindered. A region of the composition range near the eutectic point is favorable for glass formation. Here the liquid is stable down to a low temperature. The temperature interval between the liquidus and T^G is minimum, and can be passed relatively easily at a given cooling rate without crystallization. In addition, to obtain equilibrium in a two-component system, two different crystal structures have to be formed, which requires the distribution of the two types of atoms by diffusion over significant distances.

X-ray diffraction experiments show, in the case of silicate glasses, that SiO_4-tetrahedra are the main building blocks, as is the case with silicate melts. Short range order in the form of a three-dimensional network without periodicity is present. The degree of networking of the basic structure is reduced with the inclusion of metal oxides, such as Na_2O, K_2O or CaO. This creates voids, where the large metal oxide cations can be placed.

The working of glass takes place above the transformation region. For glass-blowing a viscosity of 10^5–10^8 Pa·s is useful. The slope of the log η – T curve can be different for different types of glasses. The viscosity range of 10^5–10^8 Pa·s is thus, fixed by a larger or smaller temperature interval. Glasses, where the optimal viscosity range for working is given by a narrow temperature interval, are called "short" glasses. "Long" glasses on the other hand have a less steep log η-T curve in the working region.

8.5
Metastable, Non-Crystalline Metallic Phases

As already mentioned, metallic liquids can also be transformed into non-crystalline solids. However, for this to occur, extremely high cooling rates are needed, because metallic liquid solutions have, as a rule, low viscosity in the vicinity of the liquidus temperature and have a tendency of rapid nucleation.

Glass formation in a metallic system will be discussed based on a gold-germanium-silicon alloy. Figure 8.20 reproduces the heat capacities of the phases

Fig. 8.20 Molar heat of crystalline and non-crystalline phases as a function of temperature for a Au-Ge-Si-alloy with 23.66 at.-% Au, and 9.45 at.-% Si (after [11]). C_p^K, C_p^G and C_p^L are molar heats of the mixture of crystalline of equilibrium phases, of glassy phase, respectively liquid phase T^G = temperature of glass transition, T^L = liquidus temperature

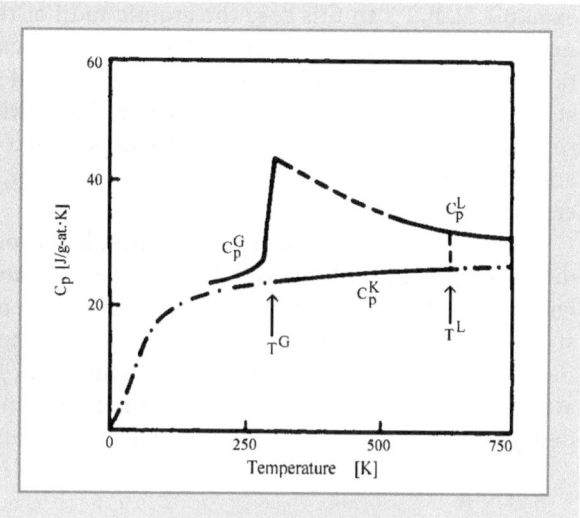

in question as a function of temperature. They are the liquid, the mixture of the equilibrium solid solutions and the glass. The heat capacity of the undercooled liquid in the stable and metastable range increases with decreasing temperature. This is probably due to an increase in short range order caused by the association to molecular species with dissimilar coordination. If the liquid is cooled sufficiently fast below the glass forming temperature $T^G \approx 295$ K, so that nucleation of the crystalline phase does not take place, a non-crystalline solid (glass) forms, whose heat capacity is close to that of the mixture of the stable crystalline phases. Heating this glass leads to the transformation into the undercooled liquid. This is connected to a large increase in the heat capacity, which is expected based on the glass transformation process discussed in Sect. 8.4 (see also Fig. 8.19).

The atoms in a metallic glass, and often in the liquid alloy, are arranged in the first coordination sphere similarly to that of the corresponding stable or metastable crystalline phase. As an example, according to Suzuki and coworkers [12], in metal glasses of the palladium-silicon system around 20 at.-% silicon, coordination relationships are present similar to those in the stable intermetallic phase Pd_3Si.

Kemény and coworker [13] found, that the local order in iron-boron glasses with 15–25 at.-% B correspond to that in the metastable phase Fe_3B.

It could be shown that, in metallic glasses, the glass temperature also increases with increasing interaction between dissimilar atoms. The range of glasses, is pushed to higher temperatures with increasing negative enthalpy of mixing of the liquid.

Many of the currently known metallic glasses have, as one component a transition metal (for example Fe, Ni, Pd) and as second component a metalloid (for example Si, B, P). In this case the atomic radii of the component, differ greatly. According to a model of Bernal, a liquid solution can be considered as a densest random packing of solid, spherical atoms of equal size. Similarly to the systematically densest packing of spheres where interstitial holes (octahedral and tetrahedral holes) are present, holes also appear in a densest random packing. In addition to octahedral and tetrahedral, holes other, of other symmetry and larger volume occur (see Fig. 8.21).

In these holes of liquid iron for example, boron atoms can easily be introduced, even if the volume required by the atom to be introduced is somewhat larger than the space available. This leads to a dense packing of atoms, and to strong interatomic interactions. The fraction of holes and thus, the fraction of the metalloid atoms to be introduced is about 20 %, referred to the number of atoms making up the basic matrix. In many systems a strong glass forming tendency is found at 20 at.-% of the metalloid component. As examples we use:

$$Pd_{81}Si_{19}$$
$$Fe_{80}B_{20}$$
$$\underbrace{Fe_{32}Ni_{36}Cr_{14}}_{T_{82}}\underbrace{P_{12}B_6}_{M_{18}}$$

T = transition metal, M = metalloid.

Evidently this phenomenon is based on the maximum occupancy of the holes in the liquid structure by fitting metalloid atoms. This densest packing of atoms with different size results in short distances, and large interatomic interactions between dissimilar atoms. This principle is, however, not an absolute condition for the formation of metallic glasses. Metallic glasses can also be formed without transition metals and without metalloids, and also with minimal difference in atomic radii of the partners. As examples we mention glasses with the composition $Ca_{73}Mg_{27}$, $Mg_{81}Ga_{19}$ or $Sr_{70}Al_{30}$. In each case, however, the condition is the existence of a preferential dissimilar coordination in the liquid.

Metallic glasses have remarkable properties. They have high mechanical strength, and are very corrosion resistant. If they contain iron as a major component they can be ferromagnetic. The industrial interest consists that with great mechanical strength they have small coercivity. The hysteresis loop is very narrow; the loss in energy by changing the magnetization direction is correspondingly small. They crystallize as metastable phases when heated to temperatures where the atomic mobility is sufficiently high. Many metallic glasses are stable much above room temperature (for example 600 K), and for a long time (order of magnitude years).

Fig. 8.21
Voids in the Bernal's closest random packing of equal rigid balls. **a** tetrahedron, **b** octahedron, **c** trigonal prism with three added half octahedrons, **d** archimedian anti-prism with two added half octahedrons, **e** tetragonal dodecahedron

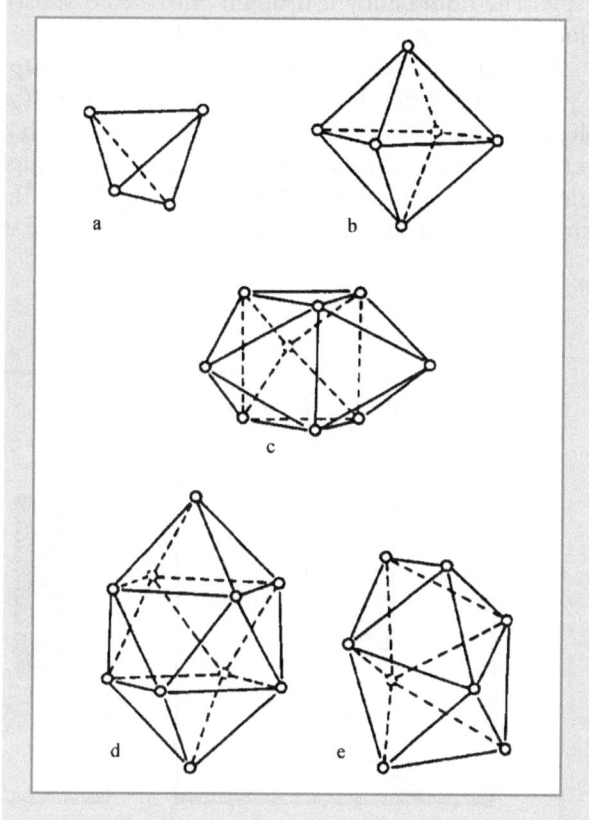

8.6
Methods to Obtain Extremely High Cooling Rates

To transform a liquid alloy into a glass, as a rule, very high cooling rates are needed. In many cases, cooling rates of more than 10^6 K/s are required to reach the glass temperature, before the crystalline phase nucleates. In addition, at these high cooling rates, if no glass formation takes place, metastable crystalline phases can be formed, which have specific properties (for example semiconductors).

A large number of methods have been developed to cool the starting liquid extremely rapidly, with the aim to obtain metastable materials, especially metallic glasses with industrially important properties. The removal of heat through convection with a cooling gas or liquid (cold helium flow around the sample, or dropping small samples into water) yields optimally cooling rates of 10^4–10^5 K/s. Heat conduction permits one to obtain significantly higher cooling rates. Thus, the mostly used methods to prepare metallic glasses are based on the latter principle. The liquid alloy is brought onto a cold substrate with high thermal conductivity, and spread out into a thin film.

The highest cooling rates obtained up to now (up to 10^{10} K/s), can be reached by the method developed by Duwez and Willens (see Fig. 8.22a). A drop of the liquid alloy is placed onto an inert substrate, which has a small hole. The surface tension prevents the small drop from falling through the hole. As a rule the liquid is heated by induction. The liquid drop forms the bottom of a tube, which is closed at the top at a certain distance with a plastic sheet. Gas pressure is applied

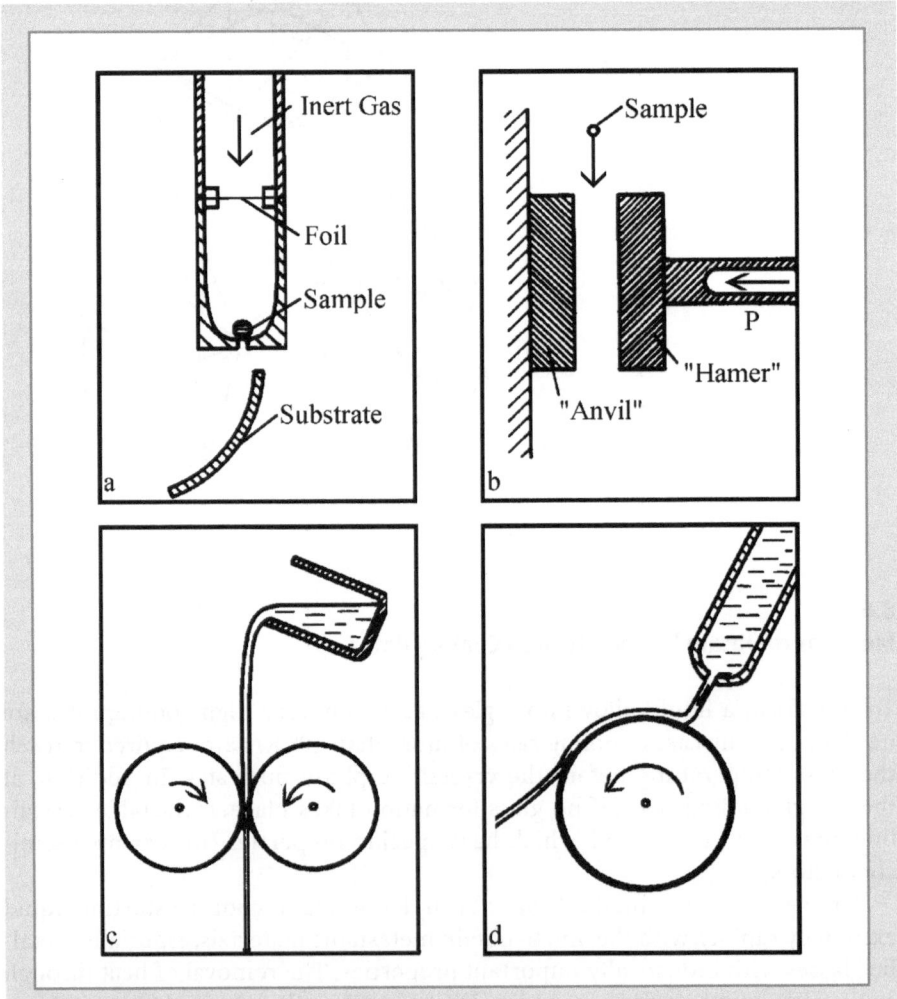

Fig. 8.22 Methods to produce extremely high velocities of cooling of metallic melt. **a** Shock wave tube after Duwez and Williams, **b** Hamer and anvil method, **c** Roller quenching, **d** Quenching of the melt at the periphery of a rotating metallic wheel

to the plastic sheet, until it breaks. The pressure wave formed pushes the liquid drop through the hole of the substrate with supersonic speed. The drop falls onto a suitable curved copper substrate and is spread out into a thin film. The sliding motion of the drop on the curved substrate produces a centrifugal force, which firmly presses the liquid onto the copper sheet during the spreading.

Only relatively small masses (approx. 100 mg) can be cooled rapidly with this method. Quenched samples with a somewhat larger mass can be obtained for example, by accelerating the liquid in a centrifuge, and thrown onto a suitable substrate. Another often used method consists to crush a drop, falling from a crucible, between two parallel metal plates (Fig. 8.22b). To ascertain that the drop spreads out rapidly, the two plates must be pushed together at high speed. This can be carried out with high pressure air, electromagnetically, with a spring or with an explosion. In addition to these, a number of other special quenching procedures have been developed.

These methods are unsuited for the preparation of glasslike alloys in industrial quantities because they yield only small samples. For the preparation of especially iron containing metallic glasses, for which cooling rates of 10^4–10^5 K/s are sufficient, the methods shown schematically in Fig. 8.22c, d were developed. As indicated in Fig. 8.22c the liquid, as a thin jet, is passed between two rapidly rotating cold cylinders. Without great difficulty a sheet 10 cm wide, and 0.1 mm thick can be obtained.

Similar bands of glasslike alloys can be obtained by transferring a liquid jet onto the edge of a rapidly rotating wheel. The glasslike solidified sheet can be removed from the wheel and can be rolled up continuously. If the flow rate of the liquid jet is coordinated with the speed of rotation of the wheel, uniform bands of any length can be obtained.

With all the methods developed up to now, the glasslike solidified material must have one dimension extremely small, because the principle of heat removal from a thin sheet was applied.

8.7
Crystallization of Glasslike Alloys

Glasslike metallic solid solutions, obtained by extreme rapid cooling of a liquid, crystallize when heated to a sufficiently high temperature. As a rule the crystalline equilibrium phase, corresponding to the composition of the solution, does not form immediately. Rather, one or more crystalline metastable phases can appear as intermediate steps. This is easily visible from a consideration of the Gibbs energies of the possible phases and the diffusion paths. As an example, Fig. 8.23 shows the Gibbs energies as a function of composition for different phases of iron-rich Fe-B solid solutions. They are the glass phase G, the α-Fe solid solution, the metastable phase Fe_3B, and the stable intermetallic compound Fe_2B. The path of crystallization of the glasslike starting phase G depends on its composition.

Fig. 8.23
Schematic presentation of
the Gibbs energy of different
crystalline phases produced
from glassy phase G in the
system Fe – B after U. Herold
and U. Köster

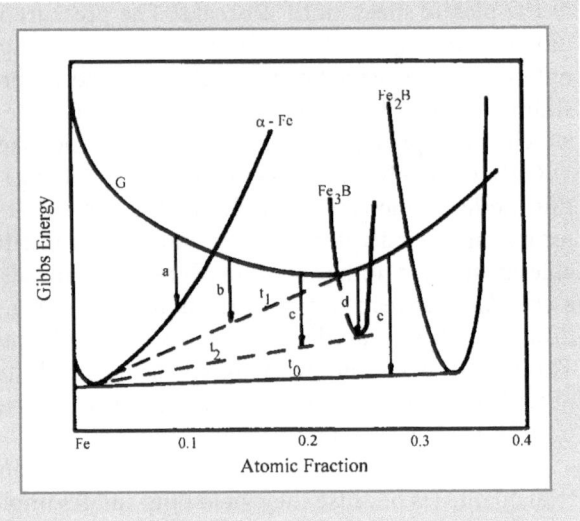

a) Crystallization without change in composition
Crystallization of a glass phase can occur without change in composition, when
the composition is close to that of a stable or metastable compound. In the case
of the Fe–B system these are reactions marked a and d (see Fig. 8.23). A supersaturated α-Fe-B solid solution, resp. the metastable compound Fe_3B forms. For
these reactions the Gibbs energy is not decreased to its absolute minimum, but
diffusion has only to move the atoms over a short distance to reach the positions in the α-phase or the Fe_3B phase. There are indications that the atomic
arrangement in the glass phase is very similar to that in the Fe_3B phase. This
may be one reason that nuclei of the stable Fe_2B phase cannot form in a simple manner by diffusion over a short path, in spite of the possible large gain in
Gibbs energy. The first formed α-Fe and Fe_3B phases can transform during further reactions into the equilibrium phases.

b) Primary crystallization of a stable phase
The crystallization of iron rich Fe-B glass alloys without change in composition takes place at the composition $x_B \approx 0.12$. At higher boron content the gain
in Gibbs energy is apparently not sufficient for rapid nucleation. At $x_B > 0.12$ an
α-Fe solid solution with low boron content can precipitate; the glass matrix becomes richer in boron (see b in Fig. 8.23). The endpoint of this crystallization is
reached when the α-Fe solid solution and glass G attain the composition of the
metastable equilibrium (α-Fe + G), defined by the tangent t_1. The thus, formed
boron rich glass can, during further temperature increase, decompose along another path than described here.

c) Discontinuous crystallization

Two crystalline phases can form, similarly to the eutectic crystallization, if a sufficiently large gain in Gibbs energy is present, and if the transport of material can proceed over larger distances (sufficient atomic mobility and time). This can produce a lamellar structure. For instance this is the case in reaction c ($G \rightarrow \alpha$-Fe + Fe$_3$B) and e ($G \rightarrow \alpha$-Fe + Fe$_2$B).

The different atoms, distributed homogeneously in the glass phase, are distributed in different proportions into the various phases. This takes place on the interface between starting matrix and the nucleated material. This reaction can occur at all compositions between $x_B \rightarrow 0$ and $x_B = 0.333$.

The discontinuous crystallization is the slowest of the reactions discussed here. It takes place three times slower at $x_B \approx 0.2$ and a given temperature than the reaction d ($G \rightarrow$ Fe$_3$B) at $x_B \approx 0.25$, which is certainly connected with the longer diffusion paths during separation into two crystalline phases.

The crystallization processes can be influenced greatly by the addition of other elements.

References

Citations

[1] M. Hansen and K. Anderko, "Constitution of Binary Alloys", McGraw-Hill Book Comp., New York (1958)

[2] B. Predel and I. Arpshofen, Viskosimetrische Untersuchungen an metallischen Schmelzen, Westdeutscher Verlag, Opladen (1978)

[3] B. Predel and T. Gödecke, Z. Metallkde., 66, (1975) 654

[4] G. Schick and K.L. Komarek, Z. Metallkde., 65, (1974) 112

[5] B. Predel, H. Bankstahl and T. Gödecke, J. Less-Common Metals, 44, (1976) 9

[6] B. Predel and H. Bankstahl, J. Less-Common Metals, 43, (1975) 191

[7] I. Arpshofen, M.J. Pool, U. Gerling, F. Sommer, E. Schultheiss and B. Predel, Z. Metallkde, 72, (1981) 776

[8] F. Sommer, D. Eschenweck and B. Predel, Ber. Bunsenges. Phys. Chem., 82, (1978) 790

[9] F. Sommer, Z. Metallkde., 73, (1982) 72

[10] H. Scholze, "Glas", F. Vieweg und Sohn, Braunschweig (1965)

[11] H.S. Chen and D. Turnbull, J. Chem. Physics, 48, 2560 (1968)

[12] K. Suzuki, T. Fukunaga, M. Misawa and T. Masumoto, Science Reports of the Research Institutes, Tohoku University, Amorphous Material Issue I, Sendai, Japan (1976)

[13] T. Kemény, I. Vincze, B. Fogarassy und S. Arajs, "Structure and Crystallisation of Fe-B Metallic Glasses", Hungarian Academy of Science, Budapest (1978)

General References

P. Duwez, "Metastable Phases Obtained by Rapid Solidification", in: "Energetics in Metallurgical Phenomena", Vol.1, Editor: W.M. Mueller, Gordon and Breach Science Publishers, New York (1965)

U. Herold and U. Koster, in: "Rapidly Quenched Metals III", Vol. 1. Editor: B. Cantor, Proceedings of the Third International Conference on Rapidly Quenched Metals, University of Sussex, Brighton, 1978; The Metals Society, London (1978)

G. Schluckebier and B. Predel, Z. Metallkde., 73, (1982)

I. Schmidt and E. Hornbogen, Z. Metallkde., 69, (1978) 221

A.R. Ubbelohde, "Melting and Crystal Structure", Clarendon Press, Oxford (1965)

"Metallic Glasses", Papers presented at a Seminar of the Materials Science Division of the American Society for Metals, 1976; American Society for Metals, Metals Park. Ohio (1978)

"Rapidly Quenched Metals III", Vol. 1 and Vol. 2, Editor: B. Cantor, Proceedings of the Third International Conference on Rapidly Quenched Metals. University of Sussex, Brighton, 1978; The Metals Society, London (1978)

CHAPTER 9

Effect of diffusion on Phase Transformations

9.1
Distribution of Components During Solidification of Liquids

In a multi-component system at the solid-liquid equilibrium the compositions of the two coexisting phases are only equal in the case of a congruently melting compound, at the melting point maximum or minimum of a solution, and in the limiting case, of the melting point of a pure component. In general, a difference in composition exists between liquid and solid. This composition difference changes during the solidification process, except at an invariant equilibrium. At a moderate solidification rate of a homogeneous liquid, a crystalline solid forms with non uniform composition.

The distribution of the components in a solidification equilibrium is described by the distribution coefficient. The equilibrium distribution coefficient is defined by

$$k_0 = \frac{c_S}{c_L} \tag{9.1}$$

where c_S is the solidus and c_L the liquidus composition at a given temperature T. The equilibrium distribution coefficient is not a constant for a given system; naturally it is independent of the solidification rate. As can be seen from Fig. 9.1 for a melting point lowering, $k_0 < 1$ and for a melting point increase, $k_0 > 1$. In addition an effective distribution coefficient is defined as:

$$k_E = \frac{c_S}{c_0} \tag{9.2}$$

where c_S is the composition of the crystal, which forms at a given moment from the liquid, which has the average composition c_0. The effective distribution coefficient, – in contrast to k_0 – depends on the solidification conditions.

The composition of the liquid changes during solidification. It can become richer (see Fig. 9.1a) or poorer (see Fig. 9.1b) in the dissolved substance B. The enrichment or depletion of the dissolved component B occurs primarily at the solid-liquid interface. This creates a concentration gradient within the liquid, which causes an equalization of the composition through diffusion within the liquid phase.

Fig. 9.1
Definition of the distribution
coefficient k_0 for equilibrium

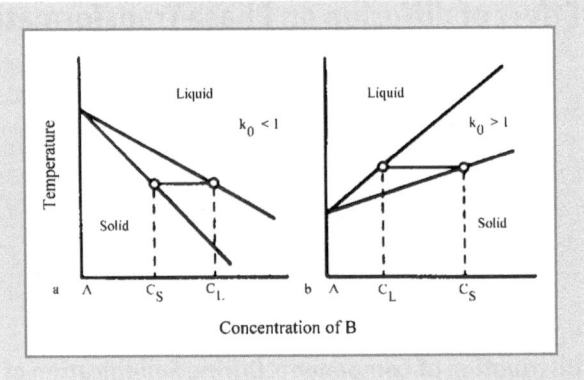

Fig. 9.2
One-dimensional solidification system

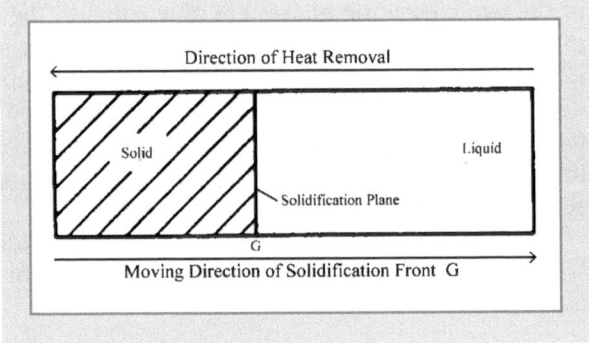

A concentration gradient also occurs in the solid being formed, because at various stages of the solidification process, solid solutions with different composition are formed.

To illustrate the effects of such diffusion, let us consider a one-dimensional solidification system (see Fig. 9.2). The interface solid-liquid G is flat. It advances with constant speed. No concentration gradients are present perpendicular to the advancing solidification front. They only appear in the direction of the advancing solid-liquid interface. An equalization of the composition takes place exclusively parallel to the axis of the system.

In the limiting case of an extremely slow solidification rate, which permits a complete equalization of the composition in the solid and liquid, the solid solution in the considered solidification state has the composition c_S, which correspond to the phase diagram at the temperature in question. Accordingly, the liquid has the composition $c_0 = c_L$. The distribution of components in the one-directional solidification system corresponds, in the presence of a melting point lowering, to that shown in Fig. 9.3a.

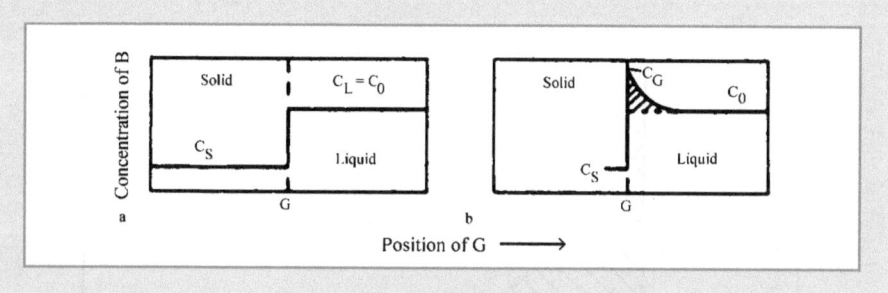

Fig. 9.3 Concentration distribution near the crystallization front G. **a** At extremely slow velocity of solidification complete exchange of concentration by diffusion, **b** deceleration of the solvent component B in the melt just before solidification front, if there is no complete distribution of concentration by diffusion at high velocities of crystallization

Figure 9.3b represent the concentration distribution for the case of a solidification rate high enough, so that a concentration equalization takes place by diffusion in the liquid, but does not permit a complete removal of the concentration gradient. Furthermore, the diffusion rate in the solid should be negligibly small. For $k_0 < 1$ an enrichment of component B takes place in the liquid in front of the solidification line, up to a maximum concentration of c_G. The concentration drops exponentially from here into the interior of the liquid, down to the average composition c_0. The concentration distribution depends on the conditions of solidification. Three different typical cases can occur: At the beginning of the solidification, at a later moment, and after the steady state started. They are illustrated in Fig. 9.4.

a) Distribution of components at the beginning of the solidification
The starting composition of the liquid is c_0. The fusion equilibria correspond to those in Fig. 9.1a. A melting point lowering is present, and thus, $k_0 < 1$. The concentration of the dissolved component B in the solid solution precipitating at the beginning of the solidification from the liquid is according to Eq. (9.1)

$$c_S = k_0 \cdot c_0.$$

The composition distribution in the vicinity of the solid-liquid interface, G, is represented in Fig. 9.4 by the line *(a)*.

b) Distribution of components after the start of the solidification, but before the onset of steady state
The dissolved substance B is enriched in the liquid in front of the interface G, because a solid solution with the composition $c_S = k_0 \cdot c_0$ ($c_S < c_0$) precipitates. After a certain amount of solid solution has precipitated, the concentration of B in the liquid at the interface has increased to c_L' ($c_L' > c_0$), without reaching steady state conditions.

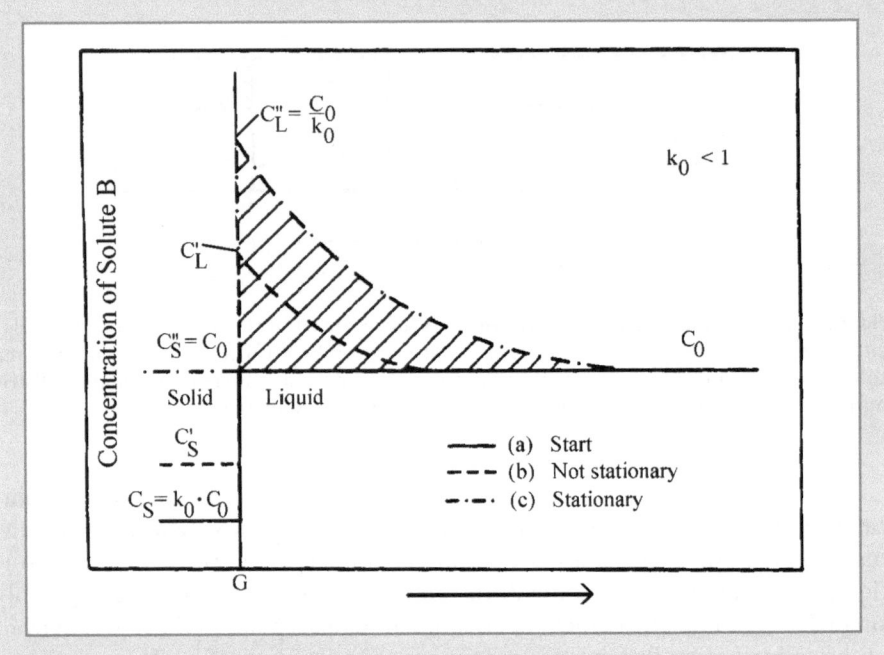

Fig. 9.4 Distribution of concentration near the crystallization front of different stages of solidification

The solid solution, precipitating now at the advancing solidification front from the adjoining liquid with the composition $c'_L > c_0$, has a concentration of B larger than at the beginning of the solidification process ($c'_S > k_0 \cdot c_0$). The existing composition distribution is represented by line (b) in Fig. 9.4. In the measure, that the solidification proceeds, the content of B at the interface G increases continuously in the liquid and in the solid. This increase in concentration occurs until after a certain quantity of solid solution, depending on the crystallization rate, is formed, and steady state is reached.

c) Steady state
A stationary state is reached when, during solidification, the amount of dissolved B liberated per unit time at the interface, equals the amount of B transported from the interface into the total volume present of the liquid. The component distribution at steady state depends on the crystallization and diffusion rates. The shape of the composition profile remains constant. In measure that the interface G moves, the complete composition profile follows.

In the stationary case a solid solution with the composition $c''_S = c_0$ crystallizes. The quantity of B, which is concentrated in front of the interface G, and which is represented by the hatched surface in Fig. 9.4, is constant. During solidification the same amount of B from the interface G enters, per unit time, the solid, as is removed by diffusion into the interior of the liquid. From $c_S'' = c_0$, one obtains with Eq. (9.1) for the concentration of B in the liquid at the interface G:

$$c_S'' = c_0/k_0$$

In the stationary state the concentration relationships are determined by the quantity

$$X = \frac{D}{R^K} \tag{9.3}$$

D is the diffusion coefficient, R^K the crystallization rate.

X represent the distance from the interface G, where the increase in concentration dc_0 in relation to the average concentration in the center of the liquid c_0, is a 1/e part of the maximum concentration increase $c_L'' - c_0$. This distance is called the "characteristic distance". In metallic systems ($D \approx 5 \cdot 10^{-4}$ cm^2/s) with a crystallization rate of $R^K \approx 10^{-2}$ cm/s is characteristic distance the $X \approx 0.5$ mm.

In the stationary state, the temperature at the interface G, is necessarily the solidus temperature at c_0, because a solid solution with the concentration c_0 is formed.

In real solidification systems, in addition to the diffusion and the composition profile, convection flows are effective in the liquid. Convection is caused by diffusion gradients, created by temperature and concentration gradients. Diffusion is decisive for the concentration profile only in a thin liquid layer, adhering to the solid being formed. This boundary diffusion layer, not touched by mechanical mixing, can, however, be so thick, that it contains the largest part of the diffusion zone. Only with strong mixing and a very small crystallization rate is the thickness of the boundary diffusion layer insufficient, to contain the complete diffusion zone, as would be the case without mixing.

9.2
Constitutional Undercooling

The dissolved component is enriched ahead of an advancing crystallization front if $k_0 < 1$, and is depleted if $k_0 > 1$. The composition is such, in both cases, that the liquidus temperature of the liquid immediately ahead of the solidification front is lower, than at greater distances from the interface G. Thus, the temperature at the interface, G, is lower than the liquidus temperature of the rest of the reservoir of liquid with the composition c_0. This can lead to an undercooling in the case when the temperature of the liquid is higher than that of the interface. The situations are represented in Fig. 9.5.

A one-dimensional solidification system is considered, where the heat removal is through the crystal being formed. The heat sink is on the left side of the system. As shown in Fig. 9.5d, the true temperature is the lowest. The temperature increases towards the interface G (see line u). At the interface the true temperature t is equal to the solidus temperature T_2 for the composition c_0

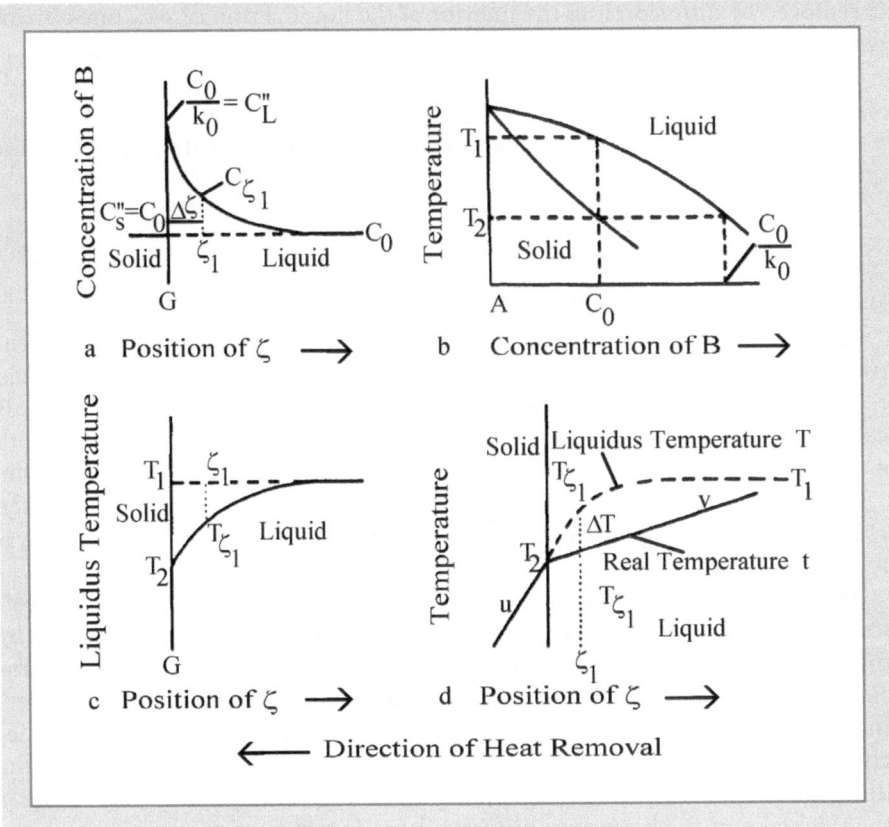

Fig. 9.5 Constitutional supercooling

(see Fig. 9.5b), because here the solid solution with the composition $c_S'' = c_0$ is formed. With increasing distance from the interface G, the temperature in the liquid increases according to line v. Thus, at every point, the liquid is at a higher temperature then the interface G. These temperature relationships around the solidification front correspond to the situation during solidification processes, when the crucible containing the liquid is cooled on one side (for example the bottom) and a portion of the material is already solidified.

In the following discussions it is assumed that the stationary state has already been reached. The compositional relationships are represented in Fig. 9.5. In the stationary situation the liquid, in contact with the solid solution being formed with the composition $c_S'' = c_0$, has the composition $c_L'' = c_0/k_0$. The composition of the liquid decreases from this maximum value through diffusion with increasing distance from G to c_0. In this diffusion range a liquidus temperature has to be assigned to every composition, according to Fig. 9.5b. Figure 9.5c shows the value of the liquidus temperature as a function of the distance from G. It increases from T_2 (solidus temperature at c_0) to T_1 (liquidus temperature at c_0).

At the interface $t = T_2$. In Fig. 9.5d the behavior of the liquidus temperature as a function of the distance from G is shown. Here the case is chosen, where the slope of the T-ζ curve at point G is greater than the slope of the t-ζ line. Thus, at small distances from G, there is $T > t$. Let the composition of the liquid be c_{S1} at point ζ_1 at a distance $\Delta\zeta$ from the interface G. The composition corresponds to the liquidus temperature $T_{\zeta1}$. The true temperature, t, of the liquid is then lower by ΔT than the liquidus temperature $T_{\zeta1}$. The liquid is undercooled, though its temperature $T_{\zeta1}$ is higher than the temperature T_2 of the solid-liquid interface G. This undercooling is called "constitutional undercooling".

9.3
Effects of the Constitutional Undercooling

The amount of attainable constitutional undercooling depends, under given cooling conditions, on the position of the liquidus and solidus lines, that is the constitution of the system. At very small temperature gradients in the liquid, $dt/d\zeta \rightarrow 0$, the maximum constitutional undercooling can almost reach $(T_1 - T_2)$ at c_0. The interval $(T_1 - T_2)$ can be very large. For example, in the copper-tin system around 7.5 at.-% Sn, $(T_1 - T_2) \approx 190$ K. In a practical situation such a large constitutional undercooling is, however, not reached in this system, because, at even smaller undercooling nucleation occurs. This causes the solidification not to take place evenly. In addition the flat solid-liquid interface becomes unstable. Protrusion can enter the constitutionally undercooled liquid, and cause periodic fluctuations of the composition in the resulting solid.

Figure 9.6 represents the formation of a cellular structure in a solid precipitating during solidification. An accidentally appearing protrusion at a point of the flat interface advances into a region of the liquid, which is constitutionally undercooled. The growth rate at the top of the protrusion is therefore greater than in the flat interface. It must be mentioned that originally there was no gradient of the true temperature perpendicular to the solidification front. A lateral expansion of the protrusion occurs only in a relatively small measure, because the solidification enthalpy which appears, increases the real temperature on the sides of the protrusion and thus, reduces the undercooling. In addition, the increase in concentration of the dissolved substance on the side surfaces of the protrusion reduces the growth rate perpendicular to the direction of growth of the protrusion top. At the junction B (see Fig. 9.6b) of the protrusion with the flat interface, the increase in the concentration of B delays the crystallization process of the flat solidification front at this place. This makes it possible for the formation of new protrusions around the first one. Finally, the solidification front advances only in the form of closely packed column shaped structures. At the boundaries of these structures (C in Fig. 9.6) the concentration of B is higher than in the interior of these structures.

These boundary regions solidify somewhat later (at a somewhat lower temperature) than the regions of the structures themselves.

Fig. 9.6
Formation of cellular structure due to constitutional
supercooling

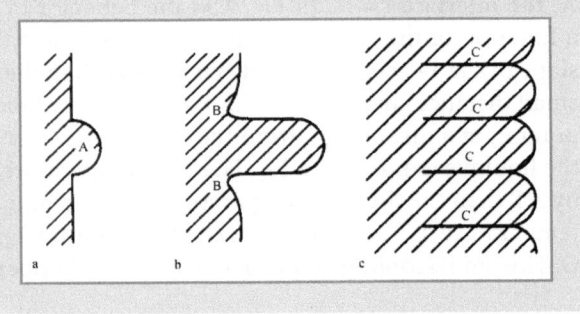

The formation of these structures can be prevented, when the crystallization rate is kept low, by using a very small temperature gradient $dt/d\zeta$. There
is then sufficient time for an almost complete elimination of the concentration
gradient near the crystallization front. In addition, the structure formation can
be prevented by creating a sufficiently large temperature gradient in the liquid.
This is also apparent from Fig. 9.7. If $dt/d\zeta$ is equal or greater than the slope of
the T-ζ curve in the immediate vicinity of the interface G, no constitutional undercooling can appear. The true temperature in the liquid is higher every
place than the corresponding liquidus temperature.

A further effect of the constitutional undercooling can be the nucleation
ahead of the crystallization front G. This can occur at a place in the liquid where
the true temperature is higher than at the interface G. This can be understood
based on Fig. 9.8. The liquidus temperature, T, is shown as a function of the distance from the solid-liquid interface G, as it may occur during the concentration
distribution in a solidification process. If the gradient, $dt/d\zeta$, of the true temperature is decreased from v_1 through v_2 to v_3, the constitutional undercooling increases. Finally, the extent of the constitutional undercooling becomes so great,
that nucleation occurs in front of the interface G. The presence of nucleation catalysts can be very helpful (heterogeneous nucleation). For the start of nucleation ahead of the crystallization front, the average composition of the liquid c_0,
the rate of advance R^K of the crystallization front, and the gradient $dt/d\zeta$ of the
true temperature in the liquid, are of importance.

Large cast blocks can consists at the edge of rod like crystals, and in the center
of a mixture of finely dispersed crystallites. First rod shaped crystals grow into
the liquid from the crucible wall, caused by the heat removal through the crucible wall. After a sufficiently large increase of the concentration of the components of the alloy in the liquid ahead of the crystallization front and corresponding constitutional undercooling, nucleation can take place in the rest of the liquid, yielding a fine grained, random structure. To avoid this nucleation, the gradient $dt/d\zeta$ of the true temperature in the liquid must be kept large. This can be
attained by cooling the crucible or increasing the pouring temperature.

To avoid an uneven structure and to prevent the formation of rod like crystals which disturb the working of the material, one must operate with a large

Fig. 9.7
Avoidance of constitutional
supercooling by a sufficient
high gradient of the true
temperature T

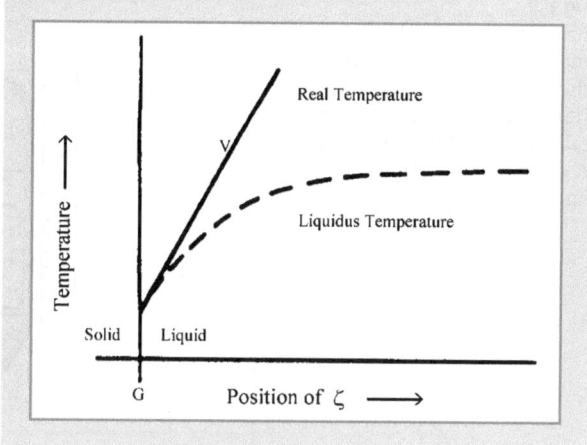

Fig. 9.8
Nucleation near the crystallization front G during constitutional supercooling

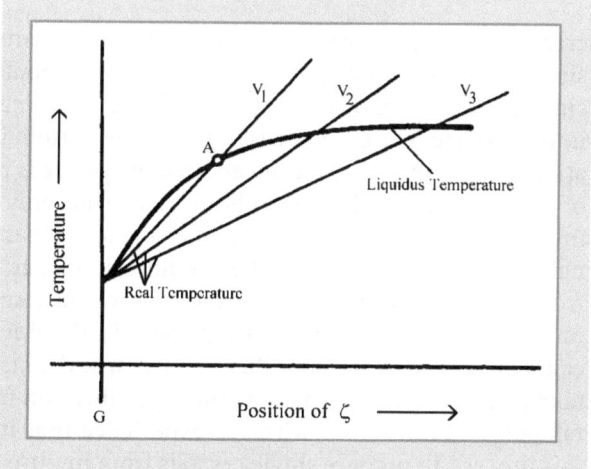

gradient $dt/d\zeta$ (see Fig. 9.8, v_1). The region G-A with constitutional undercooling is then very small. Nucleation occurs only in this small seam ahead of the
crystallization front. The liquid in the centre of the crucible remains liquid until the end of the total solidification process. The pulp like zone ahead of the solidification zone, containing the crystallites formed by the new nucleation, has
enough spaces between the crystallites filled with liquid, so that the liquid from
the center part, not yet solidifying, can flow, as is required by the volume shrinking during crystallization. This is no longer possible with a pulp zone which is
too wide. Pores are formed in the cast block, which are difficult to close with further working (rolling, forging).

For different applications, single crystals are required. As a rule they are obtained by controlled solidification of a liquid. Constitutional undercooling can

Fig. 9.9
Bridgeman-method for
monocrystal formation from
the melt. *A* locus of formation of nucleus, *E* monocrystal, *L* Liquid, *H* fixed furnace,
K direction of motion of
ampoule along the furnace

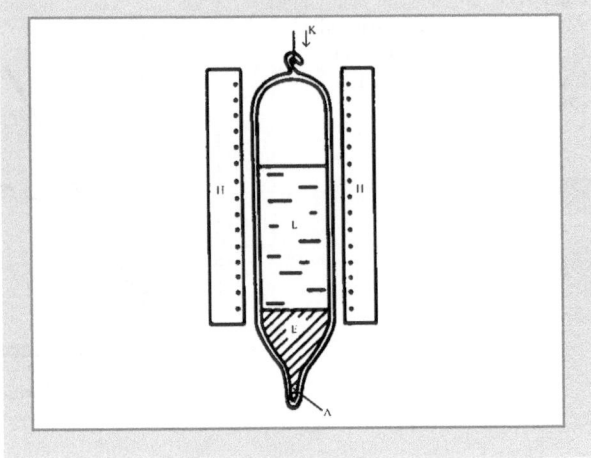

create great difficulties here. Among all the many method for the preparation of single crystals the Bridgman method shall be mentioned as an example. In the simplest case, as shown in Fig. 9.9, a liquid contained in an ampoule is lowered into a furnace. The ampoule is drawn at the bottom into a small capillary. Nucleation occurs when the solidification temperature is reached, during the lowering of the ampoule filled with the liquid. As a rule, only a single nucleus grows during further lowering of the ampoule, due to the narrowness of the capillary. Finally, during the advance of the solidification front, one single crystal is formed from the whole sample. This is the case for pure substances. If, however, a second or further components are present in the liquid, concentration gradients can form ahead of the solidification front, and constitutional undercooling can take place. Nucleation ahead of the crystallization front of the single crystal can take place. This agrees with the experience that it is generally difficult using this process, to prepare single crystals from impure substances, or prepare compound single crystals with higher content of additional components.

9.4
Purification by Zone Melting

As already discussed, from a liquid with composition c_0 and a distribution coefficient $k_0 < 1$, at the beginning of the solidification a solid crystallizes with the composition $(c_S = k_0 c_0) < c_0$ (see Fig. 9.4). With advancing solidification the concentration of the dissolved material, B, increases in the liquid as well as in the solid, until, – at a given crystallization rate – the steady state is reached. Here a solid crystallizes with composition which corresponds to the average composition c_0 of the liquid at a point very distant from the solidification front $(c_s'' = c_0)$. The solid, in a rod shape, formed by uniaxial solidification, has accordingly a composition, which increases continuously from $c_S = k\,c_0$ (immediately at the origin) to $c_S'' = c_0$ (where the stationary state begins), and remains constant until just before the end of the rod (see Fig. 9.10).

If the solidification front G is close to the end of the rod, and the reservoir of liquid is so small, that diffusion of component B away from the region where the concentration in B is increased, causes a noticeable increase in the average composition of the liquid $c_0 \rightarrow c_0'$ (where $c_0' > c_0$, see Fig. 9.11), then the composition of the liquid at the interface increases also to C_L^E, which is higher than that in the stationary state ($C_L'' = c_0/k_0$). The result is that the solid now also is formed with a higher concentration in B (C_S^E) than in the stationary state ($c_S'' = c_0$). This effect becomes more pronounced, as the interface G approaches the end of the sample (see Fig. 9.10).

Fig. 9.10
Distribution of concentration in a slab solidified in one direction at $k_0 < 1$

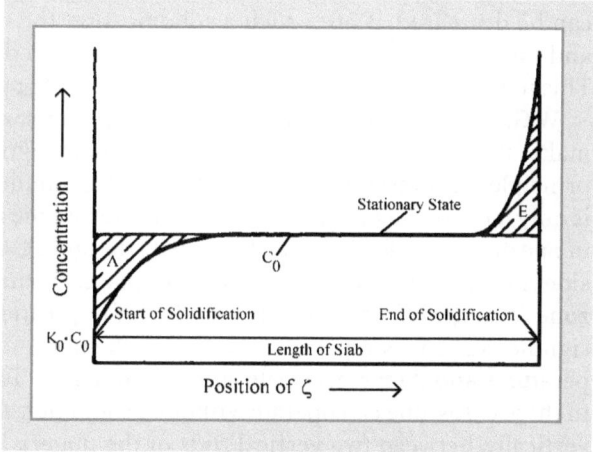

Fig. 9.11
Distribution of concentration in a melt L of a one-dimensional solidification produced slab immediately before the process of solidification ($k_0 < 1$)

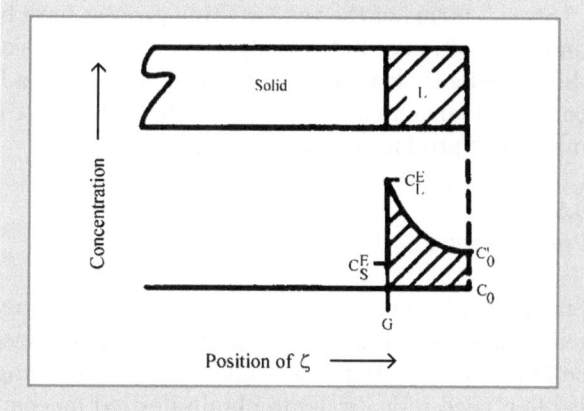

Fig. 9.12
Method of zone-melting
after W.G. Pfann. *A* newest
solidified crystal, *B* melt,
C solid not yet molten, *D* mo-
bil ring heater, *E* direction of
moving the heater

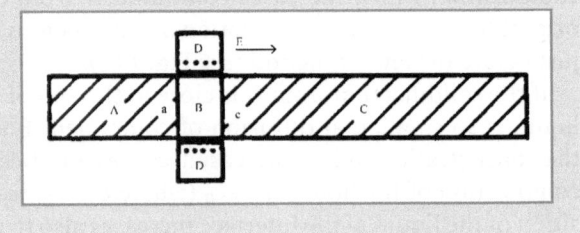

The first formed end of the uniaxial rod is poorer in B, the last formed end is richer in B than the average composition of the whole sample.

The hatched surfaces *A* and *E* in Fig. 9.10 are equal, as the quantity of B naturally remains constant during the solidification process. This effect of transporting a minority component from one end of the sample to the other can be used to eliminate impurities in a material. In principle the total amount of impurities can be decreased, if after such a solidification the end enriched in B is cut off and thrown away. The rest could again be subjected to a uniaxial solidification. This method, however, is of little use in practical applications.

W.G. Pfann has developed a practical process, called zone melting, which makes it possible to purify a substance without significant loss of material. The principle is represented in Fig. 9.12. A solid, contained in a horizontal container, is only melted in a narrow band B by a heater. As the ring shaped heater D moves in one direction (for instance *E*), it melts the solid *C* on the side c, whereas on the side *A* of the solid, crystallization takes place. During the passage of the molten zone from the left end, the introduction of impurities from the container can be significant. This is especially the case with substances which melt at high temperatures and then react easily with the container. To obtain extreme purity also in these cases, one can operate without a container. The melted zone is produced vertically, between two vertical rods of the material to be purified. The narrow melted zone is kept in its position by surface tension alone or with additional electromagnetic forces, and moved slowly through the solid rod.

Special forms of the zone melting process have been developed for the optimal purification of industrially important materials. It must be mentioned, that the opposite is also possible: Through a liquid in a long container, a cylindrical cooling arrangement is moved slowly, and thus, a narrow zone of crystallized material is produced (zone crystallization).

9.5
Precipitation Reactions in the Solid State

Starting with a solid equilibrium phase, by changing the state variables pressure, temperature or composition, it is possible to cause a transition into another solid phase or a mixture of solid phases. The change in structure of the solids is often used industrially to obtain desired mechanical, magnetic or electrical properties of materials. For practical reasons the temperature almost always is changed. A diffusionless transition of one lattice into another can serve as an atomic mechanism to transform an initial solid phase into a mixture of other phases. This is the case, for example, during the formation of martensite in the iron-carbon system. In most cases, however, the final state is reached by diffusion. Moreover, nucleation and the growth of the nuclei formed is determined by diffusion. Only the situation relating to the growth of the new phases will be considered.

The simple case of the formation of a new solid phase in a solid starting matrix is represented schematically in Fig. 9.13. An α-solid solution with the composition $x_B^{\alpha,0}$, is cooled from temperature T_1, where it is stable, to temperature T_2. Now phase β precipitates. The precipitation can take place in a continuous or discontinuous manner.

With a continuous precipitation, nucleation can occur at the grain boundaries as well as in the interior of the grains of a polycrystalline starting matrix. The atomic mechanism controlling the kinetics of growth is volume diffusion. The principle of the time dependence of the precipitation reaction is sketched in Fig. 9.14. The β-particles present in the starting matrix are separated from each other, at least at the beginning of the precipitation reaction. The concentration in these precipitated particles is x_B^β; outside, far from them, the concentration in the starting matrix is $x_B^{\alpha,0}$. In the immediate vicinity of the precipitated β-particles a diffusion zone with a concentration gradient is developed. The diffusion

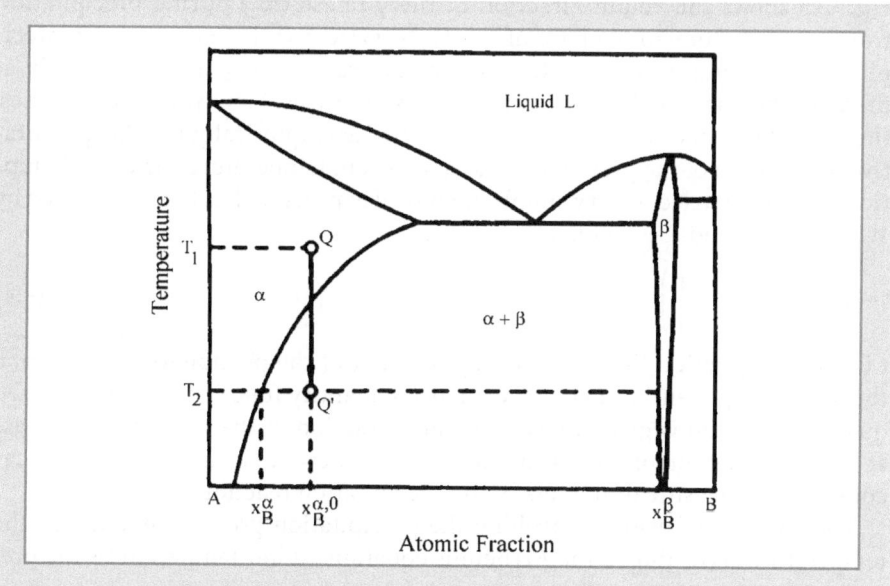

Fig. 9.13 A-rich side of the phase diagram to explain the precipitation of a new solid phase from a solid starting phase

Fig. 9.14
Distribution of concentration during continuous precipitation. In respect of the atomic fraction see Fig. 9.13. K Loci of nucleation; a distribution of concentration immediately after starting of the precipitation; b intermediate stage; c final stage after $t \rightarrow \infty$

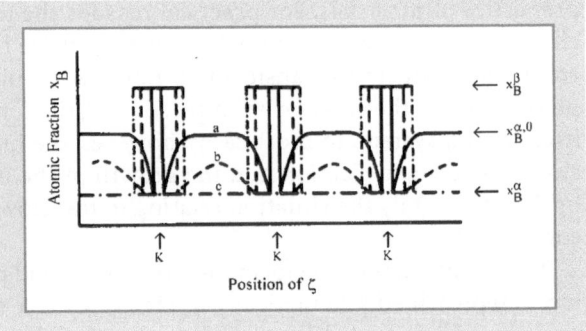

taking place here transports B to the β-particles. While the latter ones grow, the B-concentration in the matrix decreases continuously, until finally the equilibrium concentration, x_B^α, the concentration of the α-phase in equilibrium with the β-phase at T_2 is reached. The decrease of the B-content in the α-phase can, for example be continuously followed by measuring the lattice parameter using X-ray diffraction.

A quantitative analysis of the growth process can be carried out easily, by measuring the fraction of the precipitated phase as a function of time. This is possible experimentally by measuring the surface fraction of the precipitated phase microscopically on polished surfaces of the sample. As an example, Fig. 9.15 shows the volume fraction of the γ-phase (fcc) during precipitation from a supersaturated α-phase of the Fe-Ni system with 5 at.-% Ni at a precipitation temperature of 1013 K. The volume fraction of γ increases slowly at first, but increases with the growth of the γ-particles. Finally, with long times, the end value of the volume fraction is reached asymptotically. For the quantitative description of the growth process at constant temperature various assumptions have been tried. A very simple analytical expression has been presented by W.A. Johnson and R.F. Mehl:

$$y(t) = 1 - \exp\left(-\frac{t}{\tau}\right)^n \tag{9.4}$$

t is the precipitation time, τ the decay constant of the precipitation reaction, n the growth exponent, and y(t) a normalized quantity function. The normalized quantity function is given by y(t) = volume fraction of the precipitating phase at t/volume fraction of the precipitating phase at t = ∞. The inverse of the decay constant is a measure of the rate of the precipitation reaction.

The volume diffusion controlling the precipitation process depends on the temperature according to the Arrhenius equation. At low temperatures the precipitation rate can be very low.

A different way to form a new phase in a single phase starting matrix, is discontinuous precipitation. As a rule, the nucleation starts at the grain boundaries.

Fig. 9.15
Kinetics of continuous precipitation. Volume part γ is precipitated from supersaturated cubic centered Fe-5 at.-% Ni solid solution (α). This precipitated part is plotted as a function of duration of precipitation. T = 1013 K (after [1])

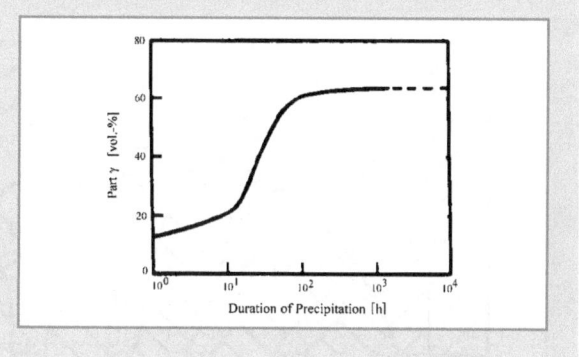

Generally a lamellar structure results, where lamellae of the newly formed β-phase lay next to the matrix, depleted in B. The lamellar precipitation structure, occuring after nucleation, and consisting of β-phase with x_B^β, and α-phase with a composition of approximately x_B^α (see Figs. 9.13 and 9.16), grows at a fixed temperature with constant rate into the starting matrix α_0 (concentration $x_B^{\alpha,0}$). The B atoms, distributed homogeneously in the α_0 matrix, are divided between the two neighboring new phases α and β. This takes place by interface diffusion between the precipitate and the starting matrix.

If continuous precipitation as a competing reaction is not effective, there is no change in composition in the part of the starting matrix not yet attacked by the precipitation. This method to form a new phase is called discontinuous precipitation, because in this case the composition of the starting matrix does not change continuously into the end composition.

As indicated in Fig. 9.16, the lattice of the α-lamellae assumes the crystallographic orientation of the grain, from where the growth of the precipitated material started, and is moving away. The rate R^A, with which the reaction front advances, depends on the mobility M and the driving force of the reaction ΔG, the gain in Gibbs energy at the completion of the reaction:

$$R^A = -\Delta G \cdot M \tag{9.5}$$

The quantity ΔG, contains, in addition to the gain in Gibbs energy due to demixing into two phases with almost equilibrium compositions, the expenditure in Gibbs energy for the formation of the lamellar boundaries.

The mobility, M, depends mainly on the atomic structure of the interface. An average lamellar distance is formed, which guaranties, under given exterior conditions (temperature, composition), the rapidest decrease of the Gibbs energy of the system. According to J. Peterman and E. Hornbogen, the rate at which the reaction front advances is given by:

$$R^A = -8 \cdot \frac{\Delta G}{RT} \cdot \frac{\lambda \cdot D_I}{l^2} \tag{9.6}$$

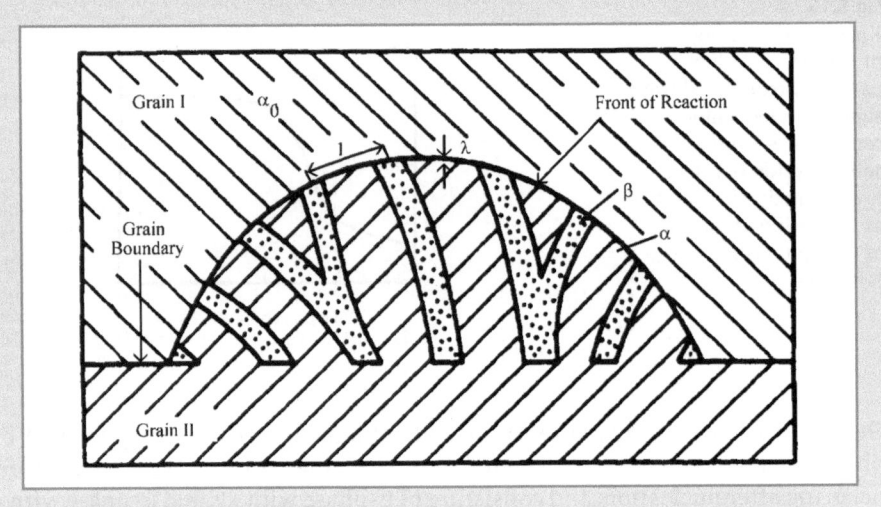

Fig. 9.16 Schematic presentation of a precipitation colony concerning lamellar discontinuous precipitation. α_0 starting matrix $(x_B^{\alpha,0})$, α α-phase (x_B^{α}) deplated by B, β β-phase (x_B^{β}), I distance between lamellae, λ Thickness of incoherent phase boundary between colony of precipitation and starting matrix

λ is the thickness of the reaction boundary between precipitated substance and starting matrix (≈ 0.5 nm), D_I, the interface diffusion coefficient in this reaction zone, l the average lamellae distance and ΔG the change in Gibbs energy during the precipitation reaction. In addition, numerous expressions were developed for the rate of advance of the precipitation front which take into account various situations.

It must be mentioned that as a rule, the equilibrium composition x_B^{α} of the α-phase is not reached immediately during the formations of precipitation cells. Rather an α-phase with $x_B > x_B^{\alpha}$ is formed. The excess in B can be decreased with volume diffusion after the discontinuous precipitation.

Furthermore, in a thin lamellar structure formed by discontinuous precipitation, a second discontinuous reaction can start. A lamellar structure is formed, with greater lamellar distance. The original finely distributed lamellar structure can be used up completely. The driving force for this secondary precipitation reaction, in an already discontinuously demixed matrix, consists of the gain in Gibbs energy due to the reduction of the total interface enthalpy and a better adjustment of the composition of the α-lamellae to the equilibrium composition x_B^{α}.

Which of the various precipitation possibilities occurs under given boundary conditions depends principally on the rate with which the system can lower its Gibbs energy. Often – depending on temperature and supersaturation – the precipitation of the equilibrium phase in the same system occurs both ways. In addition, one or more metastable phases can appear as preprecipitates, as can be observed, for instance, in duraluminum (see Sect. 8.2).

References

Citations
[1]   Predel and W. Gust, Arch. Eisenhüttenwes., 43, (1972) 657

General References
J. Burke, "The Kinetics of Phase Transformations in Metals", Pergamon Press, Oxford (1965)
B. Chalmers, "Principles of Solidification", J Wiley and Sons, New York (1964)J.
J. Manenc, "Structural Thermodynamics of Alloys", D Reidel Publ. Comp., Dordrecht, Holland (1973)
G. Matz. "Kristallisation", Springer-Verlag, Berlin (1969)
W.G. Pfann, "Zone Melting", J.Wiley and Sons, New York (1958)
H. Schildknecht, "Zonenschmelzen", Verlag Chemie, Weinheim/Bergstr. (1964)
H. Schmalzried, "Festkörperreaktionen", Verlag Chemie, Weinheim/Bergstr. (1971)
W.C. Winegard, "An Introduction to the Solidification of Metals", The Institute of Metals, London (1964)

CHAPTER 10

Organic and Polymeric Materials

10.1
Introduction

Organic materials and polymers represent important materials in our society; people studying and describing them also use thermodynamic means. The same thermodynamic conditions apply to them as to inorganic materials [1]. There are obviously structural differences between inorganic and organic materials. The building blocks of an inorganic phase diagram are molecules made up of a few (less than 10) atoms. Organic materials and polymers can contain up to 100,000 atoms. The only change we made to the basic thermodynamic equations is that the ideal Gibbs energy of mixing is represented by

$$\Delta G_{id} = R \cdot T \, (z_1 \ln z_1 + z_2 \ln z_2 + z_3 \ln z_3 \dots) \tag{10.1.}$$

where z can be mole fraction, weight fraction or volume fraction (ΔG_{id} per at.%, wt.-%, vol.-%, resp.).

The excess terms e.g. in the Hoch-Arpshofen model Chap. 6 stay the same, x is replaced by z and it has the same meaning as in Eq. (10.1). It is possible to convert data e.g. from volume fraction to mole fraction, all representations being equivalent. However, in ternary and larger systems the mole fraction representation has to be used.

10.2
Organic Materials

Figure 10.1 shows the enthalpy of mixing in the system ethylacetate-chloroform at 25°C and Fig. 10.2 shows the enthalpy of mixing in the system i-Amylalcohol-Chloroform at 25 °C. The original data are from Staude [2] given as a function of the mole fraction. The analysis and different representation of the data is from Hoch [1]. The lines are the calculated ones, using the interaction parameters obtained from the measured data. All representations are equivalent.

The molecule of Chloroform is a round ball, whereas Ethylacetate is a straight molecule and i-Amylalcohol has a side chain. In the i-Amylalcohol–Chloroform system the enthalpy of mixing is positive at high i-Amylalcohol concentrations,

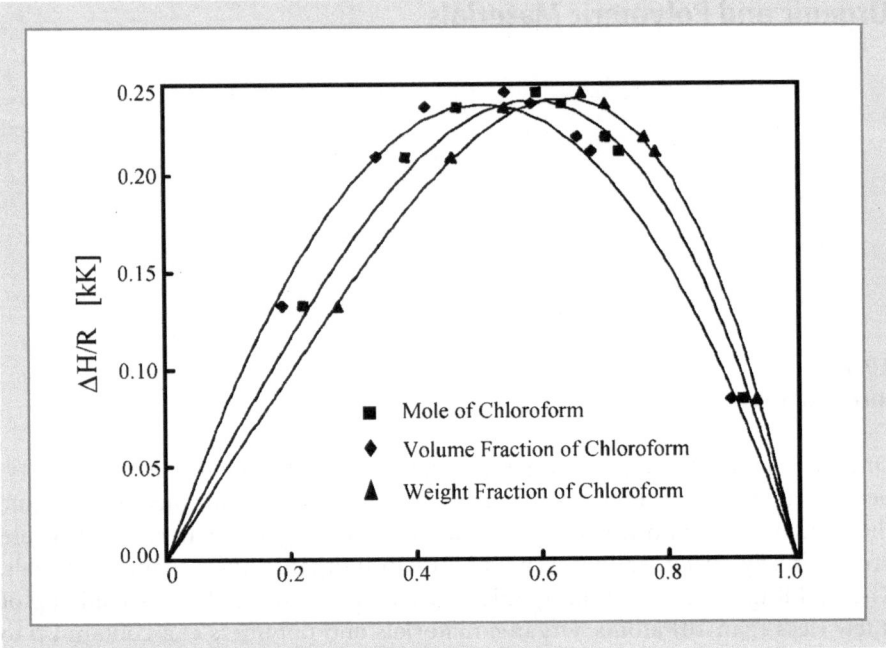

Fig. 10.1 Enthalpy of mixing of Ethylacetate-Chloroform solutions at 25 °C. Data from Staude [2]. A = Author's data: mole fraction of Chloroform, B = Volume fraction of Chloroform, C = Weight fraction of Chloroform, Lines calculated from model

and negative at high Chloroform concentrations. In the system Ethylacetate-Chloroform the volume fraction representation shows a symmetrical enthalpy of mixing.

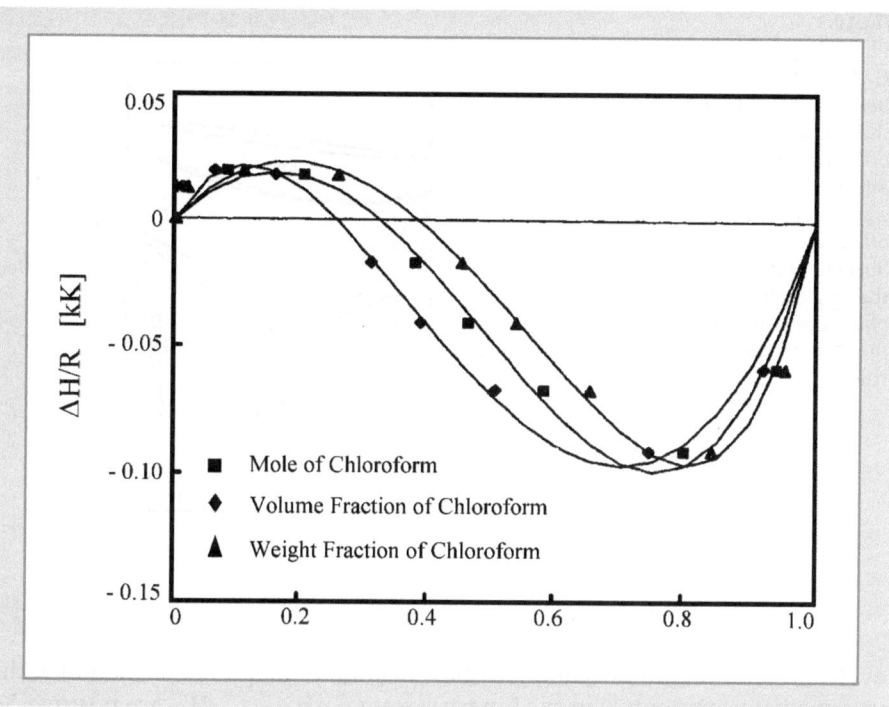

Fig. 10.2 Enthalpy of mixing of i-Amylalcohol–Chloroform solutions at 25°C. Data from Staude [2]. A = Author's data: mole fraction of Chloroform, B = Volume fraction of Chloroform, C = Weight fraction of Chloroform. Lines calculated from model

10.3
Phase Diagrams

The n-alkanes from $C_{19}H_{40}$ to $C_{26}H_{54}$ play an important role in the performance of diesel fuels; they decide the solidification temperature of the fuel. Dirand et al. [3] presented the $C_{22}H_{46}$–$C_{24}H_{50}$, $C_{24}H_{50}$–$C_{26}H_{54}$, $C_{22}H_{46}$–$C_{23}H_{48}$ and $C_{23}H_{48}$–$C_{24}H_{50}$ binary phase diagrams. The $C_{22}H_{46}$–$C_{23}H_{48}$ diagram is shown in Fig. 10.3. In Fig. 10.3 the liquidus and solidus lines indicate a system with small deviations from ideality.

The α-RII – RI transformation is possibly a first order transformation, and the + + + line between ß and RI is a second order transformation. The concave "line" separating the ß-phase from the two compounds $ß'_1$ and $ß''_1$ is surprising. In the case of metals where a solid solution forms compounds on cooling, the maximum temperature is in the center of the composition range (e.g. Cu-Au), and the boundary "line" is convex. The situation in Fig. 10.3 is caused by an entropy effect. The analysis[4] of the enthalpies of pure alkanes measured by Barbillon et al. [5], and of some binary solutions from 260–350 K by Achour

Fig. 10.3
Phase diagram of $C_{22}H_{46}$-
$C_{23}H_{48}$ [3]. The +++ line
between β and RI indicates
the beginning of the second-
order transition which
affects the lattice parameters
a and b of the phase β; the
other lines show the equilibria boundaries between
the single-phase and two-
phase domains according to
the Palatnik and Landau's
rule.

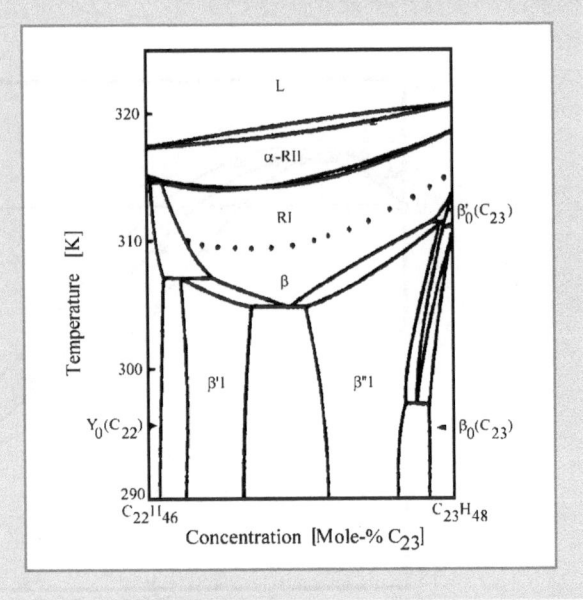

et al. [6, 7] and Sabour's [8] analysis show that the enthalpies of formation of the
intermediate compounds in the binary system $C_{22}H_{46} - C_{23}H_{48}$ are positive. To
check the entropy question, the measured enthalpy data of n-alkanes $C_{19}H_{40}$ to
$C_{24}H_{50}$ and of the binary system $C_{22}H_{46} - C_{23}H_{48}$ were analysed [4].

10.4
Polymer Blends

a) Thermodynamic properties
Enthalpies of mixing of polymer melts cannot be measured experimentally because the liquid is to viscous to obtain complete mixing of the components in a
reasonable time. Thus, model calculations are used to get information on mixing in polymeric type materials.

Figure 10.4 shows the enthalpy of mixing data published by Harris et al. [9].
The abbreviations of the organic components which are used in Fig. 10.4 are
summerized in Table 10.1.

The sign and the amount of the enthalpy of formation depends on the chemical structure of the polyesters (see Fig. 10.4). The maximal enthalpies of formation are given in Table 10.2.

The experimental DEA-DPP data show some scatter. The molecular weights
of the components are not very different. The maximum of ΔH in the system
DEA-DPP is shifted towards DPP indicating that DPP has a stronger binding
capacity than the other component of this system. This is more or less valid in
common.

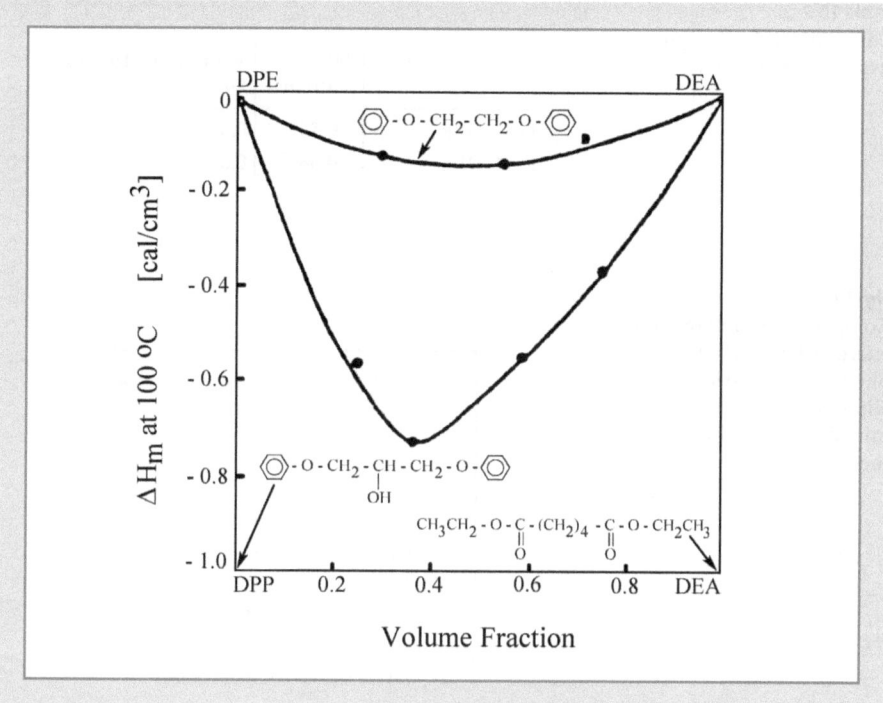

Fig. 10.4 Heat of mixing for DEA-DPE and DEA-DPP binary mixtures at 100 °C

Table 10.1 Abbreviations used in Fig. 10.4

Abbreviation	Structure formula	Melting point [°C]
DEA	CH₃—CH₂—CO(CH₂)₁—CO—CH₂—CH₃ (with two C=O above)	
DPP	⬡—O—CH₂—C H—CH₂—O—⬡ with OH	81.2
DPE	⬡—O—CH₂—CH₂—O—⬡ with OH	95.6

Table 10.2
Enthalpies of formation in
two polymer systems

Binary mixture	ΔH_{max} [cal/cm^3]	Volume fraction of DPP, Φ_{DEA}
DEA-DPP	– 0.73	0.35
DEA-DPE	– 0.08	0.5

Fig. 10.5
Comparison of interaction
energies B for blends of
phenoxy with various
aliphatic polyesters determined by three different
methods (○●▲)

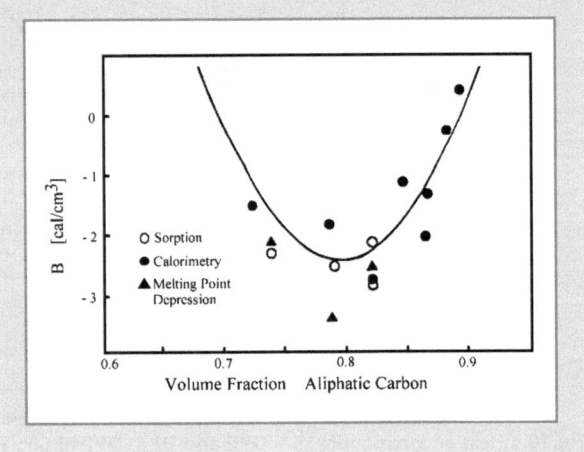

The enthalpy of formation, ΔH^P, as a function of unit volume, Φ_{DPP}, can be expressed by

$$\Delta H^P = \Phi_{DPP} \cdot (1 - \Phi_{DPP}) \cdot B \qquad (10.2)$$

where B is the interaction parameter. The dependence of B in the ester of carbonyl of aliphatic groups is shown in Fig. 10.5.

B approximately has a maximum negative value, if the volume fraction of ester carbonyls is 0.2.

The experimental enthalpy data generally are given in cal/cm^3. If one assumes a density of approximately 1 g/cm^3 and an average molecular weight of 225, the experimental data would decrease by a factor of 1.13 to represent them in kcal/g-at.

b)Phase diagram of binary polymer systems

An example of a complete phase diagram of a binary polymer system is given in Fig. 10.6 as determined by Roe and Zin [10]. It concerns mixtures containing polystyrene (molar weight 2400) and styrene/butadiene diblock copolymer (27 % styrene, molar weight 28,000). The difference of molar weights of the components is responsible for the miscibility gap in the liquid state. The gap is

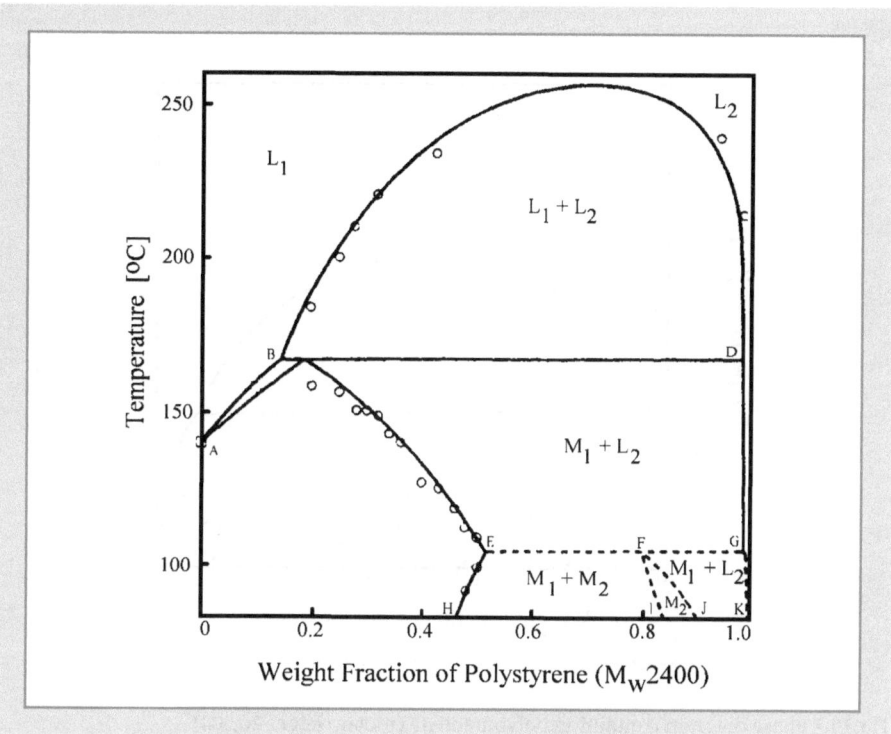

Fig. 10.6 Phase diagram of the mixture containing styrenebutadiene diblock copolymer (styrene 27 %) (molar weight 28,000), and polystyrene (molar weight 2400)

shifted to the component with the lower molar weight. In the Hoch-Arpshofen model description, polystyrene has the weaker bonding capability for it has fewer carbon atoms.

Figure 10.7 shows the binary phase diagram of styrene-butadiene (molar weight 28,000) and polybutadien (molar weight 26,000). In this case the molar weights are almost equal. Therefore the components are more soluble in each other than in the case shown in Fig. 10.6 (lower critical point). Further on, the miscibility gap is symmetrical. It should be mentioned, that the equilibria indicated by broken lines are hypothetical.

In Figs. 10.6 and 10.7, M_1, is a mesophase consisting of microdomains of the block polymer swollen with polystyrene. The mesophase M_2 contains micelle aggregates of the block copolymer suspended in the medium of polystyrene.

In the region $L_1 + L_2$ the mixture undergoes a macroscopic phase seperation into two homogeneous mixtures. In the region $M_1 + L_2$ there occurs macroscopic separation in two phases, also, in a mesophase M_1 and the liquid L_2.

An experimental method to determine the shape of the miscibility gap consists of determining the temperature at which the sample become cloudy on

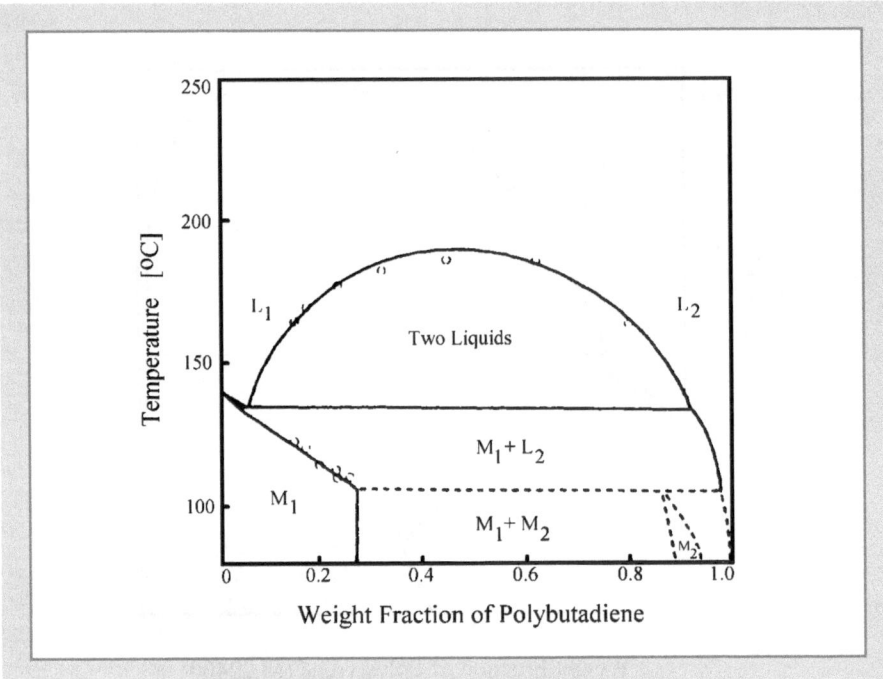

Fig. 10.7 Phase diagram containing polybutadiene (molar weight 26,000)

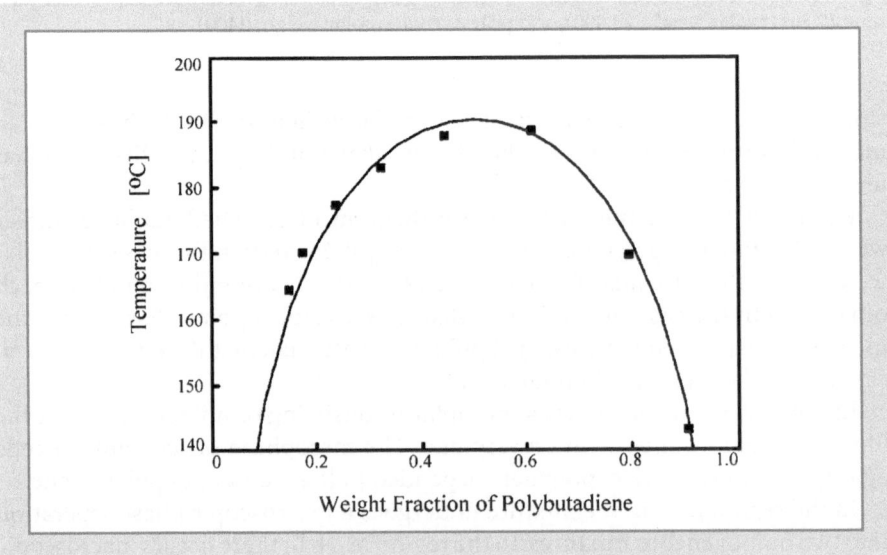

Fig. 10.8 Miscibility gap in the system styrene-butadiene block copolymer, 27 % styrene (molar weight 28,000) – polybutadiene (molar weight 26,000) [10, 1] ■ experimental,— calculated

cooling. The structures of the mesophases were determined by X-ray diffracto-graphy. The determined shape of the miscibility gap can be achieved rather exactly, as is to be seen from Fig. 10.8 [1].

References

Citations
[1] M. Hoch, Polymer Engineering and Science. 36 (20) (1996) 2485–2493
[2] H. Staude, "Physikalisch-Chemisches Taschenbuch", Akademische Verlagsgesellschaft, Leipzig (1945)
[3] M. Dirand, Z. Achour, B. Jouti, A. Sabour and J.-C. Gachon, Mol. Cryst. Liq. Cryst. 275 (1996) 293–304
[4] Z. Achour, A. Sabour, M. Dirand and M. Hoch, Journal of Thermal Analysis, 51 (1998) 477–488
[5] P. Barbillon, L. Schuffenecker, J. Dellacherie, D. Balesdent and M. Dirand, J. Chim Phys 88 (1991) 91–113
[6] Z. Achour, P. Barbillon, M. Bouroukba and M. Dirand, Thermochimica Acta, 204 (1992) 187-204
[7] Z. Achour, PhD Thesis, INPL-NANCY (1994)
[8] A. Sabour, PhD Thesis, INPL-NANCY (1994)
[9] J.E. Harris, D.R. Paul and J.W. Barlow, Polym. Eng. Sci. 23 (1983) 676–680
[10] R.J. Roe and W.C. Zin, Macromolecules 17 (1984) 189

General References
H.S. Goh, D.R. Paul and J.W. Barlow, Polymer Eng. Sci., 22 (1982) 34
C.M. Kuo and S.J. Clarson, Eur. Polym. J. 29 (1993) 661
D. Rigby, J.L. Lin and R.J. Roe, Macromolecules 18 (1985) 2269
R.J. Roe and W.C. Zin, Macromolecules 13 (1980) 1221

Subject Index

ΔG-p curve 254
ΔG-p diagram 255
ΔG-x curve 241
ΔG-x curve 264
ΔG-x curve 300
ΔG-x isotherms 240
ΔH-x curve 179
ΔSi-x curve 180
γ-loop 79
Λ-point 258

A

abnormal eutectic 44
accomodation coefficient 18, 20, 77
acid slags 129
activation energy 18, 19, 22
activity 206, 211, 221
activity coefficient 191, 212, 215
activity isotherm 216, 239
additivity 290
Ag-Te phase diagram 197
Al_2Cu 285
Al-Cu phase diagram 285
aliphatic group 336
α-quartz 14
α-tridymite 14
anharmonic contribution 204
anion 147, 148
anisotropic transformation 77
annealing 288
anomalous eutectic 45
Anorthite 128
arrest 52
associates 191, 192
association model 190, 195, 196, 198
at. % 331
atomic fraction 17, 181
atomic radii 54, 56, 175
atomic weight 18

atomic-% 17
Au-Ge-Si-alloy 305
Au-Si system 293
austenite 82, 83, 84, 85, 278, 284
austenite matrix 278
azeotropic mixtures 159

B

basic slags 129
Bernal's closest random packing of equal rigid balls 307
binary crystallization 141
binary equilibria 165
binary eutectic 140
binary eutectic crystallization 103
binary eutectic reaction 102
binary eutectic systems 104
bivalent reaction 106
bivariant reaction 137
block copolymer 337
block polymer 337
boiling curve 158
boiling diagram 164, 165
boiling point 157, 158, 159, 171
boiling point maximum 159, 159
boiling process 6
boiling temperature 158
Boltzmann constant 253
bond 199
bonding electron 199
borate 304
boundary Gibbs energy 270
bounding binary system 95
Bridgeman-method for monocrystal formation 322

E

e_1E 97
e_2E 97
effective distribution coefficient 313
elastic distortion 85
electrical resistance 41
electron microscope 84
electron microscopic investigation 39
electron microscopy 40
electroneutrality 70
ellipsoid 275
EMF measurement 222
enantiotropic transformations 10
energetic conditions 278
enthalpy 3, 4, 176, 178, 180, 183
 – of disordering 252
 – of formation 209
 – of fusion 9, 20, 21, 23
 – of melting 19, 20, 303
 – of mixing 179, 180, 184, 185, 186, 208, 234, 243, 299
entropy 178, 183, 186
entropy of mixing 179, 180, 184, 185, 199, 299
equilateral tetrahedron 143
equilibration 159
equilibria of transformation 78
equilibrium 275
equilibrium concentrations 49
equilibrium distribution coefficient 313
equilibrium point 85
equilibrium pressure 161
equilibrium state 2
equilibrium temperature 275
ester carbonyl 336
ester of carbonyl 336
etching 34
Ethylacetate 331
ethylacetate-chloroform 331
ethylalcohol-water 159
eutecoid reaction 74
eutectic 82, 163, 167
eutectic composition 167
eutectic concentration 34
eutectic crystallization 42, 43, 103
eutectic four-phase equilibrium 142
eutectic grain boundaries 37
eutectic point 95, 166
eutectic polyhedron 107
eutectic reaction 60
eutectic solidification 146
eutectic structure 44
eutectic system 60

eutectic three-phase equilibria 74
eutectic vapor mixture 163
eutectoid concentration 66
eutectoid equilibrium 66
eutectoid point 82
eutectoid reaction 68, 74
eutectoid three-phase equilibrium 82
eutectoid three-phase reaction 68
eutectoid transformation 67
evaporation 10
excess entropy 186, 237
excess entropy of mixing 202
excess free enthalpy of mixing 293
excess Gibbs energy of mixing 199, 260
excess partial entropy of mixing 236
explanation of the energetic principle of the spinodal demixing 279
explanation of the knod triangle and of the compatibility triangle 92

F

Fayalite 67
ferrite 82, 86, 278
ferrite nucleation 278
first law of thermodynamic 176
first order transformation 333
fold 278
formation of a nucleus 270
formation of cementite nuclei 278
four fields of three-phase equilibria 143
four-phase equilibrium 141, 143
four-phase plane 143
four-atom complexes 199, 199, 201
four-component system 201
fraction of coordination 288, 289
fractionating distillation 158, 159
free enthalpy 65, 178, 277
free enthalpy concerning homogenous formation of nuclei 270
free enthalpy G 179
free enthalpy of mixing 179
freezing 19
freezing point depression 23
frost 162
frost curve 161, 162
frost line 171
fusion equilibrium 34, 163, 236
fusion process 22
fusion surface projection 100
fusion temperature 20

Printing: Strauss GmbH, Mörlenbach
Binding: Schäffer, Grünstadt